Robotics and Biomimetics: Advanced Technologies

Robotics and Biomimetics: Advanced Technologies

Editor: Rowland Wilson

New York

Published by NY Research Press
118-35 Queens Blvd., Suite 400,
Forest Hills, NY 11375, USA
www.nyresearchpress.com

Robotics and Biomimetics: Advanced Technologies
Edited by Rowland Wilson

© 2017 NY Research Press

International Standard Book Number: 978-1-63238-531-4 (Hardback)

Cataloging-in-publication Data

Robotics and biomimetics : advanced technologies / edited by Rowland Wilson.
     p. cm.
Includes bibliographical references and index.
ISBN 978-1-63238-531-4
1. Robotics. 2. Robots. 3. Robots--Control systems. 4. Biomimetics. I. Wilson, Rowland.
TJ211 .R63 2017
629.892--dc23

Printed in the United States of America.

Contents

Preface

The various sub-fields of robotics along with technological progress that have future implications are glanced at in this book. This discipline is multidisciplinary in nature borrowing from fields such as mechanical engineering and computer science that deals with the design, construction, operation and use of robots. This book elucidates the concepts and innovative models around prospective developments with respect to robotics and biomimetics. The text also provides a broad spectrum of topics to its readers for a better understanding of the subject with state-of-the-art inputs by acclaimed experts of this field, this book targets students and professionals. From theories to research to practical applications case studies related to all contemporary topics of relevance to this field have been included in this book.

Various studies have approached the subject by analyzing it with a single perspective, but the present book provides diverse methodologies and techniques to address this field. This book contains theories and applications needed for understanding the subject from different perspectives. The aim is to keep the readers informed about the progress in the field; therefore, the contributions were carefully examined to compile novel researches by specialists from across the globe.

Indeed, the job of the editor is the most crucial and challenging in compiling all chapters into a single book. In the end, I would extend my sincere thanks to the chapter authors for their profound work. I am also thankful for the support provided by my family and colleagues during the compilation of this book.

Editor

Single-step collision-free trajectory planning of biped climbing robots in spatial trusses

Haifei Zhu[1], Yisheng Guan[1*], Shengjun Chen[1], Manjia Su[1] and Hong Zhang[1,2]

Abstract

For a biped climbing robot with dual grippers to climb poles, trusses or trees, feasible collision-free climbing motion is inevitable and essential. In this paper, we utilize the sampling-based algorithm, Bi-RRT, to plan single-step collision-free motion for biped climbing robots in spatial trusses. To deal with the orientation limit of a 5-DoF biped climbing robot, a new state representation along with corresponding operations including sampling, metric calculation and interpolation is presented. A simple but effective model of a biped climbing robot in trusses is proposed, through which the motion planning of one climbing cycle is transformed to that of a manipulator. In addition, the pre- and post-processes are introduced to expedite the convergence of the Bi-RRT algorithm and to ensure the safe motion of the climbing robot near poles as well. The piecewise linear paths are smoothed by utilizing cubic B-spline curve fitting. The effectiveness and efficiency of the presented Bi-RRT algorithm for climbing motion planning are verified by simulations.

Keywords: Biped climbing robots, Motion planning, Collision avoidance, Path smoothing, Rapidly-exploring random tree

Background

To release workers from tedious and dangerous high-rise tasks in truss-type environments, such as inspecting or spray-painting the frame of gymnasiums, airports and large bridges, and so on, robots able to autonomously climb poles are ideal solutions with a lot of benefits. Motivated by this, a variety of pole-climbing robots including UT-PCR [1], CPR [2], Shady3D [3], Climbot [4] and 3DCLIMBER [5] have been developed. Among them, biped pole-climbing robots (BiPCRs), whose main bodies are usually serial arms with multiple degrees of freedom (DoFs) and both ends are mounted with attaching devices, are considered to be outstanding, thanks to their high mobility in terms of multiple climbing gaits, strong ability to transit between poles and to overcome obstacles.

The ultimate goal of developing BiPCRs is to autonomously carry out high-rise tasks in place of humans.

To this end, autonomous climbing is a fundamental and essential functionality of a BiPCR. In some sense, a BiPCR can be regarded as a mobile manipulator, whose base may be changed and fixed in turn. During climbing, the robot fixes and supports itself with one of the two grippers served as the base, and moves the other one (the swinging gripper) to the target position, interchanging the roles of the two grippers in each climbing cycle. Hence to completely describe how a BiPCR climb in a spatial truss, we have to provide a series of discrete footholds and the continuous trajectories between adjacent footholds of the same swinging grippers. While the former define the gripping configurations of the BiPCR from the initial position to the destination, the latter determine the climbing motion of the robot in each climbing step. How to plan the footholds refers to climbing path planning or grasp pose planning of a BiPCR, which is out of the scope of this paper. Rather, given the footholds of the two grippers, how to plan the smooth and collision-free motion of the swinging gripper in one climbing step for a BiPCR in complex spatial trusses is an open and challenging issue and is the focus of this paper.

*Correspondence: ysguan@gdut.edu.cn
[1] Biomimetic and Intelligent Robotics Lab (BIRL), School
of Electro-mechanical Engineering, Guangdong University of Technology,
Hi-education Mega Center, Guangzhou 510006, China
Full list of author information is available at the end of the article

Climbing path planning of BiPCRs in spacial trusses has been investigated to some extent in the literature. The problem was converted into the classical traveling salesman problem considering the energy consuming during each climbing cycle in [6] and [7]. In [3], the trusses were discretized into a series of nodes and the sequence of clamping points from a given initial node to the destination one was planned by calculating the Dijkstra shortest distance and motion complexity as criteria. The above work on climbing path planning actually belongs to the category of foothold planning. However, to the best of our knowledge, single-step collision-free trajectory planning of a BiPCR climbing in complex spatial trusses has not been explored.

Climbing motion planning of a BiPCR in one climbing step is similar to that of an manipulator, since the robot is fixed and supported on a pole by the base gripper at a specific foothold, and the swinging gripper moves from its initial foothold (configuration) to the target one. Therefore, traditional algorithms for collision-free motion planning of manipulators, such as artificial potential field (APF) [8], probabilistic road map (PRM) [9], rapidly-exploring random tree (RRT) [10], and almost all the intelligent algorithms like genetic algorithm, particle swarm optimization (PSO) [11] can be utilized to generate the climbing trajectories. However, the motion planning of a BiPCR has its own features compared with that of an industrial robot. First, when the base of a BiPCR is changed and switched between the two grippers during climbing, those algorithms suitable for fixed base, such as PRM, will exhibit low efficiency. Second, some part(s) of the target pole is/are the graspable region(s) and other parts should be treated as obstacles, traditional algorithms like APF will encounter difficulties. Third, the role of a pole (target or obstacle) may interchange in different climbing cycles.

Considering the RRT algorithm has wide adaptation and good robustness to multiple degrees of freedom and dynamic environments, we address the problem of collision-free motion planning in one climbing step for BiPCRs in the spatial trusses, utilizing the Bi-RRT algorithm. The main contributions of this paper are as follows. On the one hand, the framework for climbing motion planning of BiPCRs with different DoFs is first built based on Bi-RRT. The proposed planning algorithm is adaptive to BiPCRs with different numbers of DoFs including five and six. For a 5-DoF BiPCR like Climbot-5D (hereafter we use Climbot-5D and Climbot-6D to represent the Climbot with five and six degrees of freedom, respectively) whose orientation is limited due to its special configuration, a simple but effective state expression method is presented to deal with the sampling, interpolation and

metric processes, which also adapts to Climbot-6D. On the other hand, pre-process and post-process methods are proposed in this paper to guide the swinging gripper to move away from the starting foothold and to the target foothold. In addition, cubic B-spline curves are utilized to smooth the climbing trajectories.

Theoretical analysis

Description of a BiPCR in a truss

In order to completely describe a BiPCR in a truss, we need to specify not only the position on a pole where the base gripper grasps, but also the configuration the robot achieves. Hence, two homogeneous transformation matrixes are needed—one is to locate the grasping base in the world frame (denoted as $_B^W T$), and another to indicate the swinging gripper (end-effector) with respect to the base frame of the robot (denoted as $_E^B T$), as shown in Fig. 1.

A conventional configuration description with homogeneous transformation matrix can be expressed as

$$_G^W T = \begin{bmatrix} _G^W R & _G^W p \\ 0 & 1 \end{bmatrix} \tag{1}$$

where $\{G\}$ represents the base frame attaching to the grasping gripper of the robot, $_G^W R$ and $_G^W p$ represent the rotation matrix and translation vector with respect to the world frame $\{W\}$, respectively.

On the one hand, suppose that a pole is described in the world frame by the parametric equation as

$$^W p = {}^W p_0 + t \cdot {}^W d, \quad 0 \le t \le L \tag{2}$$

where $^W p_0$, $^W d$ and L are the reference point, the unit direction vector and the length of the pole, respectively. The gripping position on the pole must thus satisfy $_G^W p \in \{^W p\}$.

On the other hand, referring to Fig. 1, using notation (α, β, γ) of $Z-Y-X$ Euler angles, the orientation of a grasp can be calculated as

$$\begin{cases} _G^W R = R_Z(\alpha) R_Y(\beta) R_X(\gamma) \\ \alpha = \arctan(n_y/n_x) \\ \beta = -\arctan(n_z/\sqrt{1 - n_z{}^2}) \end{cases} \tag{3}$$

where γ denotes the grasping direction, which restrains the rotation around the pole, $n = [n_x \ n_y \ n_z]^T$ is the unit direction vector of the pole to be grasped as shown in Fig. 1.

Problem statement

In a single climbing step, collision-free motion planning involves three adjacent footholds, one of which determines the grasping configuration of the base gripper and the other two are the initial and the target configurations

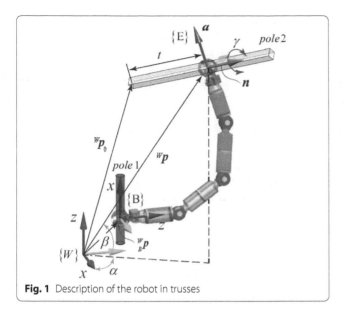

Fig. 1 Description of the robot in trusses

of the swinging gripper.[1] A feasible and collision-free trajectory is to be found between the two footholds for the swinging gripper. The problem can be described as follows.

Suppose a BiPCR grasping on a pole with one of its grippers at $_B^W T_i$, the aim is to find the feasible trajectory for the swinging gripper moving from the initial foothold $_E^W T_{i-1}$ to the target one $_E^W T_{i+1}$. There should not be any collision between the robot and the climbing environment (the truss). Let $\tau: [0,1]$ denote the trajectory and $q \in R^n$ denote the joint angles $(q_0 \ldots q_n)$ of the robot, the single-step collision-free trajectory planning of the BiPCR can be modeled as

$$
\begin{cases}
q_{\text{init}} = IK\left(_B^W T_i^{-1}\, _E^W T_{i-1}\right) \\
q_{\text{goal}} = IK\left(_B^W T_i^{-1}\, _E^W T_{i+1}\right) \\
\tau(0) = q_{\text{init}} \\
\tau(1) = q_{\text{goal}} \\
\tau : [0,1] \rightarrow C_{\text{free}}
\end{cases}
\tag{4}
$$

where C_{free} refers to the collision-free configuration space and $IK()$ represents the inverse kinematics.

Pre- and post-process for easy trajectory planning

Since the grippers of a BiPCR are usually designed to grasp objects using two fingers with V-shaped grooves, the initial and target configurations of the swinging gripper in one

climbing step are constrained with respect to the poles. As a result, the directions of the swinging gripper at the beginning and end of the climbing motion are restricted to be perpendicular to the corresponding pole. To improve the efficiency of the sampling-based algorithm, we propose a pre-process and a post-process to guide the swinging gripper to leave from the initial grasping configuration and approach the target configuration. To this end, a translation matrix is defined with respect to the gripper frame $\{E\}$ as

$$
_{P'}^E T = \begin{bmatrix} 1 & 0 & 0 & \Delta x \\ 0 & 1 & 0 & 0 \\ 0 & 0 & 1 & \Delta z \\ 0 & 0 & 0 & 1 \end{bmatrix}
\tag{5}
$$

where Δz stands for the offset along the Z axis of $\{E\}$ and Δx for the translation along the pole (X axis of $\{E\}$). $_{P'}^E T$ thus defines a new configuration of the swinging gripper with constant orientation.

With the translation matrix $_{P'}^E T$, we get a new homogeneous transformation in the pre-process describing the swinging gripper of the robot with respect to base frame $\{B\}$ as

$$
_{P'}^B T = _E^B T\, _{P'}^E T
\tag{6}
$$

And the corresponding joint angles can be found as

$$
q'_{\text{init}} = IK\left(_{P'}^B T\right)
\tag{7}
$$

In the similar manner, a new goal configuration q'_{goal} can be obtained. To ensure no collision occurs during the translation, collision detection should be conducted. After the pre-process, the original trajectory planning from P to Q is transformed to that from P' to Q', as shown in Fig. 2. In the post-process, the translational motion from P to P' and from Q' to Q is added to the planning output in turn to form a complete trajectory from P to Q.

The pre- and post-processes bring several benefits to the motion planning of a BiPCR including (1) expediting the convergence of the searching procedure with sampling-based algorithms, (2) simplifying collision check, no need to distinguish the grasped poles and the obstacle poles and (3) easy integration of collision-free trajectory planning and autonomous alignment of the gripper.

Utilization of the reachable workspace

In this paper, the reachable workspace (as shown in Fig. 3) of a BiPCR is considered to simplify the planning problem. It is clear that only those poles inside the reachable workspace, rather than the whole truss, should be considered as the target or obstacle poles in the planning. Therefore, the reachable workspace contributes to filter the poles in order to accelerate the collision detection. Moreover, it is also utilized to define the sampling area.

---

[1] The foothold of the base gripper may be at the end or in the middle of the three footholds, depending on the climbing gait—if the inchworm gait is used in the climbing step, the foothold of the base gripper is at the end, since the (front and rear) order of the two grippers are not changed; otherwise, if the turning-around gait or the flipping-over gait is employed, it is in the middle of the three footholds, since the order of the two grippers interchange [4].

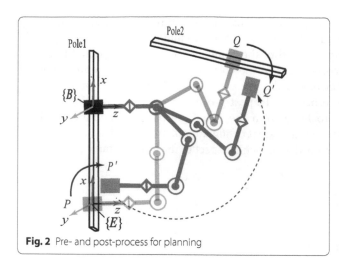

Fig. 2 Pre- and post-process for planning

Without loss of generality, taking the Climbot-5D for example, its reachable workspace can be described in polar coordinates with respect to the base frame {B}, as shown in Fig. 3,

$$
\begin{cases}
x = \rho \sin \theta \cos \varphi \\
y = \rho \sin \theta \sin \varphi \\
z = Z_0 + \rho \cos \theta
\end{cases}
\tag{8}
$$

where θ and φ are two parameters, Z_0 represents the offset along the Z axis of frame {B}, and ρ represents the radius of the workspace depending on the angle θ,

$$
\begin{cases}
\rho = l_{234} = \sum_{i=2}^{4} l_i, & \theta \in [0, \theta_{\lim}] \\
\rho = \sqrt{\rho_x{}^2 + \rho_y{}^2 + \rho_z{}^2}, & \theta \in (\theta_{\lim}, \pi]
\end{cases}
\tag{9}
$$

where l_i represents the length of the i-th link of Climbot-5D, $\theta_{\lim}$ represents the rotation angle limit of the T-type joint modules, ρ_x, ρ_y and ρ_z are obtained as

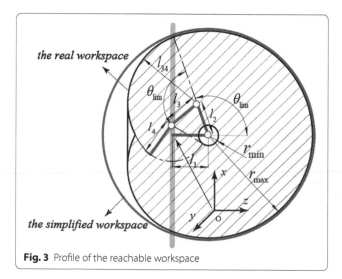

Fig. 3 Profile of the reachable workspace

$$
\begin{cases}
\rho_x = [l_2 \sin \theta_{\lim} + l_{34} \cos(\alpha + \theta - \pi/2)] \cos \varphi \\
\rho_y = [l_2 \sin \theta_{\lim} + l_{34} \cos(\alpha + \theta - \pi/2)] \sin \varphi \\
\rho_z = -l_2 \cos \theta_{\lim} + l_{34} \sin(\alpha + \theta - \pi/2)
\end{cases}
\tag{10}
$$

where $l_{34} = l_3 + l_4$ and α is an intermediate variable defined as

$$
\alpha = \arcsin \frac{l_2 \sin(\theta - \theta_{\lim})}{l_{34}}
\tag{11}
$$

The complicated mathematic expression of the robot's reachable workspace is inconvenient for application. Hence, the workspace is simplified here as a sphere with radius $\rho = l_{234}$ and center at $(0, 0, l_1)$ in the base frame {B}, as shown in Fig. 3.

Collision detection

Since the truss poles and Climbot links are cylindrical, collision detection is easily carried out through the calculation of the minimum distance between two line segments (the axes of a pole and a robotic link), which is divided into three steps as follows.

Step 1:   describing the poles and the BiPCR in the world frame {W}. Collision check can be conducted only under the condition that the robot and the obstacles are expressed in the same coordination frame. Without loss of generality, supposing "A" is an arbitrary point of the robot, its position can be calculated by the forward kinematics with respect to frame {B} as ${}^{B}\boldsymbol{p}_A$, then transformed to {W} by ${}^{W}\boldsymbol{p}_A = {}^{W}_{B}\boldsymbol{T}{}^{B}\boldsymbol{p}_A$.

Step 2:   finding the pole segments within the simplified reachable workspace of the robot. The algorithm to calculate the line segment inside a sphere can be found in [12]. This intersecting segment can be described by two points with parameters t_1 and t_2 respectively, having the form as

$$
\boldsymbol{p}_i = \boldsymbol{p}_0 + t_i \boldsymbol{d}
\tag{12}
$$

where $\boldsymbol{p}_0$ and $\boldsymbol{d}$ represent the reference point and the unit direction vector of a line segment.

Step 3:   computing the minimum distances between the remaining poles and links of the robot, and comparing with the threshold (the sum of radii of the pole and the robotic link). Collision is reported when the computed distance is less than the threshold; otherwise, there is no collision between the poles and the robot. The pseudo-codes of the algorithm are listed in Algorithm 1. In the algorithm, Seg2SegDist means the function calculating the Euclidean distance between two spatial line segments.

Algorithm 1 Collision check between two segment line

Input: l_i: the i-th link of the robot;
   $Truss$: truss segments filtered in Step2.
Output: true(collision), false(collision-free)
1: **for** $i = 1$ to 4 **do**
2:   $\delta = Seg2SegDist(l_i, Truss)$;
3:   **if** $\delta > (l_i.radium + Truss.radium)$ **then**
4:     **return true**;
5:   **end if**
6: **end for**
7: $\delta = Seg2SegDist(l_1, l_4)$;
8: **if** $\delta > (l_1.radium + l_4.radium)$ **then**
9:   **return true**;
10: **end if**
11: **return false**;

The motion planning algorithm

Bi-RRT algorithm

The RRT algorithm was first proposed by Lavalle [13] in 1998 and has been widely applied in the field of robotics since then. RRT-based algorithms may be classified into two categories: single directional and bidirectional RRTs (single-RRT and Bi-RRT). Considering the higher searching efficiency, we adopt the Bi-RRT algorithm in this paper, as shown in Fig. 4, with the pseudo-codes listed in Algorithm 2.

Algorithm 2 Finding a collision-free path for a BiPCR to transit from one pole to another.

Input: q_{start} : the initial configurations of the robot;
   q_{goal} : the goal configurations of the robot.
Output: Path: a feasible path.
1: $T_a.root \leftarrow q_{start}$; $T_b.root \leftarrow q_{goal}$;
2: **while** $time < t_{max}$ **do**
3:   $q_{rand} \leftarrow RandomState()$;
4:   **if** $Connect(T_a, q_{rand}, q_{new1}) = Advanced$ **then**
5:     **if** $Connect(T_b, q_{new1}, q_{new2}) = Advanced$ **then**
6:       **if** $Metric(q_{new1}, q_{new2}) < \varepsilon$ **then**
7:         $Path \leftarrow TracePath(T_a, T_b)$;
8:         $SmoothPath(Path)$;
9:         **return true**;
10:        $SWAP(T_a, T_b)$;
11:      **end if**
12:    **end if**
13: **end if**
14: **end while**
15: **return false**;

Constraints on grasping orientation

It is well known that a manipulator with six DoFs may reach arbitrary configuration in its workspace. The configuration of a 6-DoF BiPCR can be described by a 3-D position and a 4-D unit quaternion, similar to that of an industrial robot.

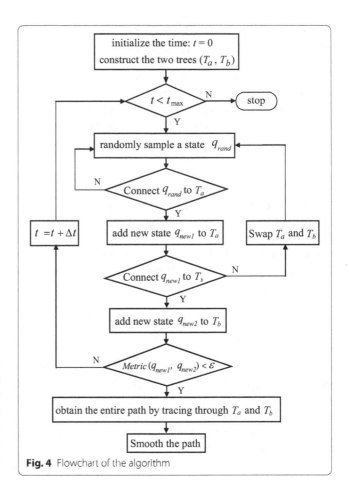

Fig. 4 Flowchart of the algorithm

Unfortunately, a 5-DoF BiPCR is unable to reach arbitrary orientation. As a result, if the Euler angles or quaternions are utilized to describe its orientation and to interpolate, the reachability of a desired configuration cannot be guaranteed. In other words, the computed configuration of the robot may not be accurate when the inverse kinematics presented in [4] is used directly.

Due to the special kinematic structure of Climbot-5D, the robot's links are always restricted in a plane (referred as "robot plane," the shaded triangle area in Fig. 5). As a consequence, the following constraints on orientation must be satisfied

$$\begin{cases} \boldsymbol{a} \cdot \boldsymbol{n} = 0 \\ \boldsymbol{a} \cdot \boldsymbol{m} = 0 \\ ||\boldsymbol{a}|| = 1 \\ \tan \beta = p_y/p_x \end{cases} \qquad (13)$$

where $\boldsymbol{m} = [-\sin \beta \quad \cos \beta \quad 0]^T$ represents the normal vector of the robot plane and p_x, p_y represent the position projection components with respect to the base frame $\{B\}$. Therefore, the grasping orientation $_E^B\boldsymbol{R}$ can be calculated through Eq. (13) with the known grasping position. In other words, once the grasping position and the direction of a pole (the vector $\boldsymbol{n}$) are given, the orientation of Climbot's swinging gripper is determined uniquely.

As a result, we may specify the configuration of a 5-DoF BiPCR using a position vector and a direction vector (six dimensions in total) and describe that of a 6-DoF BiPCR using a position vector and a unit quaternion (seven dimensions in total).

Random sampling

In order to guarantee the uniform distribution of the sampling, and to take into account the multiple gaits of a BiPCR, we sample the configuration of a BiPCR in the workspace with respect to the base frame.

On the one hand, recalling Fig. 3, the position is sampled in the robot's reachable workspace as

$$\boldsymbol{p} = \boldsymbol{p}_{\min} + r(\boldsymbol{p}_{\max} - \boldsymbol{p}_{\min}), \qquad (14)$$

where $\boldsymbol{p} \in \boldsymbol{R}^3$, $r \in [0, 1]$, $\boldsymbol{p}_{\max} = l_{234}[1 \ \ 1 \ \ 1]^T$ and $\boldsymbol{p}_{\min} = -l_{234}[1 \ \ 1 \ \ 1]^T$, ensuring that

$$r_{\min} < \| \boldsymbol{p} \| < r_{\max}, \qquad (15)$$

where $r_{\min}$ and $r_{\max}$ represent the radii of the inner inaccessible sphere and the outer reachable sphere, respectively. Considering the center of the reachable workspace has an offset to the origin of the base frame, the sampled position should be finally moved by

$$\boldsymbol{p}' = \boldsymbol{p} + [0 \ \ 0 \ \ l_1]^T. \qquad (16)$$

On the other hand, since we use vectors with different dimensions to describe the orientation of the swinging grippers of 5-DoF and 6-DoF BiPCRs, two methods are utilized to sample the orientation component, respectively. For a 6-DoF BiPCR, a simple sampling algorithm in SO(3) performs well in sampling unit quaternion [14]. For a 5-DoF BiPCR, we need to sample the direction of a virtual pole (the $\boldsymbol{n}$ component) and then calculate the grasping orientation by Eq. (13). To this end, the HEALPix algorithm [15] is employed to generate two angular parameters (θ, φ) in spherical coordinates, which is then transformed to a 3-D directional vector by $\boldsymbol{n} = [\cos\theta \ \ \sin\theta\cos\varphi \ \ \sin\theta\sin\varphi]^T$.

So far, through sampling we have achieved a 6-D random state (a 3-D position and a 3-D direction vector) for a 5-DoF BiPCR and a 7-D random state (a 3-D position and a 4-D unit quaternion) for a 6-DoF BiPCR, respectively.

Distance metric

The distance metric is very important for sampling-based algorithms. The most simple and commonly used metric can be defined as

$$\rho(\boldsymbol{q}_0, \boldsymbol{q}_1) = \omega_p \|\boldsymbol{p}_0 - \boldsymbol{p}_1\| + \omega_r f(\boldsymbol{R}_0, \boldsymbol{R}_1), \qquad (17)$$

where $\boldsymbol{p}$ and $\boldsymbol{R}$ indicate the position and the rotation components of the configuration, respectively, ω_p and ω_r represent their weighting scales, $\| \ \|$ means the standard Euclidean norm in 3-D and $f()$ stands for the measurement between two orientation matrices.

Considering that the inner product of two quaternions or vectors indicates the difference between them, one option to define the function $f()$ is

$$\psi = \arccos(\ \mathrm{dot}\ (\boldsymbol{R}_0, \boldsymbol{R}_1)), \qquad (18)$$

where $\boldsymbol{R}_0$ and $\boldsymbol{R}_1$ are quaternions or 3-D vectors.

Since the value of the rotation "distance" is limited to be less than π, we can specify the rotation distance weight as $\omega_r = 1/\pi$ to normalize the orientation distance. Correspondingly, the weight for position distance can be set as $\omega_p = 1/(2l_{234})$. Hence, the configuration distance is limited in $[0, 1]$ by

$$\rho(\boldsymbol{q}_0, \boldsymbol{q}_1) = \frac{\omega_p}{2}\|\boldsymbol{p}_0 - \boldsymbol{p}_1\| + \frac{\omega_r}{2}\arccos(\ \mathrm{dot}\ (\boldsymbol{R}_0, \boldsymbol{R}_1)).$$

Configuration interpolation

When interpolating between two configurations, it is usually separated into two parts corresponding to the position and orientation components. For the position component, a simple linear interpolation is suitable. As for the orientation component, it depends on the inner product of the two unit quaternions or 3-D vectors. If the orientations are close enough (their inner product is bigger than the pre-defined threshold), the linear interpolation algorithm is applied. Otherwise, the spherical linear interpolation algorithm is carried out, which is able to ensure the smooth interpolation between two configurations along geodesics.

Motion smoothing

Sampling-based planning may sometimes generate jerky and unnatural trajectories whose first derivatives are not continuous [16], which results in non-smooth motion or vibration of the robot. Therefore, motion smoothing is necessary. We utilize cubic B-spline fitting in this paper considering its sufficient flexibility and high-order smoothness.

The smoothing algorithm is illustrated in Fig. 6. A linear shortcut of the original piecewise linear path is first carried out to obtain a shorter path [the dash line in (b)]. The vertices of the dash line are then taken as the control points, and a non-uniform cubic B-spline is constructed as the final path [the red solid curve in (c)].

Note that we set double coincidence points at the two ends of the piecewise linear path to ensure the cubic

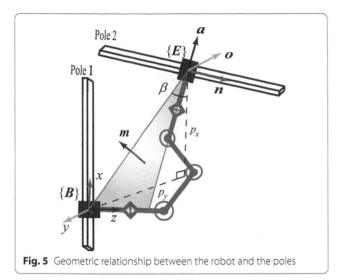

Fig. 5 Geometric relationship between the robot and the poles

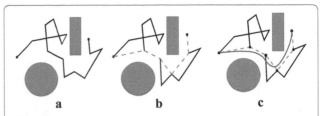

Fig. 6 Overview of the smoothing scheme: **a** the original jerky piecewise linear path; **b** path after the linear shortcut process; **c** path after fitting with the cubic B-spline

B-spline curve passing them exactly. We set also weights for each control point to adjust the shape of the cubic B-spline curve. The larger the weight is, the closer the curve gets to the control point. As discussed above, the path is composed of a series of configurations of the robot. While the position portions of the configurations are fitted with a cubic B-spline in 3-D space, the orientation portions are calculated by interpolating between every two adjacent configurations at the vertices of the shortcut path. In configuration check along of the path, the position portion may be changed by adjusting the weights of the control points of the cubic B-spline to satisfy the inverse kinematics and ensure collision avoidance.

Simulations and results

To verify the effectiveness of the theoretical analysis and the presented algorithms above, simulations are conducted in this section. The trusses are composed of cylindrical and squared poles with a diameter of 60 mm in arbitrary orientation in 3-D space. Both Climbot-5D and Climbot-6D are employed to test the proposed algorithm.

The step length for state verification in the algorithm is set to 40 mm, less than the diameters of poles, to make sure that the robot will not across a pole. The maximum node number of the two RRT trees is set to 500, and the goal-bias sampling probability is set to 0.05. Figures 7 and 8 show the simulation results with Climbot-5D and Climbot-6D, respectively.

The simulations are conducted 50 times in the same truss. A comparison between the simulations with the two BiPCRs is shown in Table 1. It can be seen from the simulations that the Bi-RRT algorithm has excellent

performance on the motion planning of the BiPCRs. The result with Climbot-5D also demonstrates the effectiveness of the processing method for sampling, distance calculation and interpolation. In addition, the simulation with Climbot-6D consumes less time than that with Climbot-5D, and has a shorter path length, owing to better dexterity with more degrees of freedom.

Conclusions and future work

Autonomous climbing is an essential function to carry out high-rise tasks with BiPCRs. Collision-free motion planning of BiPCRs in spatial trusses is an open problem, which has been addressed in this paper as a fundamental step to autonomous planning of climbing motion. A sampling-based algorithm, Bi-RRT, has been ultilized for single-step collision-free trajectory planning for BiPCRs. With appropriate description of a BiPCR in a truss, climbing motion planning has been conducted in a manner similar to that of a manipulator. The constraint on grasping orientation and the basic operations such as sampling, configuration distance calculation and interpolation have been discussed to facilitate the application of RRT. To expedite the convergence of the Bi-RRT algorithm, pre-process and post-process have been presented to deal with leaving from the starting point (the initial grasp configuration) and approaching the goal point (the final grasping configuration) of the swinging gripper. A method to smooth the piecewise linear jerky trajectory generated by the Bi-RRT algorithm has been proposed by utilizing cubic B-spline curve fitting. Simulations have verified the effectiveness of the theoretical analysis and the presented algorithm. The algorithm is general and universal for motion planning of other robots including manipulators and biped wall-climbing robots.

In the future, the algorithm will be integrated into the robot's multi-layered planner for online climbing path and motion planning. And the dynamic constraints like the limit of joint velocity, acceleration and torque will be taken into account.

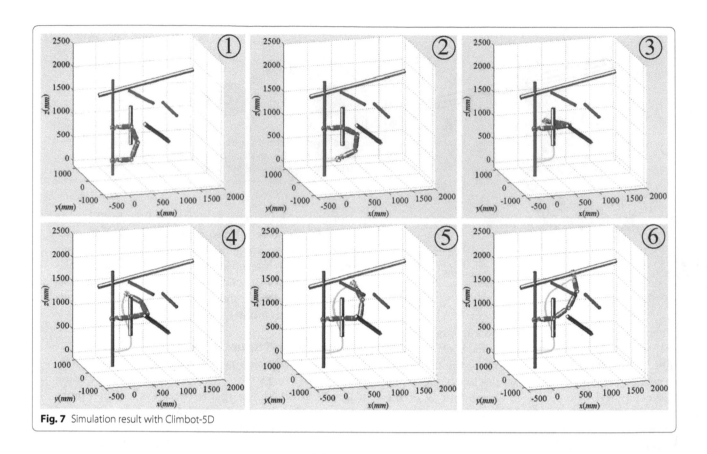

Fig. 7 Simulation result with Climbot-5D

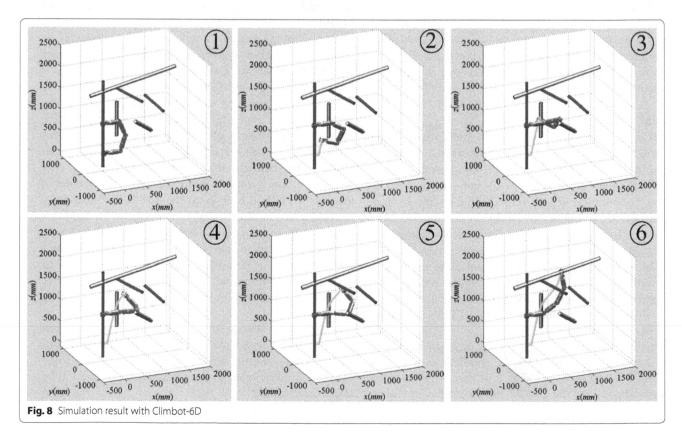

Fig. 8 Simulation result with Climbot-6D

Table 1 A comparison between the simulations

BiPCR	Time (s)	Iteration times	Tree nodes	Path length (m)
Climbot-5D	4.46	130	104	2.93
Climbot-6D	0.593	32	38	1.21

Author details
[1] Biomimetic and Intelligent Robotics Lab (BIRL), School of Electro-mechanical Engineering, Guangdong University of Technology, Hi-education Mega Center, Guangzhou 510006, China. [2] Department of Computing Science, University of Alberta, Edmonton, AB T6G 2E8, Canada.

Acknowledgements
The work in this paper has been supported by the NSFC-Guangdong Joint Fund (Grant No. U1401240), the Natural Science Foundation of Guangdong Province (Grant Nos. S2013020012797, 2015A030308011), the State International Science and Technology Cooperation Special Items (Grant No. 2015DFA11700), the Frontier and Key Technology Innovation Special Funds of Guangdong Province (Grant Nos. 2014B090919002, 2015B010917003).

Competing interests
The authors declare that they have no competing interests.

References
1. Baghani A, Ahmadabadi MN, Harati A. Kinematics modeling of a wheel-based pole climbing robot (ut-pcr). In: Proceedings of the IEEE international conference on robotics and automation. 2005. p. 2111–16.
2. Chung WK, Xu Y. A hybrid pole climbing and manipulating robot with minimum dofs for construction and service applications. Ind Robot: Int J. 2005;32(2):171–8.
3. Yoon Y, Rus D. A robot that climbs 3d trusses. In: Proceedings of the IEEE international conference on robotics and automation. 2007. p. 4071–76.
4. Guan Y, Jiang L, Zhu H, et al. Climbot: a modular bioinspired biped climbing robot. In: Proceedings of the IEEE international conference on intelligent robots and systems. 2011. p. 1473–78.
5. Tavakoli M, Marjovi A, Marques L, et al. 3dclimber: a climbing robot for inspection of 3d human made structures. In: IEEE/RSJ international conference on intelligent robots and systems. 2008. p. 4130–35.
6. Balaguer C, Gimenez A, Pastor JM, et al. A climbing autonomous robot for inspection applications in 3d complex environments. Ind Robot. 2000;18(3):287–97.
7. Chung WK, Xu Y. Minimum energy demand locomotion on space station. J Robot. 2013;2013(1):1–15.
8. Khosla P, Volpe R. Superquadric artificial potential for obstacle avoidance and approach. In: Proceedings of the IEEE international conference on mechatronics and automation. 1988. p. 1778–84.
9. Kayraki LE, Svestka P, Latombe J-C, et al. Probabilistic roadmaps for path planning in high-dimensional configurations space. Proc IEEE Trans Robot Autom. 1996;12(4):566–80.
10. Lavalle SM, Kuffner JJ. Rapidly-exploring random trees: progress and prospects. In: Proceedings of algorithmic and computational robotics: new directions. 2001. p. 293–308.
11. Lovbjerg M, Rasmussen TK, Krink T. Hybrid particle swarm optimiser with breeding and subpopulations. In: Proceedings of genetic and evolutionary computation conference. 2001. p. 469–76.
12. Ericson C. Real time collision detection. Oxford: Morgan Kaufmann Publishers; 2004.
13. LaValle SM, Kuffner JJ Jr. Randomized kinodynamic planning. In: Proceedings of international conference on robotics and automation. 1999. p. 473–79.
14. LaValle SM. Planning algorithm. Cambridge: Cambridge University Press, Cambridge University; 2006.
15. Go'rski KM, Hivon E, Banday AJ, et al. Healpix: a framework for high-resolution discretization and fast analysis of data distributed on the sphere. Astrophys J. 2005;622(2):759–771
16. Hauser K, Ng-Thow-Hing V. Fast smoothing of manipulator trajectories using optimal bounded-acceleration shortcuts. In: IEEE international conference on robotics and automation. 2010. p. 2493–98.

2

MicROS-drt: supporting real-time and scalable data distribution in distributed robotic systems

Bo Ding*, Huaimin Wang, Zedong Fan, Pengfei Zhang and Hui Liu

Abstract

A primary requirement in distributed robotic software systems is the dissemination of data to all interested collaborative entities in a timely and scalable manner. However, providing such a service in a highly dynamic and resource-limited robotic environment is a challenging task, and existing robot software infrastructure has limitations in this aspect. This paper presents a novel robot software infrastructure, micROS-drt, which supports real-time and scalable data distribution. The solution is based on a loosely coupled data publish-subscribe model with the ability to support various time-related constraints. And to realize this model, a mature data distribution standard, the data distribution service for real-time systems (DDS), is adopted as the foundation of the transport layer of this software infrastructure. By elaborately adapting and encapsulating the capability of the underlying DDS middleware, micROS-drt can meet the requirement of real-time and scalable data distribution in distributed robotic systems. Evaluation results in terms of scalability, latency jitter and transport priority as well as the experiment on real robots validate the effectiveness of this work.

Keywords: Real-time data distribution, Robot software infrastructure, Distributed computing

Introduction

Consider the following robot-assisted urban search and rescue (USAR) [1] scenario. A large-scale region should be explored after an earthquake to analyze the disaster situation and localize the victims. A team composed of several human operators, and a group of rescue robots is sent to execute this task. As shown in Fig. 1, those robots are tele-operated by the human operators through a wireless network with limited bandwidth. The human operators monitor the behavior of the robots and the video captured by them, analyze the collected data and guide the actions of the robots remotely.

This scenario involves a distributed robotic system in which various kinds of data have to be shared among participants, such as the video captured by the robots and the control commands issued by the operators. It implies some requirements to the data delivery service.

Firstly, the data distribution process should be decoupled between senders and receivers because the computing environment is highly dynamic. Secondly, multiple data streams are found in this scenario, and different data streams may have different time-related constraints. For instance, the control commands to the robots issued by the human operators should be delivered in a high priority, and the video captured by the robots can be delayed or even dropped if the network capacity is not adequate. Moreover, the data distribution scalability in terms of processing and network bandwidth overhead is a crucial issue since there are a lot of participants and the network resources are limited.

The above-mentioned requirements illustrate the concern of this paper, that is, the real-time and scalable data distribution in distributed robotic systems. Here, the term "real time" means disseminating data along with specific time-related constraints [2], such as predictable latency or a certain priority in data distribution. And the term "scalable" lays emphasis on the efficiency of this process, especially when there are a lot of participants.

*Correspondence: dingbo@nudt.edu.cn
College of Computer, National University of Defense Technology, Changsha, Hunan, China

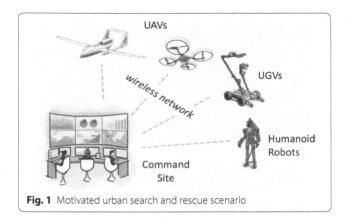

Fig. 1 Motivated urban search and rescue scenario

Realizing those two goals in a distributed robotic system is usually highly dynamic and resource limited.

Data distribution is a basic research topic in distributed computing, and there has been much progress in real-time and scalable data distribution in the traditional distributed computing systems [3]. However, little prior work has been done to adapt them into robotic settings, especially to integrate them into robotic software infrastructure. As shown in the detailed presentation of related work in "Related work" section, most of the widely accepted robot software infrastructure, such as the robot operating system (ROS) [4] and the Open Robot Control Software (OROCOS) [5], does not support real-time data distribution across computing nodes yet. This has been a serious impediment to the development and further application of distributed robotic systems.

This paper presents the design and implementation of micROS-drt, a software infrastructure that supports real-time and scalable data distribution in distributed robotic systems. A loosely coupled data publish-subscribe model for robots is proposed firstly. In this model, two kinds of message topics are defined: *general topics* without real-time assurance and *real-time topics* which support the fine definition of transport priority, latency budget, time-based filter and other real-time parameters. To reify this model, a mature data distribution standard, Object Management Group (OMG)'s data distribution service for real-time systems (DDS) [6], is adopted as the foundation of the transport layer of micROS-drt. We adapt and encapsulate the capability of the underlying DDS middleware to meet the requirements in robotic data distribution we mentioned earlier. Evaluation results in terms of scalability, latency jitter and transport priority on a test bed, as well as the experiment on real robots, show that micROS-drt can disseminate data scalably with real-time constraints in distributed robotic systems.

The remainder of this paper is organized as follows: "Research background" section introduces the research background. "Real-time data distribution model

for robots" section proposes a real-time data distribution model for robots. "MicROS-drt: architecture and implementation" section introduces the architecture of micROS-drt, as well as highlights some implementation details. "Experiments and evaluation" section focuses on the experiments on the test bed and real robots. "Related work" section presents related work.

Background

This section discusses the requirements of distributed real-time computing in robotic settings firstly and introduces two software entities highly related to our work (i.e., DDS and ROS).

Robotic distributed real-time computing

A real-time system is a system whose correctness depends not only on the logical correctness of the system but also on the time at which the results are produced [2]. Given that a robot is an autonomous agent that closely bound to the physical world, the time limit that exists in the physical space will be directly mapped to the robot software. Therefore, time-related constraints such as deadline or priority play significant roles in the logical correctness of a robot software system. The concept of robotic distributed real-time computing can be obtained by applying those real-time constraints to a complex robot made up of multiple computing nodes or a group of networked robots. The motivated scenario in "Introduction" section is a typical example of this concept.

The realization of the real-time assurance in a non-distributed environment mainly depends on appropriate scheduling of various local computing resources, such as the CPU or I/O devices. In contrast, its assurance in a distributed computing environment is much more complex. The dependencies among different processors and network resources should be considered, given that a task is accomplished by the collaboration of a set of nodes. In concrete, the realization of real-time properties in a distributed computing environment can be divided into three layers (Fig. 2):

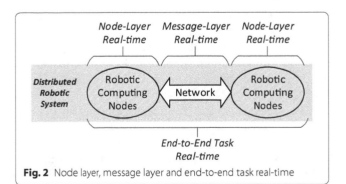

Fig. 2 Node layer, message layer and end-to-end task real-time

Layer 1 Node-level real-time is concerned with scheduling the local computing resources inside a single robotic computing node.

Layer 2 Message-level real-time is concerned with scheduling the resources on the network which connects the robotic computing nodes together. Its goal is to ensure the message-related real-time properties such as the latency or priority of data transportation.

Layer 3 End-to-end task real-time includes but not limited to the former two layers. In addition to those two layers, the real-time constraints such as task priority and deadline should be able to be propagated among all participant entities to avoid priority inversion [2].

In this paper, we mainly focus on Layer 2, that is, the timely dissemination of data among the nodes in the robotic distributed computing.

Data distribution service for real-time systems

DDS [6] is a message-oriented middleware standard proposed by OMG. It adopts a topic-based publish-subscribe communication model, in which the data producer (i.e., the publisher) does not send data directly to its consumers (i.e., the subscribers). Instead, the data are published into a channel with a specified "topic" name; all consumers who subscribe this topic receive the data without knowledge of who published them. The publisher and the subscriber are decoupled in terms of both time and space, which is suitable for highly dynamic environment such as robotic distributed computing systems.

In contrast to other message-oriented communication standards such as Advanced Message Queuing Protocol (AMQP) [7], DDS has the following two prominent features: (1) *Quality of Service (QoS) support.* DDS supports the fine control over various QoS parameters related to time constraints, including deadline, latency budgets, delivery order and priority and (2) *Scalability.* DDS adopts a peer-to-peer model with a fully decentralized architecture unlike many message-oriented middleware, which has a centralized broker. And the support of UDP/IP multicast also makes it scalable when the number of subscribers in a topic is more than one. Moreover, DDS is a mature and industrialized standard. Both commercial and open-source DDS middleware have widely been applied into many real production systems, such as aerospace, defense, industrial automation and cloud computing systems [3, 8].

The above-mentioned features are the reasons for our adoption of the DDS middleware as the foundation of micROS-drt transportation layer. The adoption of DDS middleware is the root of the scalability and the real-time capability of micROS-drt.

Robot operating system

ROS [4] is an open-source robot software infrastructure being maintained by the Open Source Robotics Foundation (OSRF). It is a meta-operating system, which means that it runs on top of the existing operating systems such as Ubuntu. By adding this additional software layer, the standardized ROS programming model in which the minimal software unit is named as "package" can be supported; thus, reuse in robot software development can be promoted. Another main feature of ROS is that it can support the distributed message publish-subscribe model (without real-time assurance), which facilitates the development of loosely coupled distributed robot software.

ROS has exerted considerable influence in the robot community, and a successful ROS-based software ecosystem has been developed. Thousands of reusable software packages and numerous tools necessary for robotic research have been accumulated on top of this platform. Thus, a major design consideration of micROS-drt is to maintain compatibility with the existing ROS packages.

Methods
Real-time data distribution model for robots

We propose a loosely coupled, topic-based data publish-subscribe model with real-time assurance as the foundation of our work. As shown in Fig. 3, there are two kinds of topics: *general topics* without real-time support and *real-time topics*. Unlike a general topic, various time-related QoS parameters can be specified in a real-time topic, and micROS-drt is expected to provide corresponding support while delivering data in this topic. Reviewing the motivated scenario enables the capture of the following real-time properties in robotic data distribution: (1) Different data streams may have different transport priorities or desired latency. (2) Since the network resource is limited, we should provide means to avoid network congestion, such as automatically dropping the out-of-date messages or time-based message

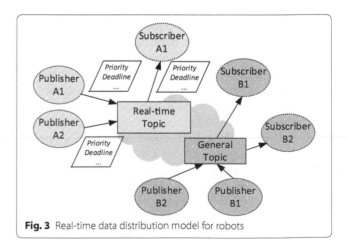

Fig. 3 Real-time data distribution model for robots

filtering. Therefore, four kinds of time-related QoS parameters are supported in our model:

Transport priority/latency budget

By specifying the message transport priority or its latency budget, the transport layer of micROS-drt can schedule its network resources accordingly and decide which one should be sent firstly. Transport priority/latency budget is the most useful real-time parameter in many real-time distributed robotic systems.

Message auto-discarding

A message can have a lifespan (i.e., valid period). It can be automatically discarded by micROS-drt when this time period has expired and the subscriber has not received it. This feature reduces the network traffic under certain circumstances. For instance, the out-of-date video frames can be automatically dropped when a network congestion takes place in the motivated scenario.

Time-based message filtering

The subscriber can specify a desired time interval of the message arriving in a topic by introducing a time-based filter. The messages that arrive ahead of the time interval are dropped to avoid wasting memory and computing resources. This filtering is useful when the software needs to handle periodic messages, such as the data from a specific sensor.

Data transport reliability

When selecting the best-effort transport means instead of the reliable one, data are delivered without arrival checks and lost data on the network are not re-transmitted. This parameter is useful when the network resource is limited and the data reliability is not a major concern.

MicROS-drt: architecture and implementation

This section provides an overview of micROS-drt, which includes its major design considerations and high-level architecture. This section also highlights some implementation details.

Design considerations

As we have stated, micROS-drt is designed to be a robot software infrastructure that supports real-time and scalable data distribution. A number of design considerations are accounted for as described below:

1. *Open source.* By adopting the open-source paradigm, we can take advantage of the community's force to improve our work and contribute to the community effectively as well. The existing achievements in the open-source society can also be reused with appropriate copyright license.

2. *Usability.* DDS has shown its potential in real-time data distribution. However, an easy target for blame is its complex, hard-to-use APIs. A design goal of micROS-drt is to seek an appropriate trade-off between usability and flexibility. "Real-time data distribution APIs" section offers more details on this design consideration.

3. *Compatibility.* As mentioned in "Robot operating system" section, there have been thousands of reusable software packages on top of the ROS platform. Keeping compatibility with these packages is a major concern in designing and realizing micROS-drt.

Architecture of micROS-drt

The architecture of micROS-drt is shown in Fig. 4. The top layer in this architecture is *the API Layer*, which provides the application programming interfaces to the robot applications. In corresponding to the two kinds of topics in the data distribution model, there are two kinds of APIs: the general data pub/sub APIs and the real-time QoS-enabled APIs. *The model layer* and *the message layer* are responsible for maintaining the data distribution model and marshaling/demarshaling the messages, respectively. The bottom layer is *the transport layer*, which consists of the negotiable transport protocol framework and a set of concrete protocols. The most important protocol is the DDS one, whose realization is based on a piece of DDS middleware.

The realization of micROS-drt is based on two open-source software entities: ROS and the DDS-compliant middleware. The topic layer and the message layer in Fig. 4 are an enhancement of the corresponding component of ROS, in which the support of the real-time topics is added. The transport layer is a mixture of the DDS Abstraction and Bridge module, the open-source DDS middleware and the existing ROS transport protocols. Theoretically, any realization of the DDS standard can be used. Currently, micROS-drt support OpenDDS (www.opendds.org) and OpenSplice DDS (www.opensplice.org).

Real-time data distribution APIs

The DDS standard, which has been denounced by the users, has hard-to-use APIs. It introduces more than 20 policies to define the QoS of the data distribution process. Furthermore, different policy combinations result in different effects. To avoid confusing the users of micROS-drt with those complex policies, micROS-drt chooses not to directly expose the DDS APIs directly and introduces a more practical and simplified real-time data distribution model (cf. Sect. "Real-time data distribution model for robots") instead. The QoS-enabled APIs of this model are an extension of the APIs in ROS, which have been widely accepted by the robotic community. The real-time

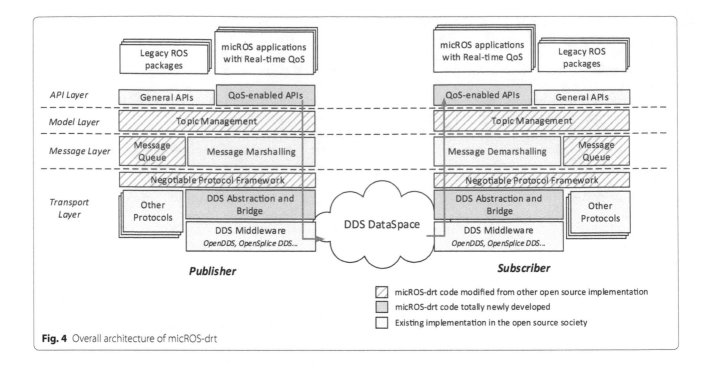

Fig. 4 Overall architecture of micROS-drt

parameters are classified into two kinds: The parameters should be set at the publisher side and the ones should be set at the subscriber side. The users can specify the parameters independently in the ROS-style APIs on both sides or just use their default values while advertising or subscribing a topic.

Table 1 presents an example of the real-time data distribution APIs in micROS-drt. It is the function to advertise a topic with QoS at the publisher side. Basically, it just adds some time-related parameters to the ROS's original *advertise()* function. In the newly introduced *advertise_qos_ops* structure, all time-related QoS parameters supported by our real-time model can be selectively specified. For example, if we want to drop out-of-date video frames automatically, we can specify the *msg_valid_time* field of this structure to an appropriate value while advertising the video topic.

DDS abstraction and bridge
The implementation of the DDS protocol in micROS-drt mainly consists of two modules as shown in Fig. 4: the DDS abstraction/bridge module and the underlying

Table 1 An example of real-time API

Function name	Overloading parameters
AdevertiseWithQoS	(topic_name, queue_size, transport_priority)
	(topic_name, queue_size, latency_budget)
	(topic_name, queue_size, advertise_qos_ops)
	(advertise_ops, advertise_qos_ops)

DDS middleware. The former can be regarded as "glue" between other parts in micROS-drt and the DDS middleware. It consists of three sub-modules: (1) *DDS capability abstraction*, which encapsulates the capability of the underlying DDS middleware as well as manages the lifetime of all DDS-related resources; (2) *Data distribution model mapping*, which maps the micROS-drt data distribution model to the DDS data distribution model, including the mapping of both topics and QoS parameters; and (3) *Message tunneling*, which encapsulates the messages that are marshaled by the message layer into a DDS message at the publisher side and extracts it from the DDS message at the subscriber side.

A challenge in the design of the DDS abstraction and bridge module is the efficiency of message handling, especially on an onboard computer of a robot which may only have limited resources. To address this issue, we strictly constrain the capability of the data distribution model as a subset of the minimum profile of the DDS standard. Thus, the underlying DDS middleware can be tailored to fit the needs of micROS-drt.

Keeping compatibility with ROS
An important design consideration of micROS-drt is its compatibility with thousands of existing ROS packages. To realize this goal, as shown in "Real-time data distribution model for robots" section, two kinds of topics are strictly distinguished in our data distribution model, and the APIs of the general topic are identical with those in ROS. Moreover, the realization of existing ROS transport

protocols such as TCPROS is retained in micROS-drt. The ROS negotiable protocol framework is enhanced to support the negotiation of the newly introduced DDS protocol. At runtime, TCPROS or other available protocols are selected transparently for the general topics instead of the DDS one when the remote node (with an official ROS) does not support the DDS protocol.

Results and discussion

A set of experiments has been conducted to evaluate micROS-drt, which include both the experiments on a dedicated test bed composed of a group of servers and the experiments on a group of real distributed robots.

Experiments on test bed

The test bed comprises of a group of servers with Intel Xeon E5-2630v3 CPU, 8 GB RAM and Ubuntu 14.04. A 1000 Mbps LAN connects the servers together. The version of OpenDDS in our experiments is 3.1.6, the version of the OpenSplice DDS middleware is 6.4, and the version of ROS is Indigo. The experiments in this subsection mainly focus on the real-time capability of micROS-drt and the scalability of our work.

Throughput and scalability

In this experiment, we test the throughput of two configurations of micROS-drt (with OpenDDS and OpenSplice DDS, respectively) as well as the throughput of ROS on the test bed which has no real-time support. As shown in Fig. 5a, the "1 publisher to 1 subscriber" model is adopted (i.e., two robots are simulated), and the message length is varied from 100 B to 50 KB. The result shows that micROS-drt is slower than the official ROS to reach the limitation of the network bandwidth. It is a normal overhead incurred by both the extra real-time assurance code and the additional field in the network message.

Although micROS-drt has no advantage over ROS in terms of throughput, it has a significant advantage in terms of scalability. To evaluate it, the above throughput experiment is re-conducted with a "1 publisher to n subscribers" model in which n is a variable from 1 to 4 (i.e., the experiment simulated 5 robots in maximum). Each message is 20 KB in length. Figure 5b shows that the throughput of ROS drops drastically when the n increases since it does not support multicast and has to send n copies of the data. In contrast, the micROS-drt performance is more stable because the underlying DDS middleware supports the UDP/IP multicast. Therefore, only one copy of the data is delivered regardless of the number of subscribers.

Latency and jitter

Latency is an important measurement to evaluate the performance of data distribution. This experiment

evaluates the message round-trip latency in micROS-drt with OpenSplice DDS, micROS-drt with OpenDDS and ROS. Each message is 100 KB in length, and 5000 messages are sent continuously. As shown in Fig. 5c, although the average latency in the three configuration/products is about the same, the latency standard deviation is totally different. The ROS latency without real-time support has a large deviation, and the latencies in the two configurations of micROS-drt are both significantly low. The snapshot of the latencies of 1000 messages in Fig. 5d also shows this trend. It indicates that the behavior of micROS-drt is more predictable, which is an important property of the real-time software.

Transport priority

This experiment validates a major real-time capability of micROS-drt, that is, the message transport priority management. Two servers are connected by a virtual network over the 3G cellular network. Three topics are advertised with different priorities at the publisher side, and each of them sends messages with 5 KB length continuously. Sending many messages of this kind in a short time window results in network congestion, especially with a cellular network that has limited bandwidth. The experiment is terminated when 300 messages have been received at the subscriber side, and the arrived messages in each topic are counted. Figure 5e shows that with real-time assurance support, the topic with high priority delivers 173 messages and the low priority one delivers only 17 messages successfully. As a comparison, in the same experiment on ROS which has no transport priority support, the arrived messages are distributed evenly among those three topics.

Experiments on real robots

MicROS-drt is the software infrastructure of the distributed robotic research platform in our laboratory. Figure 6a shows a compact-size three-wheel robots in this platform, which we designed mostly based on the commercial off-the-shelf hardware. The onboard computer is an ODROID XU3, a credit card-size embedded development board with a Samsung 1.8GHz ARM-based CPU. Other parts on this robot including a mbed microcontroller, two brush-less motors with encoders, an highly integrated IMU, sonar and IR sensors, and an optional general or RGB-D camera.

To maximally exploit the science research and education potential of this robot, micROS-drt is ported to it, and a software stack is constructed for this robot mainly based on existing ROS packages. With thousands of existing ROS packages, distributed or swarm robotic experiment settings (e.g., multi-robot coverage path planning [9]) can be quickly constructed. Figure 6b shows

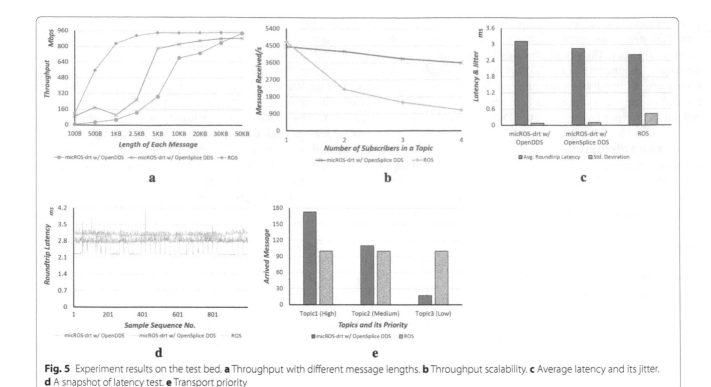

Fig. 5 Experiment results on the test bed. **a** Throughput with different message lengths. **b** Throughput scalability. **c** Average latency and its jitter. **d** A snapshot of latency test. **e** Transport priority

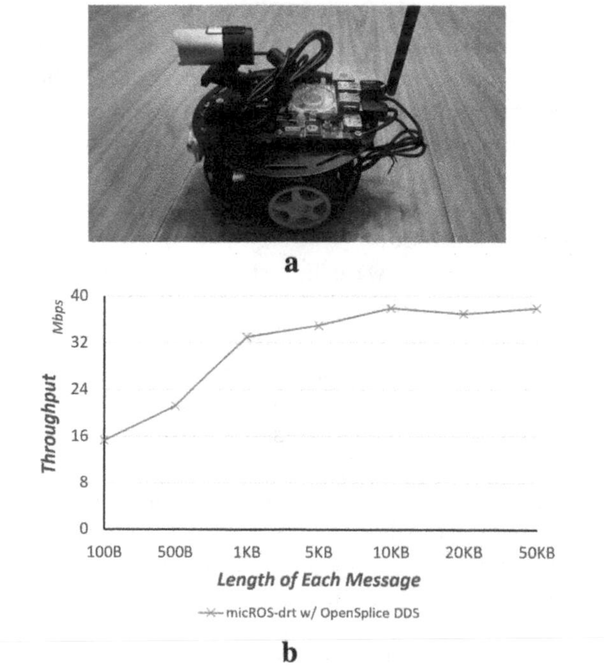

Fig. 6 Experiment on the real robots. **a** MicROS-drt test robot (ARM CPU and Wi-Fi network). **b** Throughput on the test robots

the throughput test results between two robots which are connected by an 802.11n Wi-Fi wireless network.

Related work

In the early days, many robotic software infrastructures adopt the real-time Common Object Broker Architecture (CORBA) [10] to support robot cooperation or teleoperation, such as the work in [11] and [12]. Real-time CORBA provides a solution on real-time distributed computing. However, it is based on a tightly coupled client/server model. In contrast, our work adopts a loosely coupled publish-subscribe model, which is more suitable for dynamic scenarios. Another similar work is the Dynamic Data eXchange (DDX) project [13], which enables runtime data sharing for distributed robotic systems. DDX is realized through an efficient shared memory mechanism. However, this centralized paradigm hinders the scalability of the data distribution process. In contrast, the DDS technology we adopt has a fully decentralized architecture.

The use of publish-subscribe paradigm in robotic data distribution has recently been given increasing attention. The message transfer mechanism in ROS [4] is a typical example, which is without real-time support. Attempts have been made to integrate DDS into robot software infrastructure. In [14], DDS has been adopted to improve the performance of RoboComp, a robot software framework that originally relies on the Internet Communications Engine (ICE). A QoS-enabled middleware, Nerve, intended for networked robots and based on DDS is introduced in [15]. Since ROS has been widely accepted by the robotic community, there are also some similar attempts. In [16], the ROS–DDS proxy is introduced to support the interaction of multiple robots. It validates the feasibility to make ROS and DDS work together. However, the proxy realization is message specific. In other words, a corresponding proxy has to be developed manually for each kind of message. In contrast, our solution is general for all messages, and the DDS middleware is fully transparent to the upper layer applications. Another undertaking effort is ROS 2.0 [17], the next big step of ROS, which is expected to be released in the near future. It also adopts DDS as its underlying transport means. However, according to the information that has been disclosed such as its preview APIs [18], the real-time assurance in distributed computing environment is not its major concern.

Conclusion

This paper presents micROS-drt, a robot software infrastructure, which supports real-time and scalable data distribution in distributed robotic computing. It is based on a loosely coupled data publish-subscribe model that supports both the topics with real-time QoS and the ones without real-time support. To reify this model, micROS-drt adopts a mature real-time data distribution standard, the OMG's DDS, as its underlying transport means. A set of experiments on both the dedicated test bed and the real robots has validated the effectiveness of its real-time assurance capability and its scalability. In our future work, we will enhance micROS-drt to support the real-time property propagation from network resources to local computing resources.

Authors' contributions
BD and HW designed the architecture of micROS-drt as well as participated in the implementation of this work. ZF, PZ and LH participated in the implementation of micROS-drt. All authors read and approved the final manuscript.

Acknowledgements
This work was supported by the special program for the applied basic research of National University of Defense Technology (No. ZDYYJCYJ20140601) and the National Natural Science Foundation of China (No. 61202117).

Competing interests
The authors declare that they have no competing interests.

References
1. Liu Y, Nejat G. Robotic urban search and rescue: a survey from the control perspective. J Intell Robot Syst. 2013;72(2):147–65.
2. Stankovic JA. Real-time computing. Byte. 1992;17(8):155–62.
3. Pérez H, Gutiérrez JJ. A survey on standards for real-time distribution middleware. ACM Comput Surv (CSUR). 2014;46(4):49.
4. Quigley M, Conley K, Gerkey B, et al. ROS: an open-source robot operating system. In: ICRA workshop on open source software, 2009.
5. Bruyninckx H. OROCOS: design and implementation of a robot control software framework. In: Proceedings of IEEE International Conference on Robotics and Automation, 2002.
6. Pardo-Castellote G. OMG data-distribution service: architectural overview. In: Proceedings of international conference on distributed computing systems, 2003.
7. O'Hara J. Toward a commodity enterprise middleware. Queue. 2007;5(4):48–55.
8. Hoffert J, Schmidt DC, Gokhale A. Adapting distributed real-time and embedded pub/sub middleware for cloud computing environments. In: Proceedings of ACM/IFIP/USENIX international conference on middleware, 2010.
9. Galceran E, Carreras M. A survey on coverage path planning for robotics. Robot Auton Syst. 2013;61(12):1258–76.
10. Fay-Wolfe V, DiPippo LC, Cooper G, et al. Real-time CORBA. IEEE Trans Parallel Distrib Syst. 2000;11(10):1073–89.
11. Song I, Karray F, Guedea F. A distributed real-time system framework design for multi-robot cooperative systems using real-time CORBA. In: Proceedings of IEEE international symposium on intelligent control, 2003.
12. Yoo J, Kim S, Hong S. The robot software communications architecture (RSCA): QoS-aware middleware for networked service robots. In: Proceedings of SICE-ICASE international joint conference, 2006.
13. Corke P, Sikka P, Roberts J, et al. DDX: a distributed software architecture for robotic systems. In: Proceedings of Australasian conference on robotics & automation, 2004.
14. Martínez J, Romero-Garcés A, Manso L, et al. Improving a robotics framework with real-time and high-performance features. In: Simulation, modeling, and programming for autonomous robots, Springer Berlin Heidelberg, 2010, p. 263–74.
15. Cruz JM, Romero-Garcés A, Rubio JPB, et al. A DDS-based middleware for quality-of-service and high-performance networked robotics. Concurr Comput: Pract Exp. 2012;24(16):1940–52.
16. Bich M, Hartanto R, Kasperski S, et al. Towards coordinated multirobot missions for lunar sample collection in an unknown environment. J Field Robot. 2014;31(1):35–74.
17. ROS 2.0. https://github.com/ros2.
18. Thomas D, Woodall W, Fernandez E. ROS 2.0: developer preview. In: ROS developer conference, 2014.

3

Microdesign using frictional, hooked, attachment mechanisms: a biomimetic study of natural attachment mechanisms

Bruce E. Saunders[*]

Abstract

Part 1 completed the studies of five long-shafted, cellulose, frictional, hooked probabilistic fasteners. Part 2 identified three substructures prevalent in the natural world for probabilistic fasteners and detailed the collection of voxel dataclouds while measuring from the natural fluorescence of their composing chitin and cellulose under the laser illumination of a confocal microscope. In this part 3, consideration is given to the development of a behaviour-optimised bioinspired probabilistic attachment system that is thermodynamically inert due to attachment substructures, such as interlocking setae, that act as arrestors and temporary interlocking devices. The three devices of part 2 are considered for their relative merits, and one part is modelled for a rapid prototyping device. If one is considering the question of shape versus material, then it is at this stage that it is a very important issue since one is considering fundamental, simple shapes and the materials used to form them are of finite variety. Hence, the final design will hinge upon design for manufacture and component material qualities, in this case copper.

Keywords: Microadhesion, Miniaturisation, Cellulose, Chitin, Hook, Self-assembly, Scaling effects, Biomimetic, Bioinspire, Probabilistic, Friction, Fields, *Arctium minus*, *Apis mellifera*, *Omocestus viridulus*, Biosensor, Dimensionless groups, Copper

Background

It is strange to think that one can find a biomimetic principle on demand. This study, however, has found at least one in the consideration of its objective, to take all known studies of hooks and to compare them and others in order to define a new hook that is advantageous to design for a purpose that precludes all known uses so far, i.e. that is intended for a use that has not so far been defined. However, it is such that it can be assumed that all the new designs are going to be on the purpose/possibility frontier, namely in this case, microdesign and the design of micron-ranged size structures that do not altogether behave in a "normal" way under use, such as the tarsal hooks of an insect and their manner of sticking or adhering to a surface. It is somewhat of a mystery yet but which can be considered and mimicked to find its own properties that reflect their sizes (the hooks).

It is through the design of these autonomous structures and their variants that microdesign such as those parts used by computer-aided medical devices are possible. But it is a mystery yet in the scientific literature, how these marvels are designed such that it can be studied. There are new ways of doing things that need chronicling so that their progress can be chartered and modified according to new discovery.

The title of the thesis proposal was "The Functional Ecology and Mechanical Properties of Biological Hooks in Nature". This tells us that each hook must be studied as a part of a system; hence, the complete set of mechanical properties will only be revealed once the hook has been brought into interaction with a substrate of some properties of its own. Attention is now drawn to the publication of a paper that is significant to this study, namely [1] that is recently published where in it is described how a process of electro-chemical deposition is used to draw the bead of the new substance, namely copper, to draw in the "cubic pixels" or voxels, that then make up the shape

*Correspondence: brucesaunders23@hotmail.co.uk
University of Bath, 54 Ballance Street, Bath, UK

required or drawn, or in this case, scanned on a confocal microscope as executed in paper 2 of this research upon three specimens which were all measured under the microscope [2]. This enables the process to continue, of evaluating the progress such that it can be concluded how to best design a hook made of copper at the size of 100 micron span say, manufacturing it in a macrofashion and predicting its behaviour in an absolute fashion.

Being considered here is a necessarily obscure manner of approaching the issue, to predict what will be the manner of making these structures and how the design will process the data of the new part such that it will behave in a fashion that will be predictable or not depending on the way it is fashioned or the material itself and its issues with solidarity and maintaining shape. How to draw a free-body diagram of such a structure, for instance, would be quite specific and yet very different from predicted Newtonian behaviour. There is a way to get all of the parts of the design into one sphere and that is to generate a curve that shows their behaviours under different variants, according to different criteria. It is not going to be easy to explain the use of dimensionless groups to produce fluid-flow performance curves, but these make the use of criteria that may come into the design such as Brownian motion or molarity of particles in self-assembly or electro-chemistry. Each will affect the solution of the entropy of the system but will also preclude the discovery of any new criteria unless they are included in the symmetry of the equation that makes all the forces equal at any one state or stage. Otherwise, we will have movement, and this is what we are trying to arrest with a probabilistic fastener. In short, all forces must be balanced.

Saunders [3] describes the discovery of an apparent tensile shape/size-scaling effect while completing a study of long-shafted cellulose hooks started by Gorb. The *Arctium minus* hook was found to be exceptionally strong, an effect answered by calculation as being an illustration of the hook's propensity to be strong through either composition or shape and therefore span and fibre content. This is of little importance since we are simply warned that there are these effects present in Nature.

There is however a consequence of our study which is that we have a selection of five long-shafted hooks of cellulose from which to choose and the most frequent and strongest, namely *A. minus* was chosen, as per George de Mestral and his Velcro. It is then that we undertook to scan the hook under the confocal microscope together with two other samples noted for their frequent occurrence in British wildlife, namely the tarsi of the British common wasp and grasshopper (*Apis mellifera* and *Omocestus viridulus*). This led to the conclusion that we had the route to three evolutionary pathways plotted on

our confocal microscope and preserved for further analysis or manipulation in the form of .tiff files [2]. Each is a permanent reusable fastener, and yet each does equip its owner without the need for a sample to identify it and make it from, i.e. they are without a pattern, possessing only a genetic code for growth and form.

Functional ecology

When considering the system of the hook-shaped structure, going back to the Cambrian Era, one is struck by the fact that there is no apparent control system. Natural Laws prevail. The first hook shapes appearing on the fossil record were of primitive cellulose and chitin.

When studying a biological specimen the prescribed ethos is to study its interaction within its system since it is the system that is of interest to a biologist and to a designer. The entire system needs to be considered, of a frictional fastener such as the probabilistic (i.e. non-random) fastener which the hooks *A. minus* form, and how the fastener shall be designed is based upon the performance of these hooks and their substrates. In the case of the growth and formation of hook shapes in Nature, their biochemistry will be of equal importance and must have been fundamental to the very first organisms that appeared with shapes of biomaterial.

It is possible to break up a system into discrete segments or sections, and this enables the identification of biodesign indicators such as scaling effects that predict the alignment of microfibrils for instance or the percentage volume required of a matrix cavity in a gel.

A free-body diagram is such an example and is selected and constructed along subjective lines for the purposes of force analysis. A free-body diagram can enclose an entire system or can be used to analyse part thereof.

Botanical hook structures and biodesign indicators for a reusable, frictional, silent, probabilistic attachment mechanism

What we need to consider is that we have passed step one, the choice of a cellulose representative of a single-hook fastener type, exposed on the end of a shaft where behaviour is more isolated from the base or surface from which it originates. This led us to a discussion above. There is then the choice of three, which has been presented [2]. Now it must be established which of the three is most suitable for pre-production analysis and study, and a cellulose single hook has been chosen. This is because

(a) It is a non-assembly.
(b) It is long and therefore accessible to the head of a 3-D printer for its overhang if necessary.

For this work there are special qualities that may be included in the design which may or may not increase the quality of the attachment such as the inclusion of setae-like protrusions from the shaft or hook itself. These might ease adhesion.

So it is a conclusion of the previous research that we must produce a sample hook in .stl data file format that can be used as a basis or datum structure and that can morph into variations according to need and application. It is then understood that all the variations will be available to research, as would be those of the other two samples which have yet to be deciphered onto Solidworks, but it may not be deemed necessary in the light of [1] where direct control of the data transfer is allowed, and thus, the treatment of the structures under testing will be available in copper only which is a very conducive, malleable substance. From a point of view of application, what needs be considered is the requirements of application of the attachment and what bioprinciple we can derive from this structural mimicry and that includes the obvious—a tactile manner of carrying a load up a direct incline of 90°. It also includes the obvious use of copper's varying impedance under stress which is used in strain gauges and can now be used in the biomedical sensor field.

It is with this in mind that all thought of concluding with a 3-D representation of all three when they are quite readily available through the microscope's own image-ware is not economical. Instead, a single sample is included that is to scale and leads to the following development of the design for the purposes of manufacture and testing. It is noted here that all the sundry tests have been carried out through the Solidworks software itself such as FEA, but we do not know the true directions from which forces are applied when considering analysis of the hook. A point load seems inappropriate as does a limited forcing being applied through the length of the shaft in sheer. It should be under pressure throughout, not as it seems here, as the cross-stresses are prevalent throughout the real-life loading of the hook.

The point of using Solidworks, in spite of its problems with an analogous material, is that it converses with a rapid prototyping device and produces a file in .stl format, but its FEA capabilities which are vastly insufficient.

Aim

To transfer the data from SEM to 3-D digitised form of an *A. minus* long-shafted hook, one of the three specimens selected and measured in [2] so that it can be used as a base design for the morphing of shape into variants for testing, in the understanding that the ultimate bioinspired principle being studied is microadhesion; therefore, the test specimen is at the limits of manufacturing capability. It is to be drawn and then analysed using the latest available software (circa 2004). All information pertaining to this subject area is to be noted.

Methods
Morphological recording: 2-D digitising

2-D digitising an object is, as it implies, limited by the effect of distortion by the planar image on the 3-D object. This makes it inaccurate, but for the dimensions being considered it was considered sufficient. In the pursuit of the .stl file and the form of 3-D computer graphics output that could be altered, the final conclusion of [2] was to make a direct reconstruction of a model in Solidworks. To gather the data 2-D digitising was applied. Thereafter, a model was constructed of a field of hooks that could be rapid prototyped and manipulated for testing purposes using the new method outlined in [1]. This is the only way of creating these structures on record. Each sample can now be rapid prototyped in copper with s suitable device and tested for its features as needed. This is for further research.

Using Fig. 1 it was possible to digitise two splines for the inner and outer profiles. Diameters were measured perpendicular to the inner spline and used to reconstruct the hook using the loft feature, the result of which is shown below in Fig. 2.

The hook and shaft are drawn in proportion, and it was observed at this point that given Nature's reputation for energy efficiency and balance, the shaft walls are near parallel, making it clear that they are hiding some secret with regards their form. And that is the inner form, the non-homogenous fibrous, cellular flesh of the hook and shaft and their inner control system or genetic instructions that led the flower, mauve and bright and well suited to the pollinators of the area to transform into the perfect seed dispersal mechanism using the same structures, though desiccated. The flattened flange is optional, but follows from the fact that there are further structural options available, such as flanges or setae, substructures that could play a part in anchoring more firmly in the substrate or providing a brake for adhesion and detachment, mimicking the attachment mechanism of the dragonfly head arrestor mechanism [4]. At this stage a universal foot is being considered for a one-size-fits-all attachment mechanism that can anchor to a wall or vertical flat surface irrespective of the size of the object, and then, it shall be scaled down according to need and ability to adhere and detach repeatedly.

Finite element analysis

Committing this image to FEA was in fact superfluous but analogous to the execution of the package. The material analogue is very unsuitable. But it illustrates one further point to be made about FEA, and that is it is wildly

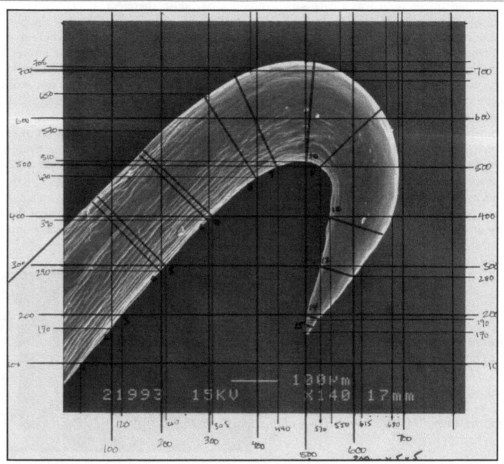

Fig. 1 Electron micrograph with grid superimposed, of an *A. minus* hook in profile. Each *bar* represents an interval of 100 microns. This is one of the three species studied in [2] and measured under a confocal microscope set for cellulose then chitin. Both materials fluoresced

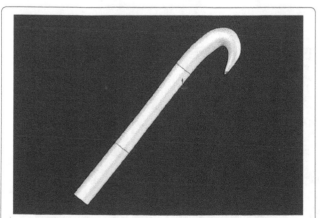

Fig. 2 3-D image of reconstructed *A. minus* hook. *Note* the material analogue is uniform and homogenous. Only the hook is an obvious stress concentration. Otherwise, there are no obvious indicators of conservation of material due to lack of applied stress. Hook span = 200 microns

unsuitable for composite modelling of any form. And further it is known that the cuticle surrounds the cellular content of the hook and this forms a sheaf much like a thin-walled pressure vessel. The shaft resembles a long thin tube, and it deforms under bending like one too. Therefore, the FEA in this package is not advantageous to analysing the cellulose behaviour. However, it is good for measuring the distortion under loading of a micron-sized copper hook, but one must get the loading correct.

Assembly of the probabilistic, frictional, multiuse, silent, attachment mechanism morphology for a universal foot based upon cladistic studies

This was the first thing that was considered: How to hold the detached figure of the hook between two solid handles to manipulate it when complete. It was decided that it was of no purpose whatsoever to make do with a size-limited sketch and that only the real size specimen and model would do, and therefore, it was a necessary feat to be able to image the specimens completely so as to be able to visualise their capabilities in a virtual reality medium or in a model for an attachment mechanism for manufacture. However, [1] has presented feat of engineering that must be used.

Figure 2 shows the configuration of a test specimen, for a single evolutionary path, that of the *A. minus*. There are two more specimens in part 2 [2] that give the same result, but have not been reproduced in Solidworks for the purposes of economy. These two own the second and third position in the ranking of all possible manufacture-able attachment mechanisms, second only by nature of their complexity, one being of two parts, each identical, and the third made up of three parts, two of which are identical. Again with chitin, one wonders whether a thin-walled pressure vessel calculation under bending would be more suitable than the FEA of a homogenous material that Solidworks allows for.

Results

The standard form of a result table is omitted here because it has yet to be performed. This is a way of getting through to the reader that it is about the morphology or shape of the characteristic attachment mechanism and not its FEA. There are no FEA results although there are indicators on the drawings. This is for the purposes of illustration only.

Describing the reconstruction from 2-D of the *A. minus* hook

It was once impossible to take this work past the hypothetical allusion to a possible solution through the invention of some material that could be used to manufacture at this scale. This changed directly as a result of [1]. This material is now selected as being copper. It is with a certain use that we associate copper, namely conductivity of electricity and heat, not only that it is "green" material that it is relatively easily reclaimed.

Thus, we are arriving at a solution to the problem of how to attach the smallest of electrodes to a motherboard for instance, using friction. It is not necessary to say more, just to apply the testing array of formulae to show that it is feasible to get to this end solution through the manufacture of these hooks using this method.

As was discussed earlier in the "Background", it is the choice that matters, of the shape that is most appropriate to suit the purpose as illustrated by Nature. Here we have dictated that it should be the long-shafted hook of the *A. minus* that is the model, the decision made as a basis of the results of part 2 of the study when we discovered that the evolutionary sparkle had gone into the development of a cellulose hook that was fibrous and therefore

non-homogenous. Therefore, it can be predicted from shape efficiency constraints that all of the hook will not be used to absorb stress in copper where it is homogenous in structure as laid down by the layered manufacturing device. So it is true that there will be some use in FEA but only if specified correctly for forces and anchorage points and surfaces that are accurate. It will be the same for both of the new acts called the error note of the state, namely the way in which it uses a lot of its energy to absorb sound and noise to stop it from being a friction-destroyed device and rather reusable without destructive tendencies. This will be shown as heat of friction and not available to be used as noise or light as an ignition spark.

Figures 3 and 4 below show the way that the hook was drawn or constructed using the software. It must be done

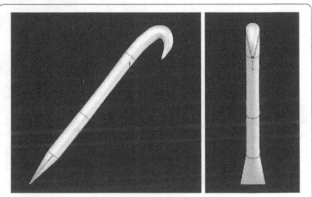

Fig. 3 Front and side views of hook with tapered shaft as per the *A. minus* hook. At this stage we are looking towards a universal probabilistic fastener that can be morphed from this structural example, or one of the other two or a combination or two or more forms/assemblies

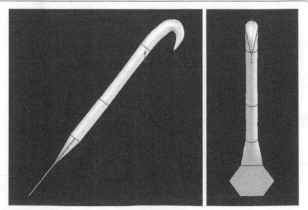

Fig. 4 Front and side view of *A. minus* hook with added flat hexagonal flange. Basic assumptions are made here about the forming of the structure. The configuration depends entirely on the configuration of the layered manufacturing device and its pattern of electro-deposition of copper and its jig

like this for the first principle of the test; thereafter, it is within the possibilities of Nature itself that it will form a new hook in the future when it is able to adapt to the new shape of the substrate. Then, it will adapt through malleability, and thus, a new hook shall be developed that does not forego the use of new materials but rather dictates how they should be formed to have properties to endure and make it feasible.

Figure 5 below shows the loading and mode of distortion of the hook, flexing and absorbing the extension and movement of the tip of the hook.

Hook field structure

It has been noted in the *Nachtigal* [5] that hooks in nature are assembled in different configurations and numbers. Gorb compared the mass of the fruit with the contact separation force of single hook in order to assess the hook performance for each species and the number of hooks required to support the fruit which could be viewed as a measure of design efficiency. This is of course, Nature's design. Instead, this research continues to avail itself of the known thing that is the scaling effect and the fact that hooks in a field with a modular design could help fulfil a number of roles or permutations and in fact is necessary in order to consider these tiny hooks at all, to consider them collectively.

With the modelling process with a development of a modular hook with a supporting flange analogue, the opportunity exists to experiment with:

1. Field configurations, densities and numbers.
2. Flange shapes, i.e. square, rectangular, circular and octagonal

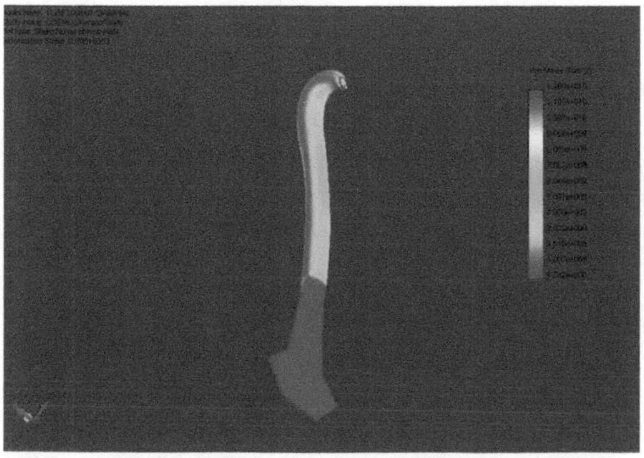

3. Flange attachment mechanisms for both attaching bracts to each other to form composite fields as well as for attachment to a structure with a further attachment mechanism as a substructure.
4. Flange shapes also offer the opportunity to manufacture the hooks in flattened rows.

Attention is drawn to Fig. 6 for extensions to the model:

Discussion

The 3-D rendering of the model is hoped to be probabilistic, but it is impossible as of yet to produce a model of the scale required using a rapid prototyping device, to produce a product of some variant of the 3-D model as a testable product at the scale to take advantage of the hook-span scaling effect.

The collective effect of the hooks at a small scale in size should be markedly different from that of the arithmetic sum. Under the static load of its own (small) mass it seems it is largely friction that holds the attachment in place, but atomic force microscopy illustrates the strength of atomic forces such as covalent bonding which occurs with the surface molecules as two surfaces approach each other, particularly if they are both hydrophobic such as dry wool and a seasonally desiccated seed pod. Local, internal, turgor pressure will have an effect upon the surface area that forms the bond as will any adhesives, chemical or otherwise. The loss of friction through moisture exposure and thin-film flow must

Fig. 5 The maximum deformation under loading. A point load at the tip, constrained at the base along the flange. There is nothing unexpected about the mode of deflection

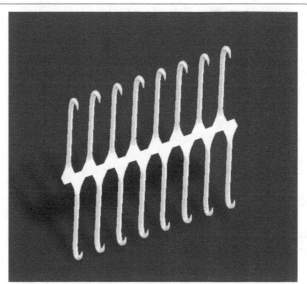

Fig. 6 A zipper configuration in isometric view. This illustrates the possibilities of a composite formation of long-shafted hooks acting a coordinated fashion. The point being illustrated here is that although we are seeking a universal "foot", it is as unlikely to look like a foot as a drone looks like a hummingbird

be studied too, for the opportunity to perfect the on–off release mechanism of the foot, short of using a shape memory material.

The desire is for a 3-D computer graphics model of a probabilistic fastener that does not destroy itself or the substrate and that is able to attach and detach to a variety of them and that can be analysed using computational analysis, technologies and techniques. It will be friction based and in the realm where masses and reactions to them are small compared to the frictional and other forces generated.

Walking the line bridging engineer and biologist, with knowledge in an exchange, it should be noted here that the Solidworks software is not of sufficient capacity for the demands of working with a complex biomaterial.

Conclusion

As part 3 of this series of three papers this culminates from the consideration of the results of research into five species of long shaft hook including the *A. minus* species, various microscopy techniques including confocal microscopy and the design of a bioinspired model for hooked structures of the order of 200 microns span, and two further attachment mechanisms of similar size, of insect chitin.

There are limitations to what can be achieved by the pull of technology. Excited at first by the discovery of the cubic voxel result of part 2 [2], it was soon dampened by the result that it was impossible to work with the voxels in C++. It is hoped that this will yield further work until a proper procedure is established for the study of these attachment mechanisms at their natural scale and order of size in various materials.

The 3-D model produced in Solidworks 2004 is suitable for application to fields of long-shafted hooks at a micron-size, and its qualities have to be assessed in a testing/manufacturing circumstance. It is expected that these attachments will be probabilistic and frictional in behaviour. The rate of change will vary from hook to hook as the size/shape changes, and this can only effectively be measured using the new way of manufacturing the hooks through electro-deposition such that they are of the size but of a different material, thereby mimicking the scaling effects and the biological principal that guides their behaviours governing the size/volume, strength/ weight, friction coefficient (μ)/size and other relations. It must be remembered that the scaling effects associated with a burdock hook may not be related directly to Young's modulus. More important could be the shape and surface area versus frictional coefficient. Similarly, this could apply to forces like surface tension, hygroscopic forces and others.

There are other problems to overcome before manufacturing these hooks *en masse*, but it does point to a method of effectively producing magnetic hooks that have been used to conduct electricity commercially as well as heat. Attachment methods are important as well as the way they all come out, which can be distorted from all shape and needs to be considered too. Destructive testing is the only way of assessing this.

This form of data collection and transfer is regarded as another form of mathematics, abstract and applied character. It makes the statement that all members of the set of hooks can be made under the conditions of the microscopy settings given in paper 2 of the set [2]; therefore, a part of the experimentation is already complete. All that is left is manufacture and testing. Only samples that are completely translucent are used, so the intensity of the light is high and able to be differentiated from the background light.

This problem never would have been encountered had the topic been fasteners in general since a straight attachment mechanism would have been sufficiently simple to analyse and model but not manufacture either. It is the mechanical properties that are sought, and the only way of achieving this is to make them from a new material that does not get a lot of attention these days, copper, but through the act of forced self-assembly by applied voltage, it may be possible to arrive at an alternative manufacturing technique or deposition pattern as well as alternative materials such as silicates or gold.

Acknowledgements
With thanks to the publishers, Springer-Open, for their patience and assistance in making these papers possible.

Author's information
This HESRC research was a funded PhD was intended for a Biologist, but conducted by a Mechanical Engineering graduate. Completed in 2005/6, publication has been slowed by ill health. This is the final part in the series of 3. Bruce E. Saunders attended the University of Bath for a period of time (June 2002 to March 2005) before a dispute over the value of the work led to a break in relations. This is when the data were collected. Further, he was suffering from PTSD- and trauma-induced psychosis for which he is registered disabled.

Competing interests
None.

References
1. Hirt L, Ihle S, Pan Z, Dorwling-Carter L, Reiser A, Wheeler JM, Spolenak R, Vörös J, Zambelli T. Template-free 3D microprinting of metals using a force-controlled nanopipette for layer-by-layer electrodeposition. Adv Mater. 2016;. doi:10.1002/adma.201504967.
2. Saunders B. Biomimetic study of natural attachment mechanisms—imaging cellulose and Chitin part 2. J. Robot. Biomim. 2015;2:7. doi:10.1186/ s40638-015-0032-9.

3. Saunders B. Biomimetic study of natural attachment mechanisms—*Arctium minus* part 1. J. Robot. Biomim. Special issue on Micro-/Nanorobotics. 2015;2:4.

4. Gorb E, Gorb SN. Contact separation force of the fruit burrs in four plant species adapted to dispersal by mechanical interlocking. Plant Physiol Biochem. 2002;40:373–81.

5. Nachtigall W. Biological mechanisms of attachment, the comparative morphology and bioengineering of organs for linkage, suction and adhesion, translated by Biederman-Thorson MA. Heidelberg: Springer, ISBN 3-540-06550-4.

4

Hydrodynamic study of freely swimming shark fish propulsion for marine vehicles using 2D particle image velocimetry

Mannam Naga Praveen Babu[1*], J. M. Mallikarjuna[2] and P. Krishnankutty[1]

Abstract

Two-dimensional velocity fields around a freely swimming freshwater black shark fish in longitudinal (XZ) plane and transverse (YZ) plane are measured using digital particle image velocimetry (DPIV). By transferring momentum to the fluid, fishes generate thrust. Thrust is generated not only by its caudal fin, but also using pectoral and anal fins, the contribution of which depends on the fish's morphology and swimming movements. These fins also act as roll and pitch stabilizers for the swimming fish. In this paper, studies are performed on the flow induced by fins of freely swimming undulatory carangiform swimming fish (freshwater black shark, $L = 26$ cm) by an experimental hydrodynamic approach based on quantitative flow visualization technique. We used 2D PIV to visualize water flow pattern in the wake of the caudal, pectoral and anal fins of swimming fish at a speed of 0.5–1.5 times of body length per second. The kinematic analysis and pressure distribution of carangiform fish are presented here. The fish body and fin undulations create circular flow patterns (vortices) that travel along with the body waves and change the flow around its tail to increase the swimming efficiency. The wake of different fins of the swimming fish consists of two counter-rotating vortices about the mean path of fish motion. These wakes resemble like reverse von Karman vortex street which is nothing but a thrust-producing wake. The velocity vectors around a C-start (a straight swimming fish bends into C-shape) maneuvering fish are also discussed in this paper. Studying flows around flapping fins will contribute to design of bioinspired propulsors for marine vehicles.

Keywords: Carangiform swimming, Caudal fin locomotion, Flow visualization, Propulsor hydrodynamics, Particle image velocimetry, Pectoral fins, Reverse von Karman vortex street, Wake

Background

Aquatic animal propulsors are classified into lift-based (e.g., penguins, turtle forelimb propulsion and aerial birds), undulation (e.g., fishes, eels), drag-based (e.g., duck paddling) and jet mode (e.g., jelly fish, squids). Fishes use a combination of lift-based and undulating modes mainly using its undulating body, pectoral and caudal fins to achieve propulsive forces. Fishes also generate thrust by using its tail fin, paired fins and its body. Certain combinations of flapping motions and angles of body achieve greater speed and better maneuvering capabilities. Flapping foil propulsion systems, resembling fish fin propulsion mode, are found to be much more efficient than the conventional screw propellers [1, 2]. The application of fish propulsion to water crafts is found to have higher propulsive efficiency, better maneuvering capabilities, less vibrations, low emissions and more eco-friendly. Biological aquatic animal locomotion, its mechanism and their successful application to marine vehicles are being studied by different researchers. Muller [3] used 2D PIV to visualize the flow around the aquatic animals and to demonstrate the creation of vorticity and their contribution to thrust generation. Muller et al. [4] studied the water velocity near fish body using PIV and described the wake mechanism behind it. Drucker and Lauder [5] studied the bluegill sunfish pectoral fins 3D wake structures using PIV. Sakakibara et al. [6] used stereoscopic

*Correspondence: mpraveenmn@gmail.com; oe13d006@smail.iitm.ac.in
[1] Department of Ocean Engineering, Indian Institute of Technology Madras, Chennai 600 036, India
Full list of author information is available at the end of the article

PIV for capturing three components of velocity distribution on live goldfish along with particle tracking velocity in order to determine spatial velocity, acceleration and vorticity. Past researchers [7–18] carried out experiments on hydrodynamic studies of fish locomotion as well as maneuvering by using PIV system. In the present study, a shark fish which belongs to the sub-carangiform is kept in a glass tank (Fig. 1) and the water particle kinematics around its tail and fins are observed using a two-dimensional PIV system while the fish try to swim forward. In sub-carangiform of locomotion, last one-third aft length of body muscle is used for generating thrust in addition to its caudal fin, whereas thunniform fishes' the caudal peduncle and tail fin are responsible for thrust production. The sub-carangiform fishes can move its caudal fin at a higher amplitude compared with form of fishes, resulting better thrust generation. That is the reason for choosing this form of fish for the present study. It moves forward by flapping its caudal fin and body undulation. The pressure distribution around the body and the caudal fin is shown in Fig. 2. There are positive and negative pressure regions along the body. The fluctuations of these pressure distributions result in a propulsive force, pushing the fish forward. The shape of the caudal fin reduces the amount of displaced water during oscillation of tail fin, thereby reducing turbulence and frictional drag on the body without the loss of propulsive power. The velocity diagram of sub-carangiform fish caudal fin is shown in Fig. 3. In sub-carangiform swimming fish, the thrust is developed by the rear part of the body and the tail fin. The thrust generated by the tail fin is given by Eq. (1). The caudal fin is moving normal to free-stream velocity, V_o, with a transverse (sway) velocity equal to V_N. It is possible for caudal fin to attain a thrust component which provides a forward propelling force. The rotational component (yaw) is not considered in this case. The above equation is the simplest case of pure translation motion normal to a free stream V_o [19].

$$T = \left(\frac{1}{2}\right)\rho V_R^2 \left(\frac{dC_L}{d\alpha}\right) S\alpha \cos \beta \qquad (1)$$

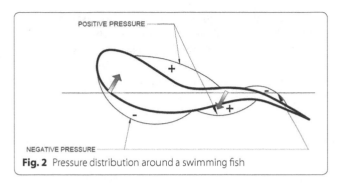

Fig. 2 Pressure distribution around a swimming fish

where

$$\beta = \cos^{-1}\left(\frac{V_N}{V_R}\right) \qquad (2)$$

where ρ represents density of fluid in kg/m³, V_R is resultant velocity in m/s, S is surface area of fin in m², α is angle of attack in (rad) and $(dC_L/d\alpha)$ is slope of lift curve for caudal fin.

In the present study, flow visualization experiments are carried out to visualize the flow pattern around the caudal, pectoral, anal and dorsal fins of a freely swimming fish using two-dimensional (2D) particle Image velocimetry (PIV) system.

Methods

Freshwater black shark (Labeo chrysophekadion) with a body length of 26 cm is used for the present experimental study. The fish is placed inside a glass tank of size $L \times B \times D = 75$ cm $\times$ 29 cm $\times$ 37 cm, with water level at 28 cm, and it is allowed to swim freely in the tank. The fish swims across the tank length, and the PIV measurement is taken at the steady phase of its movement, which is observed to be in the middle one-third portion of the tank. In this experiment, the laser pulse is operated

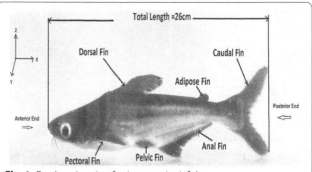

Fig. 1 Freely swimming fresh water shark fish

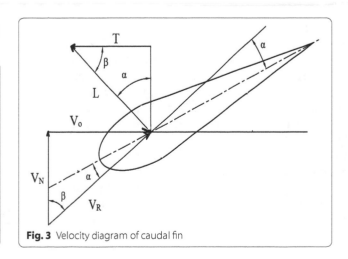

Fig. 3 Velocity diagram of caudal fin

continuously and fish will always cross this laser plane in multiple times with same time interval. Then, a range of images is selected for processing velocity fields. From the visual observations, based on the recorded video, the Strouhal number of the freely swimming shark fish used in this experiment is approximately 0.23, where the tail fin oscillation frequency is 0.6 Hz, amplitude is 0.1 m and the fish swimming speed is 0.26 m/s.

Experimental setup

Two-dimensional PIV technique is used to study the flow around a swimming fish. This helps in clearly understanding the instantaneous velocity vector fields of the flow field around the fish. It is a non-intrusive experimental technique which can measure the whole flow field with high spatial and temporal resolution at any instant.

The PIV technique involves the introduction of tiny particles called 'seeder particles' into the fluid path. The size and density of seeder particles are chosen such that they follow the flow path faithfully at all operating conditions. Hollow glass spheres with a mean diameter of 10 μm are used as the tracer particles. The seeding particles in the plane of interest are illuminated by a laser sheet of appropriate thickness 0.5–2.5 mm. Two images (an image pair) of the illuminated flow field are obtained within a separation time 'Δt' by means of high-resolution camera. The displacement of the tracer particles during the time interval 'Δt' gives velocity of the fluid particle. The experiments are performed at three different time intervals, $\Delta t = 300$, 620 and 900 ms. If the Δt is less than 300 ms, no swirl of velocity vectors is observed. Then, the Δt gradually increases from 300 to 900 ms, and velocity fields around the fish body are observed. The Reynolds number (Re) of swimming fish is in the range of 10^5, and at low Re number, the 2D velocity fields do not affect.

The PIV setup used in the present study is shown in Fig. 4. The PIV system used in this work consists of (1) a double-pulsed Nd-YAG (neodymium-doped yttrium aluminum garnet) laser with 200 mj/pulse energy at 532 nm wavelength, (2) a charge-coupled device (CCD) camera with a 2048 by 2048 pixels and an image capturing speed of 14 frames per second (fps), (3) a set of laser and a camera controllers and (4) a data acquisition system. The laser sheet is aligned with the longitudinal vertical (XZ) and transverse (YZ) planes. The camera is positioned in front of the test section at 90° to the laser sheet (see Fig. 4). The size of seeding particles is very important in obtaining proper images. The particles should scatter enough light, and too large particles may not follow the flow path. Measurement of the velocity field using PIV is based on the ability of the system to accurately record and measure the positions of small traces suspended in

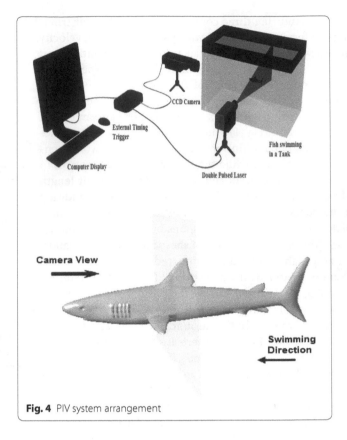

Fig. 4 PIV system arrangement

flow as function of time. The PIV system measurement scheme is shown in Fig. 5.

In PIV measurement scheme, the images are divided into a number of small sections called interrogation windows or regions. The corresponding interrogation regions in frame 1 and 2 are correlated using cross-correlation method. The maximum of the correlation corresponds to the displacement of the particles in interrogation window. The displacement gives the vector length and direction in interrogation zones. Small interrogation windows

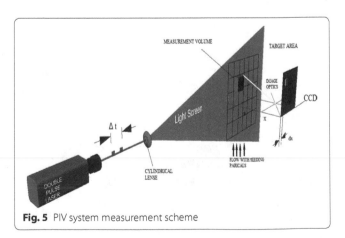

Fig. 5 PIV system measurement scheme

give more vectors but contain less particles. The main advantage of the cross-correlation approach is displacement that can be obtained with directional ambiguity.

In the experimental setup, care should be taken to make the laser sheet, camera axis and test object lie in the same plane. The laser sheet should be aligned perfectly vertical to the calibration plate (Fig. 6). In PIV calibration, the images obtained are focused and the scale factor (calibration constant) necessary for further processing of the images is obtained. This calibration plate is placed parallel to the light sheet and approximately in the middle of calibration sheet. The sheet consists of a grid of dots with a large central dot surrounded by four small dots. The distance between two dots on the calibration sheet is 5 mm. Images are captured on the calibration sheet when it is placed in the light sheet. These images are analyzed by the computer software 'Davis' [20], and the scale factor is obtained by comparing the apparent distance between the dots provided by CCD camera and the actual distance between the dots of 5 mm. Once the images are captured, the camera is focused on the sheet such that all dots appear sharp.

The velocity fields obtained by PIV are used to determine the pressure fields. The pressure and velocity are linked by the Navier–Stokes (NS) equations, and the pressure can be measured indirectly by measuring the velocity field. There exist two methods to measure the pressure field indirectly. The first method is direct spatial integration of the momentum equations [21, 22]. The second method is solving a Poisson equation for the pressure field [23]. The present study does not include the comparison aspects of velocity field to pressure field.

PIV results and discussion

Fishes generate propulsive forces, are able to maneuver rapidly and stabilize its body motions using its fins

such as pectoral, dorsal, pelvic, anal and caudal fins (see Fig. 1). By using its fins, fishes can control roll, pitch and yaw motions. The paired pectoral fins (one on each side) are used for maneuvering as well as for instantaneous stopping (braking) [24]. The median dorsal fins act as keels, used for directional stability and to prevent from spinning or rolling. Pelvic fins and anal fins are used as stabilizers. Caudal fin is used for propulsion, maneuvering and braking. The flow visualization experiments are carried out on a freely swimming sub-carangiform mode shark fish in longitudinal vertical (XZ) plane and transverse (YZ) plane by using two-dimensional particle image velocimetry.

The flows around the fins of freely swimming fish are analyzed, and the velocity vector fields are presented here. In this analysis, we are presenting a raw CCD (charge-coupled device) image and the processed image at different time intervals (Figs. 7, 8, 9, 10, 11, 12, 13, 14, 15, 16, 17, 18). The white boundary line represents the body of fish. The primary vortex regions are marked as V_1, V_2 in images. Figure 7 shows CCD image and velocity vector field around caudal fin at $\Delta t = 900$ ms. The caudal fin possess thrust-producing wake, resembling a reverse von Karman vortex street. In reverse von Karman vortex street, upper row vortices rotate in anticlockwise and the lower row vortices rotate in clockwise direction. Fishes are able to generate thrust depending on the amplitude and frequency of oscillation of the caudal fin. By varying the frequency and amplitude of fin oscillation, fishes can achieve the fin oscillation in the Strouhal number range of 0.2–0.5 to attain a propulsive force. The fish can move the caudal fin in both translational (sway) and rotational (yaw) modes for its efficient propulsion. The flow around the caudal fin of steadily swimming fish with counter-rotating vortices in vertical plane is shown in Fig. 7. The center of fish tail-shed vortices appears to be about 45 deg inclined to the centerline. During steady swimming, fishes orient the body at an angle to the flow. The propulsive force generated by caudal fin movement is directed to body through center of mass. Figures 8 and 9 show CCD image and velocity vector field around adipose and anal fins at $\Delta t = 900$ ms. In the PIV experiments, the images are taken with a time difference between two consecutive images, $\Delta t = 300$ and 900 ms. The image qualities are found to be acceptable in both the cases. The adipose and anal fins generated vortices that pass downstream, interacting with caudal fin vortices, while flapping its tail from starboard side to port side, and are found to form stronger vortices, thus helping in the generation of improved propulsive force.

Figure 10 shows CCD image and velocity vector field around adipose and anal fins at $\Delta t = 300$ ms. Orientation of the caudal fin in this figure shows the flexibility present

Fig. 6 Calibration plate

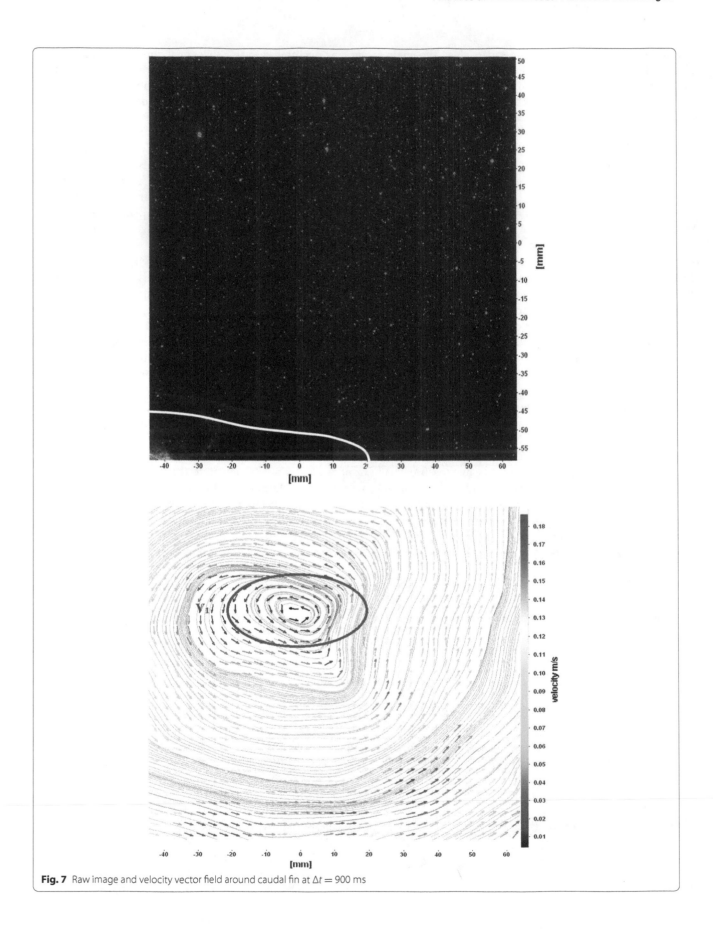

Fig. 7 Raw image and velocity vector field around caudal fin at $\Delta t = 900$ ms

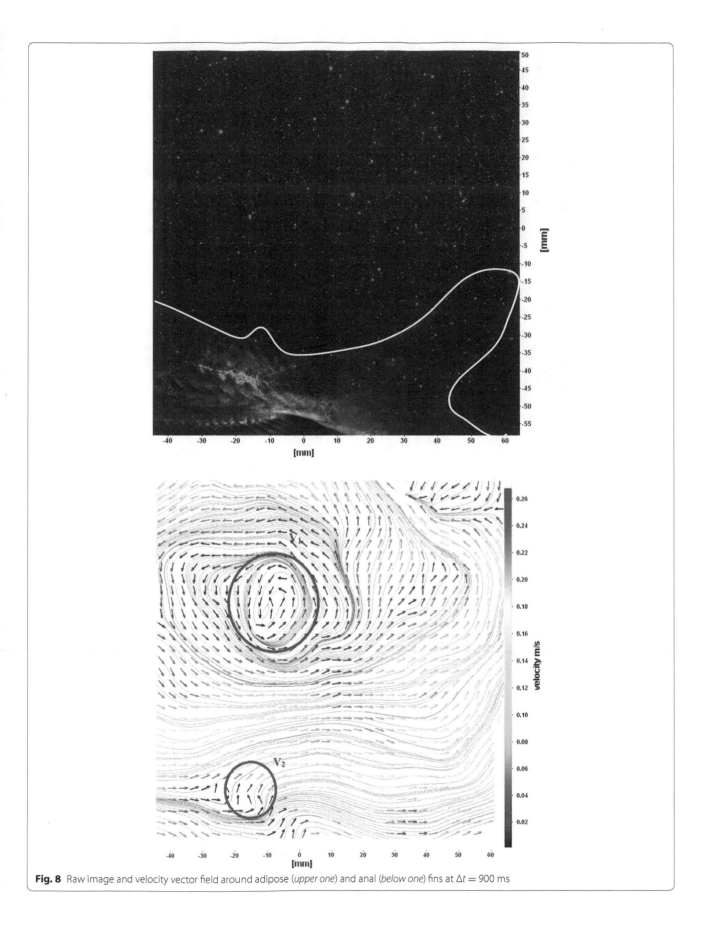

Fig. 8 Raw image and velocity vector field around adipose (*upper one*) and anal (*below one*) fins at $\Delta t = 900$ ms

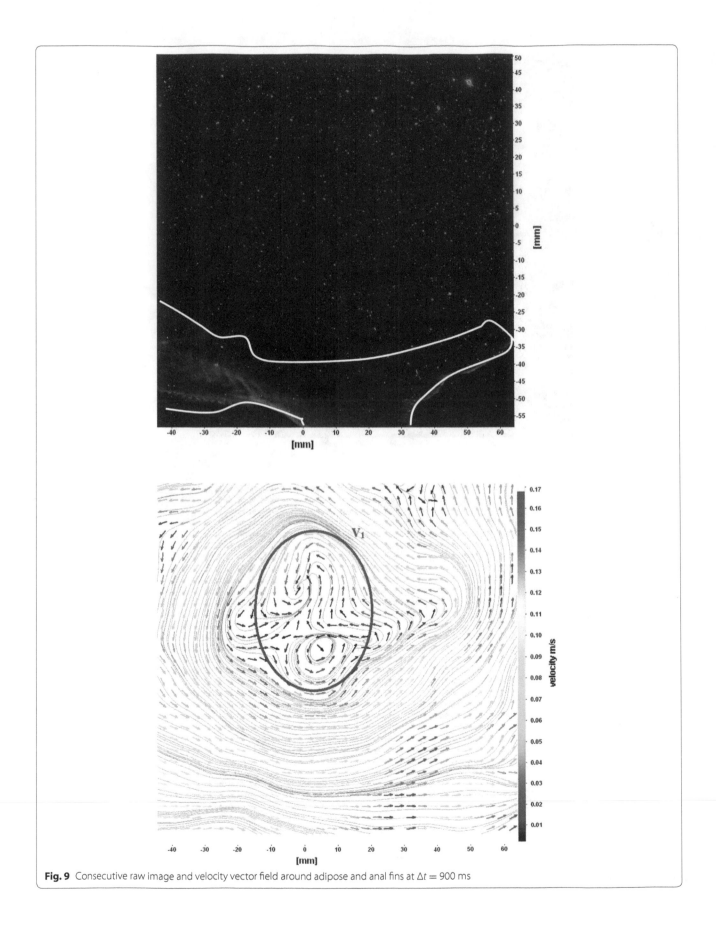

Fig. 9 Consecutive raw image and velocity vector field around adipose and anal fins at $\Delta t = 900$ ms

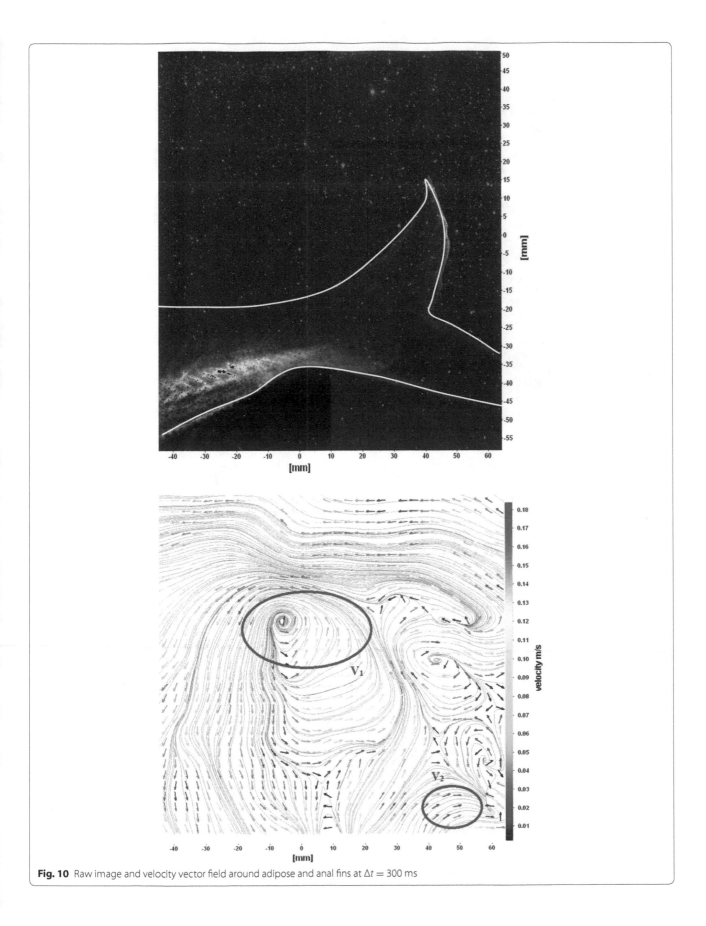

Fig. 10 Raw image and velocity vector field around adipose and anal fins at $\Delta t = 300$ ms

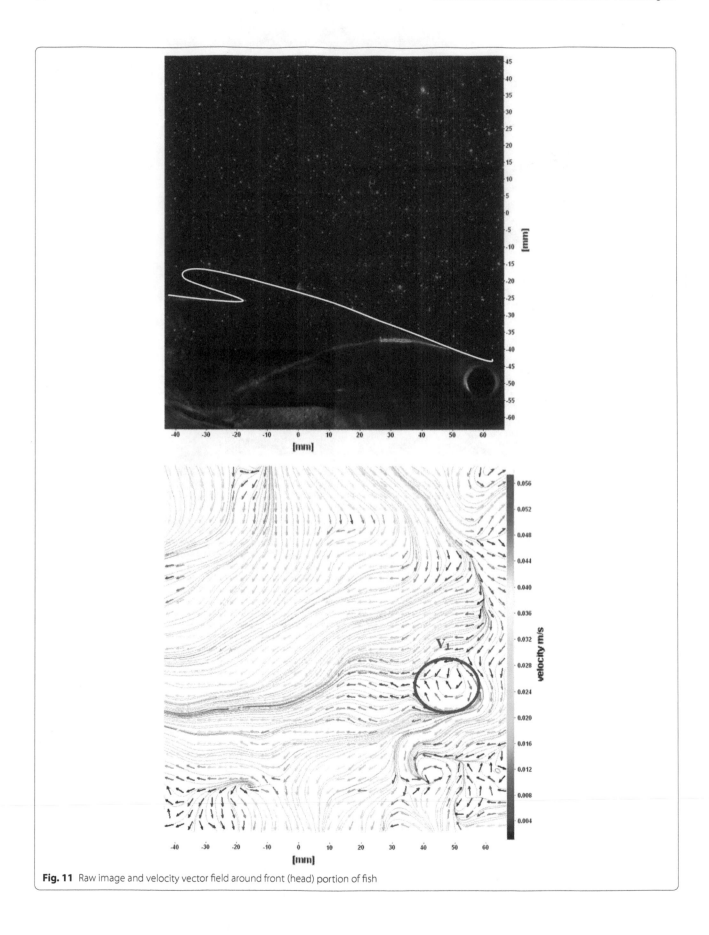

Fig. 11 Raw image and velocity vector field around front (head) portion of fish

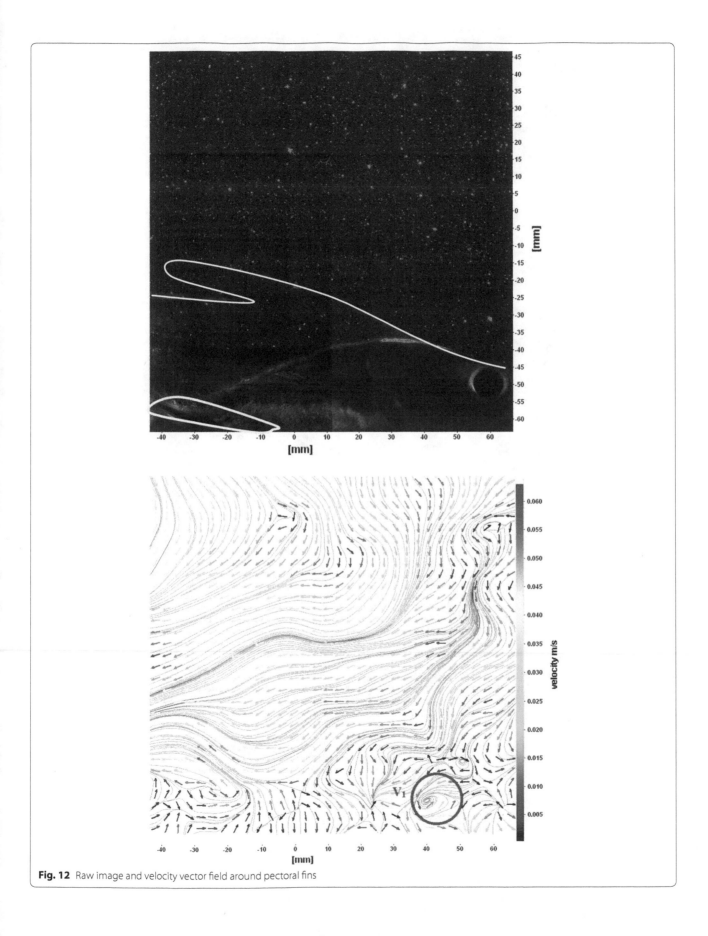

Fig. 12 Raw image and velocity vector field around pectoral fins

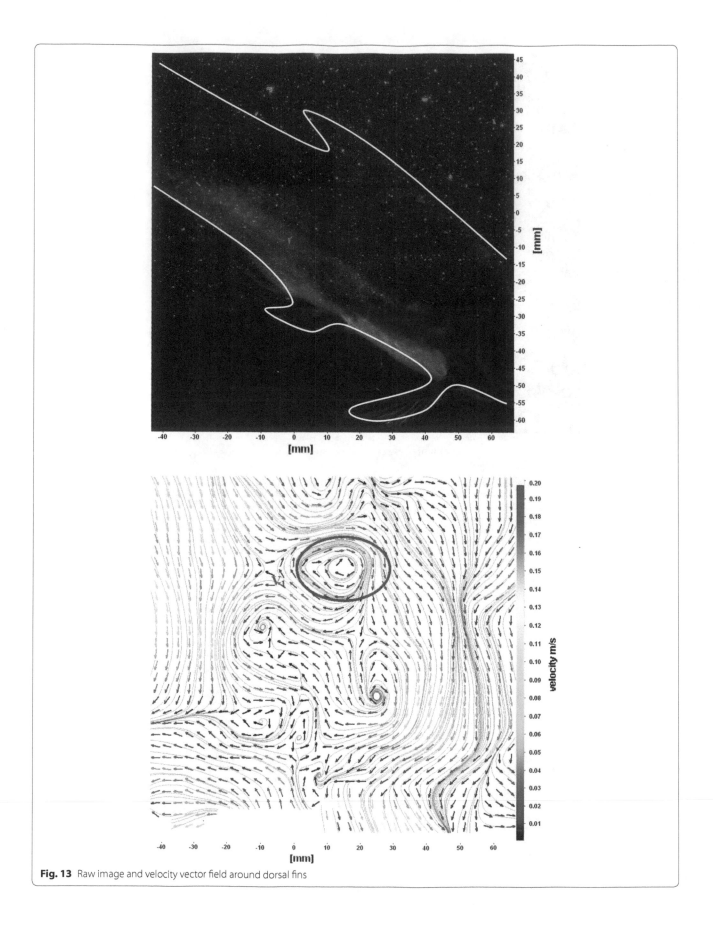

Fig. 13 Raw image and velocity vector field around dorsal fins

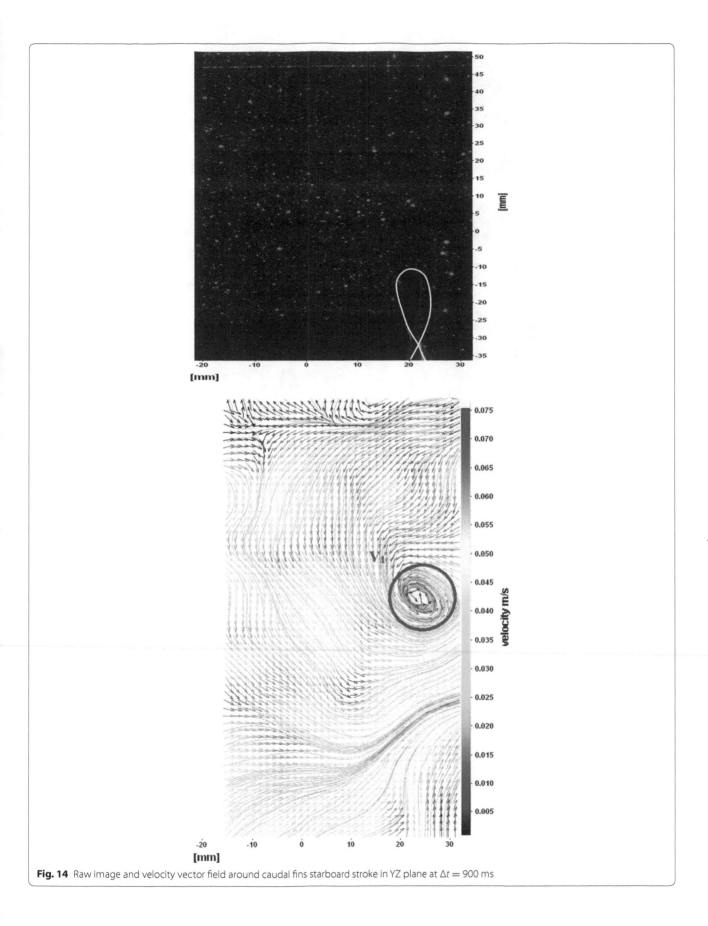

Fig. 14 Raw image and velocity vector field around caudal fins starboard stroke in YZ plane at $\Delta t = 900$ ms

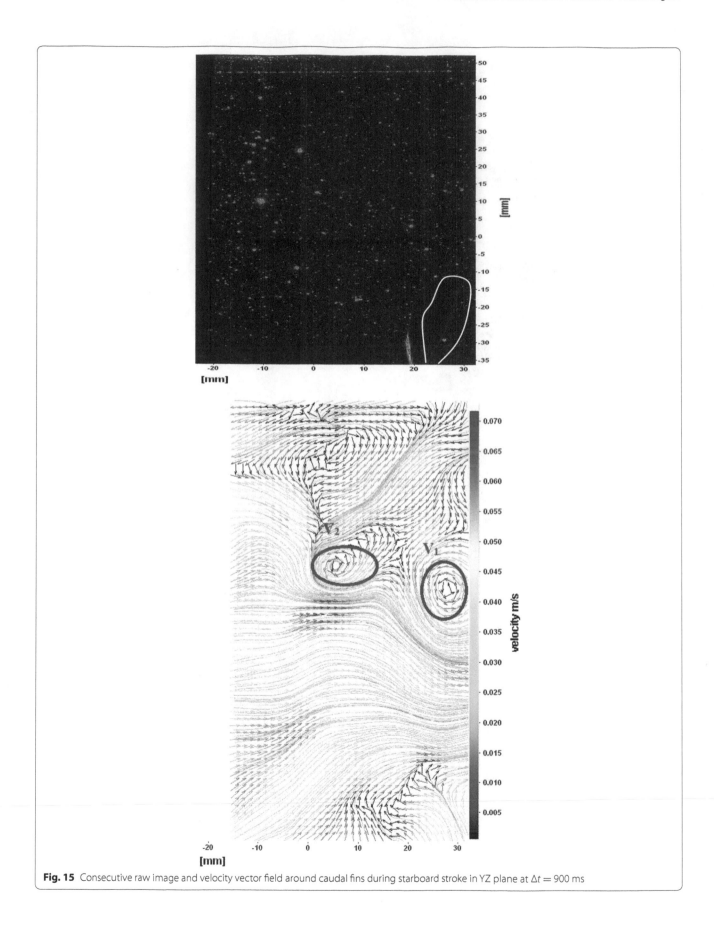

Fig. 15 Consecutive raw image and velocity vector field around caudal fins during starboard stroke in YZ plane at $\Delta t = 900$ ms

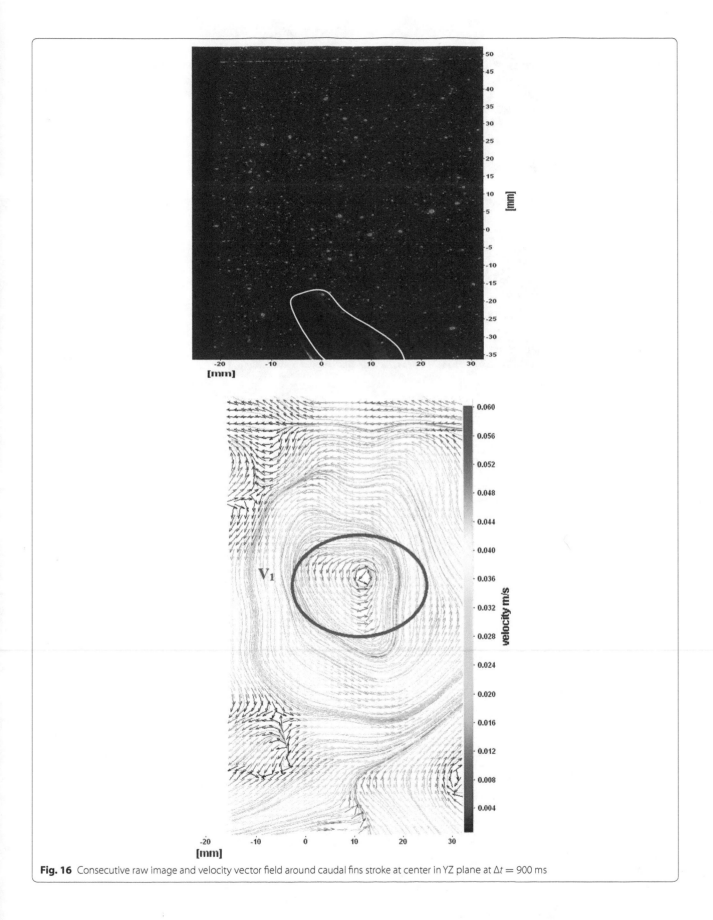

Fig. 16 Consecutive raw image and velocity vector field around caudal fins stroke at center in YZ plane at $\Delta t = 900$ ms

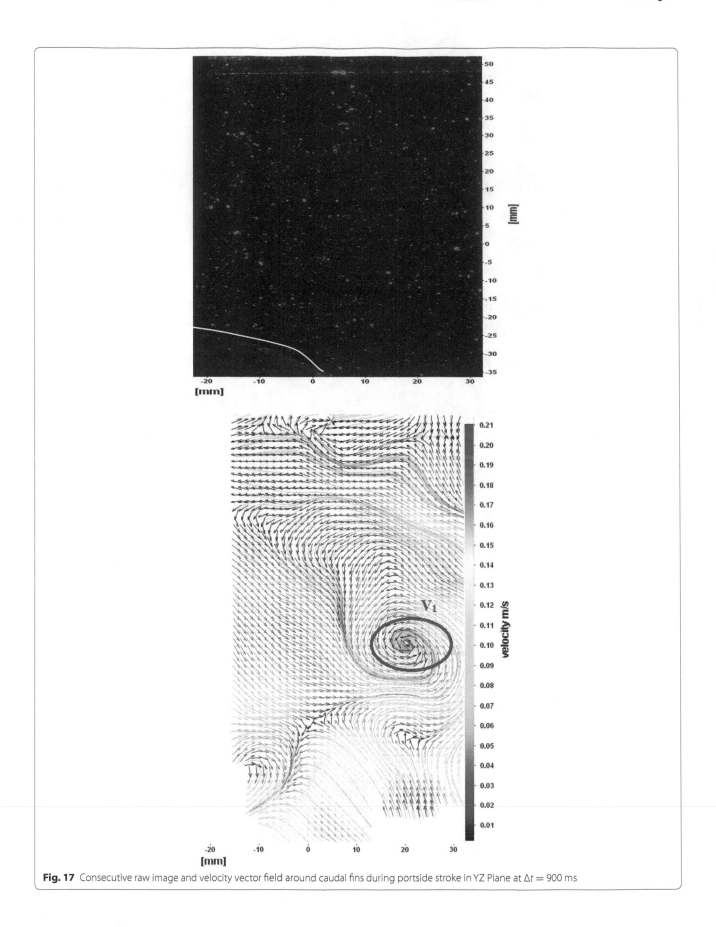

Fig. 17 Consecutive raw image and velocity vector field around caudal fins during portside stroke in YZ Plane at $\Delta t = 900$ ms

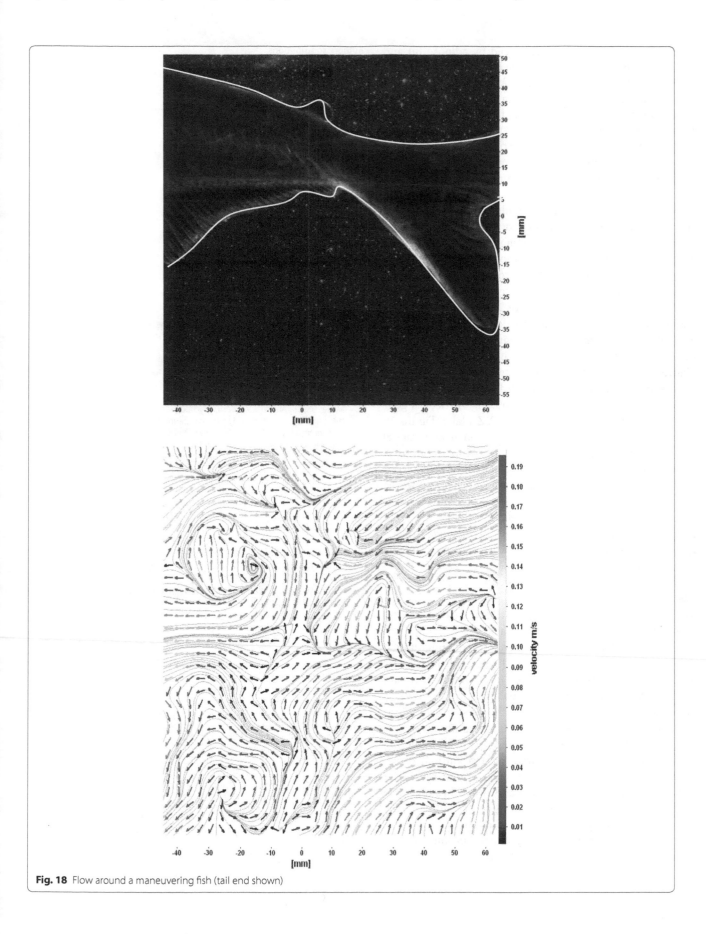

Fig. 18 Flow around a maneuvering fish (tail end shown)

in its movements. The jets produced by adipose fin and anal fins are observed in the peduncle region (region containing the tail and the body). At the posterior end of the fish, a pair of counter-rotating vortices is observed. Figure 11 shows CCD image and velocity vector field around anterior portion of fish. At low amplitudes and frequencies of caudal fin, when Strouhal number (st) is less than 0.2, the vortices become inward and thus the fish experiences drag due to these vortices. This wake resembles like a von Karman vortex street shown in Fig. 11. Figure 12 shows CCD image and velocity vector field around pectoral fins. These paired pectoral fins undergo deformation during their flapping cycle. It undergoes chordwise and spanwise deformations as well as twisting. During power stroke and return stroke, the effective angle of attack of flow with fin increases, thereby producing thrust in both the strokes. Figure 13 shows CCD image and velocity vector field around dorsal fins. Dorsal fins generate strong vortices. Flow leaving the dorsal and anal fins rolls up and then interacts with caudal fin vortices. Figures 14 and 15 show CCD image and velocity vector fields around caudal fin in starboard stroke in YZ plane at $\Delta t = 900$ ms. A pair of counter-rotating vortices is generated around the caudal fin in the YZ plane. Figure 16 shows CCD image and velocity vector field around caudal fin stroke in YZ plane at $\Delta t = 900$ ms, while the fin is at the center plane. A jet with high velocity flow is observed at the top of the caudal fin. Figure 17 shows CCD image and velocity vector field around caudal fins, during port-side stroke, in YZ plane at $\Delta t = 900$ ms. Figure 18 shows flow around a maneuvering fish. During maneuvering of a fish, jets are observed at the side of fish causing a turning moment instantaneously.

Summary and conclusions

The flow visualization experiments are carried out on a freely swimming freshwater black shark using two-dimensional particle image velocimetry in longitudinal vertical (XZ) and transverse (YZ) planes. The velocity vector fields show that both paired fins (pectoral fins) and median fins (dorsal, anal and caudal fins) produce reverse von Karman vortices resulting in the flow jets and consequent thrust (propulsive force). It is also observed that the fin flexibility in chordwise and spanwise direction substantially improves the thrust generation and direction control of the fish. The fish anal fin and caudal fin vortices are also presented here and show that they also contribute to the fish propulsive force. By studying the nature flow velocity distribution around fish fins propulsion systems, one can design flapping foil propulsion systems for ships and underwater vehicles.

Authors' contributions
All authors were equally involved in the study and preparation of the manuscript. All authors read and approved the final manuscript.

Author details
[1] Department of Ocean Engineering, Indian Institute of Technology Madras, Chennai 600 036, India. [2] Department of Mechanical Engineering, Indian Institute of Technology Madras, Chennai 600 036, India.

Acknowledgements
The authors would like to thank Department of Ocean Engineering, IIT Madras, India, for providing support for doing this project.

Competing interests
The authors declare that they have no competing interests.

References
1. Politis GK, Belibasakis KA (1999) High propulsive efficiency by a system of oscillating wing tails. In: CMEM, WIT conference.
2. Anderson JM, et al. Oscillating foils of high propulsive efficiency. J Fluid Mech. 1998;360:41–72.
3. Müller UK, Van Den Heuvel BLE, Stamhuis EJ, Videler JJ. Fish foot prints: morphology and energetics of the wake behind a continuously swimming mullet (*Chelon labrosus* Risso). J Exp Biol. 1997;201:2893–906.
4. Muller UK, Stamhuis EJ, Videler JJ. Hydrodynamics of unsteady fish swimming and the effects of body size: comparing the flow fields of fish larvae and adults. J Exp Biol. 2000;203(2):193–206.
5. Drucker EG, Lauder GV. Locomotor function of the dorsal fin in teleost fishes: experimental analysis of wake forces in sunfish. J Exp Biol. 2001;204(17):2943–58.
6. Sakakibara J, Nakagawa M, Yoshida M. Stereo-PIV study of flow around a maneuvering fish. Exp Fluids. 2004;36(2):282–93.
7. Stamhuis E, Videler J. Quantitative flow analysis around aquatic animals using laser sheet particle image velocimetry. J Exp Biol. 1995;198(2):283–94.
8. Videler JJ. Fish swimming, vol. 10. Berlin: Springer; 1993.
9. Wakeling JM, Johnston IA. Body bending during fast-starts in fish can be explained in terms of muscle torque and hydrodynamic resistance. J Exp Biol. 1999;202(6):675–82.
10. Webb PW. Hydrodynamics and energetics of fish propulsion. Bull Fish Res Board Can. 1975;190:1–159.
11. Webb PW, Blake RW. Swimming. In: Hildebrand M, editor. Functional vertebrate morphology. Cambridge: Harvard University Press; 1983.
12. Weihs D. A hydrodynamic analysis of fish turning manoeuvres. Proc R Soc Lond B. 1972;182:59–72.
13. Westerweel J. Fundamentals of digital particle image velocimetry. Meas Sci Technol. 1997;8(12):1379.
14. Wolfgang MJ, et al. Near-body flow dynamics in swimming fish. J Exp Biol. 1999;202(17):2303–27.
15. Wu YT. Hydromechanics of swimming propulsion part 1: swimming of a two-dimensional flexible plate at variable forward speeds in an in viscid fluid. J Fluid Mech. 1971;46:337–55.
16. Blake RW. Fish locomotion. Cambridge: Cambridge University Press; 1983.
17. Breder CM. The locomotion of fishes. New York: New York Zoological Society; 1926.
18. Babu MNP, Krishnankutty P, Mallikarjuna JM. Experimental study of flapping foil propulsion system for ships and underwater vehicles and PIV study of caudal fin propulsors. In: Autonomous underwater vehicles (AUV), 2014 IEEE/OES, p. 1, 7, 6–9 Oct 2014.
19. Gero DR. The hydrodynamic aspects of fish propulsion. Am Mus Novitates. 1952;1601:1–32.
20. http://www.lavision.de/en. Accessed 20 May 2014.

21. van Oudheusden B, Scarano F, Roosenboom E, Casimiri E, Souverein L. Evaluation of integral forces and pressure fields from planar velocimetry data for incompressible flows. Exp Fluids. 2007;43:153–62.

22. de Kat R, van Oudheusden B, Scarano F. Instantaneous planar pressure field determination around a square-section cylinder based on time-resolved stereo-PIV. In: 14th international symposium on applications of laser techniques to fluid mechanics; Lisbon, Portugal. 2008.

23. Gurka R, Liberzon A, Hefetz D, Rubinstein D, Shavit U. "Computation of pressure distribution using PIV velocity" data. In: 3rd international workshop on PIV. Santa Barbara, CA, US, p. 101–6; 1999.

24. Drucker EG, Lauder GV. Wake dynamics and fluid forces of turning maneuvers in sunfish. J Exp Biol. 2001;204:431–42.

5

On the development of intrinsically-actuated, multisensory dexterous robotic hands

Hong Liu, Dapeng Yang*, Shaowei Fan and Hegao Cai

Abstract

Restoring human hand function by mechatronic means is very challenging in robotics research. In this paper, we first make a brief review on the development of dexterous robotic/prosthetic hands, and then detail our design philosophy of several robot hands. We make a concentration on a type of intrinsically-actuated robot hands, wherein the driving, transmission, and control elements are totally embedded in the hand. According to different application scenarios, we develop robot hands in two parallel lines, dexterous robotic hand and anthropomorphic prosthetic hand. In both, the hand's actuation, sensing, and control subsystems are highly integrated and modularized. This feature endows our robot hands with compact appearances, simple integration, and large flexibilities. At last, we give some perspectives on the future development of dexterous hands from the aspects of structure, functionality, and control strategies.

Keywords: Robotic hand, Prosthetic hand, Intrinsic actuation, Modular design

Background

As a powerful tool, a large variety of robotic systems has been applied to help human beings explore unknown or hazardous areas such as outer space, deep sea, or contaminated nuclear plants. To achieve effective explorations, a dexterous end-effector with superior operation and perception capabilities is an urgent need. Although traditional grippers can deal with some simple, fixed tasks (grasping and transferring workpieces), their low commonality, humble perception and insufficient flexibility make them hardly competent to complex operations in unstructured environment. Then, dexterous robotic hands (DRHs) with multiple degrees of freedom (DOFs), superior operational and perceptional capabilities arouse great attentions in the robot society [1]. Currently, although a large progress has been made, the DRHs available on the market still cannot compete to biological hands due to current technical constraints

on actuators, sensors and control means. It is indicated that, rather than simply imitating the human hand, the research should switch to fully exploiting the robot hand's advantages, while considering specific requirements (manipulative dexterity, grasp robustness, or human operability) that allow for successful, fluent, and dexterous operations [2].

As a branch of robotic hand research, the anthropomorphic prosthetic hand (APH) is a type of biomechatronic device used to restore hand motions for amputees or paralyzed patients. On this topic, great efforts have been made from both robotics and biomedical engineering. However, current prosthetic hands still cannot compete to a human hand in respect of structure, sensing, and control strategy. Only a few of prosthesis products can obtain their commercial success. Because of unintuitive control feelings, lack of sensory feedback, and poor hand functionality [3], a large portion of users often refuse to use their prosthesis. After analyzing human hand's activities of daily life (ADL's), a study reveals that a superior hand prosthesis should have more controllable functions, faster response/shorter reaction time, and an intuitive control and feedback strategy [4].

*Correspondence: yangdapeng@hit.edu.cn
State Key Laboratory of Robotics and System, Harbin Institute
of Technology, HIT Science Park, No.2 Yikuang Street, Nangang District,
P.O. Box 3039, 150080 Harbin, People's Republic of China

The advanced prosthetic hand systems are then characterized by its anthropomorphic appearance, congenital dexterity (including both mechanical structure and sensors), and high-level mechatronic integration. As for the hand's manipulation capability, it is normally held that the hand dexterity improves as the number of active joints increases. However, studies also shows that, as the number of the active joints increases, the dexterity of a prosthetic hand may even decrease due to the intensified control complexity. Therefore, the prosthetic hand design should consider more comprehensive factors, such as the compromise between dexterity and controllability, the suitability and adaptability of the sensory feedback, as well as essential neural rehabilitation principles [5].

After briefly reviewing some representative studies, in this paper, we detail our development process of several DRH and APH prototypes. From a view of biomechatronics, we also prospect some directions on the development of advanced robot hands, after fully acknowledging the challenges in front of us.

Representative studies

So far, over hundreds of robot hands, including DRHs and APHs, have been developed in academic colleges, research institutions and companies. Among these hands, the DRH is a special topic aiming to reproduce human hand's manipulation dexterity by mechatronic means. According to drive position (inside or outside the hand), the DRH can be mainly divided into two categories: intrinsic actuation pattern (IAP) or extrinsic actuation pattern (EAP). Some representative DRHs with specifications of number of fingers, number of active DOFs, actuation configuration, and transmission mechanism are shown in Table 1.

The design of robotic hands on the early stage, such as Stanford/JPL Hand and Utah/MIT Hand, are generally focused on the hand's anthropomorphism and multi-sensory integration. The DLR-I Hand [7] is a representative of the first generation DRH featured with independent actuation. To enhance the hand's appearance and operating flexibility, the DLR-II Hand [8] further introduces an extra DOF between the thumb and the ring finger for offering palm curling. Driven by air muscles, the shadow hand has more than 20 DOFs that endows the hand with noticeable grasp functions. Besides pure IAP and EAP, many DRHs (such as the iCub hand [15]) also adopt a hybrid driven pattern, wherein multimodal sensors (tactile, position, and force) are also integrated for providing more proprioception information.

Some DRHs are developed for special space and military applications, such as the NASA's Robonaut Hand I, II, and the DLR's Dexhand. Nowadays, the Robonaut hand II has been tested successful in the International Space Station (ISS) to assist astronaut. Meanwhile, the Dexhand also has some special design, such as its transmission system (Dyneema tendon plus harmonic reducer), control system (totally integrated into the hand), and communication system (CAN Bus and VxWorks controller), for properly working in the space environment. In addition, the DLR's HASy Hand is a new type multi-finger DRH to be used in the bionic Hand-Arm system [10]. It has a similar size, weight, and even behavior to a human hand. To reproduce the dynamic characteristic of the human hand, joints of the DLR HASy Hand are integrated with a special variable stiffness actuation system (VSA, consisting of servo modules and elastic elements) [33]. All actuation and electronic systems are embedded in the forearm, making it easy to be integrated in any concrete applications.

Together with DRH, the development of APH also gets a large promotion. During the last decades of 20th century, many multi-DOF prosthetic hands come into being, such as, Southampton prosthetic hand [34], Oxford Intelligent prosthetic hand [35], Stanford prosthetic hand [36], and NTU prosthetic hand [37]. Due to the actuation techniques and manufacturing level at that time, these prosthetic hands are generally large, heavy, and provided with very limited number of sensors. Upon entering 21st century, the development of APH shows a diverse tendency where the design guidelines are no longer simply "reforming" an existing DRH or totally "reproducing" the human hand. Both scope and depth of interdisciplinary fusion with relevant to mechanic, electronic, biology and control are getting strengthened in the APH development. Today, an ideal APH should possess a human-like appearance, as well as high-level dexterity. As well, it should be comfortable to wear and, more importantly, easy to control. We list a collection of representative APH prototypes, as Table 2 shows.

In particular, the DARPA extrinsic hand adopts a special actuation mechanism named Cobot [49]. It consists of one power motor and 15 steering motors that is able to output 15 channels of motions. According to specific needs, the continuously variable transmission (CVT) device (including operating motor, position sensor, power transmitting ball, operating roller and synchronizing gear sets) is able to output varying torque moment and speed.

Comparing with IAP hands, the EAP prosthetic hands are superior in compactness, dexterity, actuation manner and power. Tendon actuation is usually adopted in EAP hand since there is sufficient space in the palm allowing for more active DOFs and sensors. In addition, the actuation components outside the hands are not limited by space anymore, by which motors with greater power can be used. On the other side, considering the overall volume and weight, IAP prosthetic hands

Table 1 Dexterous robotic hands (selected)

Name	Fingers	Degree of freedom	Actuation configuration	Transmission mechanism	Ref.
Okada hand	3	11	Extrinsic	Tendon + Pulley	[6]
High-speed hand	3	8	Intrinsic	Harmonic reducer	[7]
Pinching hand	5	18	Intrinsic	Gear + Pulley	[8]
Ultrasonic hand	5	20	Intrinsic	Ultrasonic motors + Elastic elements	[9]
DLR-I hand	4	12	Intrinsic	Tendon	[10]
DLR-II hand	4	13	Intrinsic	Belt + Linkage + Gear	[11]
Dexhand	4	12	Hybrid	Tendon	[12]
DLR HASy hand	5	19	Extrinsic	Tendon	[13]
UB-II hand	3	11	Extrinsic	Tendon	[14]
UB-III hand	5	16	Extrinsic	Tendon	[15]
DIST hand	4	16	Extrinsic	Tendon	[16]
ARTS hand	5	11	Hybrid	Tendon + Gear + Worm	[17]
iCub hand	5	9	Hybrid	Tendon	[18]
Shadow hand	5	24	Extrinsic	Tendon	[19]
Stanford/JPL	3	9	Extrinsic	Tendon	[20]
Utah/MIT hand	4	16	Extrinsic	Tendon	[21]
Robonaut hand	5	14	Extrinsic	Tendon	[22, 23]
Extrinsic hand	5	11	Extrinsic	Tendon	[24]
Intrinsic hand	5	15	Intrinsic	Belt + Ballscrew	[25]
Gifu II hand	5	16	Intrinsic	Linkage + Gear	[26]
Gifu III hand	5	16	Intrinsic	Linkage + Gear	[27]
NAIST hand	4	12	Intrinsic	Linkage + Gear	[28]
NAIST hand 2	5	16	Extrinsic	Tendon + Gear	[29]
TWENTY-ONE	4	13	Intrinsic	Linkage + Gear	[30]
KIST hand	4	9	Intrinsic	Spatial linkage	[31]
ZJUT hand	5	20	Extrinsic	Flexible pneumatic actuator	[32]
DLR/HIT I	4	13	Intrinsic	Linkage + Gear	–
DLR/HIT II	5	15	Intrinsic	Belt + Tendon	–

usually use small-power direct current (DC) motors and tend to adopt a pre-tightening-free mechanism in actuation. Besides, the number of the sensors and embedded chips (CPU, ROM, etc.) are largely restricted such that sufficient information about the manipulation cannot be instantly processed. This problem could be solved along with the development of advanced electronic/computer engineering. One big merit of the IAP prosthetic hands is their application flexibility for different amputation degree. The individual difference of patients requires less re-design or re-configuration procedures for IAP prosthetic hand. Thus, these hands are more likely to be standardized, commercialized and maintained.

DLR/HIT dexterous robotic hands

The HIT-I hand (Fig. 1a) is our first dexterous hand prototype developed under a collaborative effort between Harbin Institute of Technology (HIT) and German Aerospace Center (DLR) in 2001. The HIT-I hand adopts a modular design concept that all four fingers (little finger excluded) are driven by embedded motors with tendons, because of which the degree of system integration is greatly improved and the size of the hand is well controlled at that time. Position sensor and force/torque gauges are embedded thus that the hand can accomplish some multisensory hand operations. However, due to the quality of the tendons and digital level of that time, the hand only promises a comparably low compatibility and robustness.

Based on HIT-I hand, refinement work for improving the mechanic/electronic reliability and human-like appearance is proposed in the design of DLR/HIT Hand I (Fig. 1b) [50]. Each finger is modularized as three joints with three active DOFs, wherein the metacarpophalangeal (MCP) joint has two active DOFs and the proximal interphalangeal (PIP) and distal interphalangeal (DIP) joints had one active DOF (coupled through a four-bar linkage). In the metacarpal joint (TM) of the thumb, an

Table 2 Anthropomorphic prosthetic hands (selected)

Name	Year 20~	Fingers joints DOF	Force velocity	Motors and configuration	Transmission mechanism	Size weight	Ref.
Cyber hand	03	5/15/16	70N 45°/s	6/DC Extrinsic	Tendon	95 % 360 g	[38]
Manus hand	04	5/10/4	60N 90°/s	3/BLDC Intrinsic	Tendon	120 % 300 g	[39]
IOWA hand	04	5/15/5	–	5/DC Intrinsic	Tendon	100 % 90 g	[40]
Fluid hand	04	5/8/8	110N 57°/s	1 Gear pump Intrinsic	8/Fluid actuator	100 % 350 g	[41]
Tokyo hand	05	5/15/12	0.4 Nm 200°/s	7/SM Extrinsic	Tendon	– 584 g	[42]
UB III	05	5/15/16	70N 250°/s	16/DC Extrinsic	Tendon	120 % –	[15]
SMA hand	08	5/15/7	– 41°/s	7/SMA Extrinsic	Tendon	50 % 250 g	[43]
Dong-Eui hand	08	5/15/6	14N –	6/DC Intrinsic	Tendon	– 400 g	[44]
Vanderbilt hand	09	5/16/17	–	5/GA Extrinsic	Tendon	– 580 g	[45]
Intrinsic hand	09	5/15/19	4.7 Nm 360°/s	15/BLDC Intrinsic	Motor	–	[46]
xtrinsic hand	09	5/11/21	– 360°/s	1/Cobot Extrinsic	Tendon	–	[47]
EA hand	09	5/16/5	80N 225°/s	5/DC Extrinsic	Tendon	100 % 580 g	[48]

DC direct current motor; *BLDC* brushless DC motor; *SM* servo motor; *SMA* Shape memory alloy actuators; *GA* gas actuator

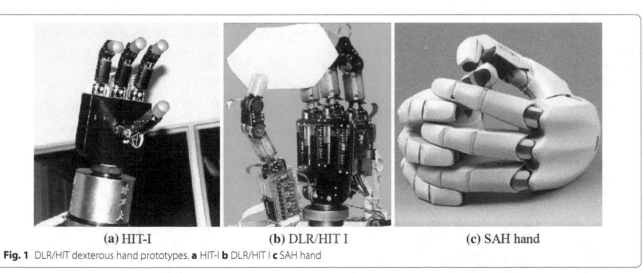

(a) HIT-I   (b) DLR/HIT I   (c) SAH hand

Fig. 1 DLR/HIT dexterous hand prototypes. **a** HIT-I **b** DLR/HIT I **c** SAH hand

extra DOF is provided for realizing thumb opposition, thus that the relative position between thumb and four digits can be ensured in various grasping tasks. Gears, harmonic reducer, and linkages constitute to the hand's transmission system. The actuation and control system is totally embedded and digitalized as much as possible. This design minimizes the hand's weight and reduces the number of tendons used for driving the joints. With collaboration of Schunk company, a commercialized version of the dexterous robot hand, SAH (Fig. 1c), is also available on the market and receives many success applications. The package design of the SAH largely improves its appearance and effectively protects the electronic system and cables within the hand.

In 2008, our newest generation DRH prototype, DLR/HIT Hand II [51], is presented with five identical fingers and a human-like curved palm, as Fig. 2 shows. To improve the hand's manipulation dexterity and operation intelligence, a total of 15 active DOFs and 140 sensors (position, force, temperature, and tactile) are integrated based on IAP. Each finger can be divided into two modulated units, 2-DOF basic joint unit and 1-DOF finger unit, within each the motors, reducers, sensors and electronic systems are totally built-in. By adopting micro brushless DC motors, timing belt, harmonic drive, and tendon coupling, the size and shape of DLR/HIT hand are significantly reduced (similar to an average mature hand). To further save space, sensors are tried to be integrated with the hand's mechanical structures, such as, the torque gauge is a transmission linkage in the basic joint, and the 3D tactile grid is an embedded layer on the finger pads. Attributed to its multimodal sensors, the DLR/HIT Hand II has a superior perception capability during various operation tasks.

Multi-fingered prosthetic hands

Since 2001, five prototypes of HIT-DLR anthropomorphic prosthetic hands (HITAPH) have been developed, as Table 3 shows. Designed based on DLR/HIT hand II, the HITAPH I–III has five fingers, and each is composed of three knuckles (2 knuckles in the thumb of HITAPH I and II). The HITAPH I–III are actuated by three DC motors, which are installed at the TM joint of the thumb, MCP joint of the index finger, and MCP joint of the middle finger, respectively. The middle finger, ring finger and little finger are co-actuated through torsional springs and linkages. Taking advantaging of the underactuation principal [52, 53], the inter-finger actuation of the hand is realized through elastic components, which provides the hands with an adaptive grasp to various objects. Among them, the HITAPH III [also called anthropomorphic robot (AR) hand III] [54] makes an improvement on its human-like appearance and grasp power.

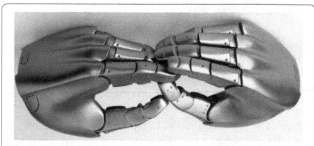

Fig. 2 DLR/HIT Hand II

To further improve the hand's dexterity, the HITAPH IV [5] is developed with five active DOFs (that is, all five fingers are individually actuated). Attributed to advanced actuation techniques, the volume of the IAP hand is only 85 % of that of HITAPH III. The total hand weight is about 450 g. The output force at the fingertip can reach up to 10N. Curved palm and scattered finger configuration endow the hand with more anthropomorphic characters in appearance and grasping. Reducing the number of the non-standardized mechanical and electronic elements is one critical request in the design of HITAPH IV, aiming to improve the hand's interchangeability and maintainability. Meanwhile, the packaging design of the hand is also considered in the design, in connection with its actuation capacity, thermal and life design.

To further improve the thumb finger's mobility, an additional DOF is provided at the TM joint for realizing opposition. Attributed to this extra DOF, the thumb can reach to each fingertip of the other four fingers. Another big revision is the reduced number of knuckles, (two, instead of three in HITAPH IV [55]), for briefing the mechanical structure while keeping the hand's functionality and reliability. A total of six DC motors are embedded, while the actuation force at the fingertip can reach up to 12N. The worm gear, instead of bevel gears, is adopted in the MCP joints, while tendon coupling mechanism is maintained in the DIP joints. With addition to the position and force sensors, a 3-D tactile sensor [56] is designed that can measure one perpendicular force and two tangential forces applied on the fingertips. The number of the parts of the HITAPH IV is largely reduced compared with former prototypes, making it very promising in commercialization.

Challenges and future work

There are mainly two trends for developing DRHs, one is anthropomorphism-oriented and the other is task-oriented. For the first one, the robot hands are devised with much more human-like properties, as in its kinematics (hand structure, DOFs, grasping functionality, etc.), dynamics (stiffness, damping, friction, etc.), and perception capabilities (position sensors, force/torque sensors, tactile grid, slipping sensors, etc.). While for the task-oriented trend, the DRHs are designed according to some specialized tasks or environments, such as the Robonaut 2 hand and Dexhand, both for extravehicular activities on the ISS. Generally speaking, the word "anthropomorphism" is a very complicated concept including numerous influential factors that lacking anyone of them may lead to an underperformance design. Under current conditions, how to make a compromise between the hand's appearance and functionality, or how to establish a proper performance index [57] to compare design and

Table 3 Specifications of the HIT-DLR anthropomorphic prosthetic hands

Picture	Name	Year	Fingers/ joints/ motors	actuation- transmission	Sensors	mass size	Force velocity
	HITAPH I	2003	5/14/3	Underactuated linkage + Belt 1-1-2	Position, torque	500 g 130 %	10N –
	HITAPH II	2005	5/14/3	Underactuated linkage + Belt 1-1-3	Position, torque	500 g 110 %	10N –
	HITAPH III	2007	5/15/3	Underactuated linkage 1-1-3	Position, torque	500 g 120 %	10N 72°/s
	HITAPH hand IV	2009	5/15/5	Coupling linkage 1-1-1-1-1	Position, torque	450 g 85 %	10N 89°/s
	HITAPH hand IV	2013	5/11/6	Tendon + Worm 2-1-1-1-1	Position, torque, 3D-tactile	475 g 80 %	12N 45°/s

thus make a suitable choice among alternatives, is still an urgent work in the robot hand research.

Attributed to its thumb's dexterity, the human hand can achieve so many versatile grasps and delicate operations. Decoding the thumb's movement in respect of DOF configuration is a big challenge in robotics research. We devote to analysis the thumb's movement based on human hand anatomy and biomechanics theory, and then to give a set of appropriate DOF configurations (TM-MCP-DIP, 2-1-1, 2-2-1 [58], 3-1-1 [59]) that can be used in DRHs for achieving a variety of dexterous operations. From the point of directional dexterity, we attempt to analyze these configurations on specific manipulation tasks, which can further facilitate our selection according to different application scenarios. How to arrange the thumb on the hand is another challenge. For achieving

versatile and effective grasp patterns, an efficient method needs to be proposed for appropriately positioning the thumb on the palm.

The dexterous manipulations requested by the DRHs are not only promised by its anthropomorphic shape and motion, but also its high-speed processing system (sensor measurement, data analysis, kinematic calculation, etc.) and real-time control algorithms (task interpretation, motion planning, sensory feedback, etc.). Because of massive data calculation and interchange, selection of an appropriate control structure and platform, highly-integrated hardware and software hierarchy and suitable communication protocol are critical for DRH realizing a real-time manipulation. Currently, based on EtherCAT, a real-time control design and validation platform [60] has been developed, on which a large variety of algorithms, such as the impedance control strategy with coordinated multi-finger manipulation and optimized grasping forces [61], can be been verified.

For achieving delicate manipulations, the DRHs need to know necessary information about its outer environment (obstacles) and the object (stiffness, size, shape, and weight) to operate. The information is generally provides by our proprioception experiences (body schema) sensed by our central neural system through a long-term, multi-sensory stimulation (visual, tactile, force, tension, etc.). The tactile sensor [62] has been now widely integrated into the fingertip to acquire such information as the contact status, position, and some other physical properties (stiffness, texture, etc.) about the object. Even a primary haptic sense (object shape and category) can be reconstructed by using the tactile sensor and position sensors integrated on the robot hand. However, how to effectively fuse these information into the control scheme, thus to improve the hand's operation compliance, precision and intelligence, is still an open question [63].

The main task of a prosthetic hand is to to help physically disabled people restore hand functions in living environment (ADLs). General APHs should have three main features, as human-like appearance (size, weight, textures, compliance, etc.), mobility, and perception. Besides, state-of-the-art APHs request even more dexterous operations, given very limited choice on the actuation styles and DOF configurations. How to realize a large portion of human hand functions in very low actuation cost is very ambitious in APH research. Besides, high-precision position control (such as, to nip a needle) and accurate force control (such as, to grasp a fragile cup) are both frequently required in the daily use of APHs. How to devise a smart control strategy properly working on different conditions is another question. For controlling the prosthetic hands, the surface myoelectric signals (sEMG) collected from the residual neuromuscular system (stump) are widely accepted. Traditional mode-switching methods established on EMG amplitude only give very limited functions, discrete robot-like finger movements, and unintuitive control feelings. By introducing the pattern recognition method [64], a large progress has been made; however, there is still a big gap between the research and its real application [65, 66]. Intrinsic timing-varying characters of the EMG signals, environmental change (electromechanical status, temperature, moisture, sweating, etc.) of the bio-machine interface, and confounding factors (body postures, contraction strength variations, involuntary EMG activations) largely affect the long-term usage of clinical APHs. In this case, the control of APHs should consider other alternative peripheral nervous signals, such as ultrasonic signal [67], mechanomyography [68], near-infrared spectroscopy [69] and electrocorticography [70], to be used in the control channel, and multi-sensory means [71], such as vision [72, 73] and tactile sense [74, 75], to be used in the feedback channel. With the big progress of the worldwide scientific research on artificial cognition and brain-computer interface, a fully-embodied hand avatar controlled by our brain with utmost ease will come soon.

Conclusions

This study focuses on the introduction of the development route of intrinsic actuation dexterous hands and prosthetic hands, giving a brief overview on the current artificial dexterous hands and prosthetic hands. With the progressing of science and technology, robotic hands are gradually approximating to human hands in dexterity and perception, based on which they can finish various complicated operations in the manufacturing process, activities of daily life, and exploration of unknown environment.

Authors' contributions

HL conceived the manuscript, participated in the data collection and analysis, and helped revise the manuscript. DPY carried out the survey study and drafted the manuscript. SWF helped draft the manuscript. HGC helped supervise the study and revise the manuscript. All authors read and approved the final manuscript.

Authors' information

Liu H received the B.S. and the Ph.D. degrees from the Harbin Institute of Technology (HIT), Harbin, China, in 1986 and 1993, respectively. In 1991, he joined the German Aerospace Center (DLR) as joint Ph.D. candidate and worked on three finger robot hand. Since 1993, he has worked as a Key Researcher on the DLR four-finger Hand and the HIT/DLR five-finger Hand. In 1999, he received the Chang Jiang Scholars Program and received the Professorship of the HIT. Currently he is the head of School of Mechatronic Engineering, HIT and the director of the State Key Lab of Robotics and Systems. His research interests include space robotics, dexterous robot hands, and prosthetic hand systems.Yang DP received the B.S. degree in mechanic engineering from the Harbin Institute of Technology, in 2004, and the M.S. and Ph.D. degrees in mechatronic engineering from the Harbin Institute of Technology (HIT), in 2006 and 2011, respectively. Since 2011, he has been with the State Key Laboratory of Robotics and System, HIT, where he holds the position of an associated professor. His research interests include dexterous robotic/prosthetic hand systems, biomedical signal processing and control, human–machine interaction, and machine learning techniques in robotics.Fan SW received his B.S., M.S. and Ph.D. degrees in mechatronics engineering from Harbin Institute of Technology, Harbin, China, in 2001, 2005, and 2010, respectively. Since 2005, he has been with the State Key Laboratory of Robotics and System, Harbin Institute of Technology, where he currently holds the position of an associate researcher. His current research interests include anthropomorphic robot dexterous hand, prosthetics hand, tactile sensor, and grasping.Cai HG received the B.S. degree in mechanic engineering from the Harbin Institute of Technology, in 1958. Since 1985, he has been a professor at the School of Mechatronic Engineering, Harbin Institute of Technology. Since 1997, he has been an academician of the Chinese academy of engineering. Prof. Cai successfully developed China's first welding robots and spot welding robots, and solved critical problems in the robots' trajectory control accuracy and path predictive control. His current research interests include industrial robot, space robot and control algorithms.

Acknowledgements

This work is in part supported by the National Program on Key Basic Research Project (973 Program, No. 2011CB013306), and the National Natural Science Foundation of China (No. 51205080).

Competing interests
The authors declare that they have no competing interests.

References

1. Pons JL, Ceres R, Pfeiffer F (1999) Multifingered dextrous robotics hand design and control: a review. Robotica 17(6):661–674
2. Bicchi A (2000) Hands for dexterous manipulation and robust grasping: a difficult road toward simplicity. IEEE Trans Robot Autom 16(6):652–662
3. Atkins DJ, Heard DC, Donovan WH (1996) Epidemiologic overview of individuals with upper-limb loss and their reported research priorities. J Prosthet Orthot 8(1):2–11
4. Peerdeman B, Boere D, Witteveen H, in't Veld RH, Hermens H, Stramigioli S, Rietman H, Veltink P, Misra S (2011) Myoelectric forearm prostheses: state of the art from a user-centered perspective. J Rehabil Res Dev 48(6):719–737
5. Liu H, Yang DP, Jiang L, Fan SW (2014) Development of a multi-DOF prosthetic hand with intrinsic actuation, intuitive control and sensory feedback. Ind Robot Int J 41(4):381–392
6. Okada T (1979) Object-handling system for manual industry. IEEE Trans Syst Man Cybern 9(2):79–89
7. Namiki A, Imai Y, Kaneko M, Ishikawa M (2004) Development of a high-speed Multifingered Hand System. In: Proceedings of the international conference on intelligent manipulation and grasping, Genova, pp 85–90
8. Hoshino K, Kawabuchi I (2005) Pinching with finger tips in humanoid robot hand. In: Proceedings, 12th international conference on Advanced robotics ICAR, IEEE, 2005, pp 705–712
9. Yamano I, Maeno T (2005) Five-fingered robot hand using ultrasonic motors and elastic elements. In: Proceedings of the 2005 IEEE international conference on robotics and automation ICRA, IEEE, pp 2673–2678
10. Hirzinger G, Fischer M, Brunner B, Koeppe R, Otter M, Grebenstein M, Schäfer I (1999) Advances in robotics: the DLR experience. Int J Robot Res 18(11):1064–1087
11. Butterfaß J, Grebenstein M, Liu H, Hirzinger G (2001) DLR-Hand II: Next generation of a dextrous robot hand. In: Proceedings 2001 IEEE international conference on robotics and automation ICRA. IEEE, pp 109–114
12. Chalon M, Wedler A, Baumann A, Bertleff W, Beyer A, Butterfaß J, Grebenstein M, Gruber R, Hacker F, Kraemer E Dexhand (2011) A Space qualified multi-fingered robotic hand. In: IEEE international conference on robotics and automation ICRA, IEEE, pp 2204–2210
13. Grebenstein M, Albu-Schäffer A, Bahls T, Chalon M, Eiberger O, Friedl W, Gruber R, Haddadin S, Hagn U, Haslinger R The DLR hand arm system (2011) In: IEEE international conference on robotics and automation ICRA, IEEE, pp 3175–3182
14. Melchiorri C, Vassura G (1994) Implementation of whole-hand manipulation capability in the UB hand system design. Adv Robot 9(5):547–560
15. Lotti F, Tiezzi P, Vassura G, Biagiotti L, Palli G, Melchiorri C (2005) Development of UB hand 3: early results. In: Proceedings of the 2005 IEEE international conference on robotics and automation ICRA, IEEE, pp 4488–4493
16. Caffaz A, Cannata G (1998) The design and development of the DIST-Hand dextrous gripper. In: Proceedings. 1998 IEEE international conference on robotics and automation, IEEE, 1998, pp 2075–2080
17. Controzzi M, Cipriani C, Jehenne B, Donati M, Carrozza MC (2010) Bio-inspired mechanical design of a tendon-driven dexterous prosthetic hand. In:2010 annual international conference of the IEEE on engineering in medicine and biology society (EMBC), IEEE, pp 499–502
18. Schmitz A, Pattacini U, Nori F, Natale L, Metta G, Sandini G (2010) Design, realization and sensorization of the dexterous iCub hand. In: 2010 10th IEEE-RAS international conference on humanoid robots (humanoids), IEEE, pp 186–191
19. Shadow Dexterous Hand™—Now available for purchase! http://www.shadowrobot.com/products/dexterous-hand/ Accessed 01.08.2016
20. Mason MT, Salisbury JK Jr (1985) Robot hands and the mechanics of manipulation. The MIT Press, Cambridge, pp 3–93
21. Jacobsen SC, Wood JE, Knutti D, Biggers KB (1984) The UTAH/MIT dextrous hand: work in progress. Int J Robot Res 3(4):21–50
22. Lovchik C, Diftler MA (1999) The robonaut hand: A dexterous robot hand for space. In: Proceedings. 1999 IEEE international conference on robotics and automation, IEEE, pp 907–912
23. Bridgwater LB, Ihrke C, Diftler MA, Abdallah ME, Radford NA, Rogers J, Yayathi S, Askew RS, Linn DM (2012) The robonaut 2 hand-designed to do work with tools. In: 2012 IEEE international conference on robotics and automation (ICRA), IEEE, pp 3425–3430
24. Mitchell M, Weir RF (2008) Development of a clinically viable multifunctional hand prosthesis. Proceedings of the 2008 myoelectric controls/ powered prosthetics symposium, Fredericton. IEEE, New Brunswick, pp 45–49
25. Weir R, Mitchell M, Clark S, Puchhammer G, Haslinger M, Grausenburger R, Kumar N, Hofbauer R, Kushnigg P, Cornelius V (2008) The intrinsic hand–a 22 degree-of-freedom artificial hand-wrist replacement. Proceedings of myoelectric controls powered prosthetics symposium, IEEE, New Brunswick, 233–237
26. Kawasaki H, Komatsu T, Uchiyama K (2002) Dexterous anthropomorphic robot hand with distributed tactile sensor: gifu hand II. IEEE/AMSE Trans Mechatron 7(3):296–303
27. Mouri T, Kawasaki H, Yoshikawa K, Takai J, Ito S (2002) Anthropomorphic robot hand: Gifu hand III. Proceedings of the 2002 IEEE international conference on autonomic and autonomous systems, Jeonbuk. Korea, IEEE, pp 1288–1293
28. Ueda J, Kondo M, Ogasawara T (2010) The multifingered NAIST hand system for robot in-hand manipulation. Mech Mach Theory 45(2):224–238
29. Kurita Y, Ono Y, Ikeda A, Ogasawara T (2011) Human-sized anthropomorphic robot hand with detachable mechanism at the wrist. Mech Mach Theory 46(1):53–66
30. Iwata H, Sugano S (2009) Design of anthropomorphic dexterous hand with passive joints and sensitive soft skins. In: SII 2009. IEEE/SICE International Symposium on System Integration, IEEE, pp 129–134
31. Kim E-H, Lee S-W, Lee Y-K (2011) A dexterous robot hand with a bio-mimetic mechanism. Int J Precis Eng Manufactur 12(2):227–235
32. Wang ZH, Zhang LB, Gzj Bao, Qian SM, Yang QH (2011) Design and control of integrated pneumatic dexterous robot finger. J Central South Univ Technol 18:1105–1114
33. Schäffer A, Eiberger O, Grebenstein M, Haddadin S, Ott C, Wimböck T, Wolf S, Hirzinger G (2008) Soft robotics, from torque feedback-controlled lightweight robots to intrinsically compliant systems. IEEE Robot Autom Mag 15(3):20–30
34. Kyberd PJ, Chappell PH (1994) The southampton hand: an intelligent myoelectric prosthesis. J Rehabil Res Dev 31(4):326–334
35. Kyberd PJ, Evans MJ, Winkel S (1998) An intelligent anthropomorphic hand with automatic grasp. Robotica 16:531–536
36. Dashy R, Yen C, Leblanc M (1998) The design and development of a gloveless endoskeleton prosthetic hand. J Rehabil Res Dev 35(4):388–395
37. Huang HP, Chen CY (1999) Development of a myoelectric discrimination system for a multi-degree prosthetic hand. Proceedings of the 1999 IEEE international conference on robotics and automation, detroit. Michigan, IEEE, pp 2392–2397
38. Carrozza MC, Cappiello G, Micera S, Edin BB, Beccai L, Cipriani C (2006) Design of a cybernetic hand for perception and action. Biol Cybern 95:629–644
39. Pons JL, Rocon E, Ceres R (2004) The MANUS-HAND dextrous robotics upper limb prosthesis: mechanical and manipulation aspects. Autonomous Robots 16:143–163
40. Yang J, Pitarch EP, Abdel-Malek K, Patrick A, Lindkvist L (2004) A multifingered hand prosthesis. Mech Mach Theory 39(6):555–581
41. Kargov A, Werner T, Pylatiuk C, Schulz S (2008) Development of a miniaturised hydraulic actuation system for artificial hands. Sens Actuators, A 141(2):548–557
42. Arieta AH, Katoh R, Yokoi H, Wenwei Y (2006) Development of a multi-DOF electromyography prosthetic system using the adaptive joint mechanism. Appl Bion Biomech 3(2):101–111
43. Andrianesis K, Tzes A (2008) Design of an anthropomorphic prosthetic hand driven by shape memory alloy actuators. In: Proceedings of the 2nd Biennial IEEE/RAS-EMBS international conference on biomedical robotics and biomechatronics, biorob, pp 517–522
44. Jung SY, Moon I (2008) Grip force modeling of a tendon-driven prosthetic hand. In: 2008 international conference on control, automation and systems, ICCAS, Seoul, pp 2006–2009

45. Fite KB, Withrow TJ, Shen X, Wait KW, Mitchell JE, Goldfarb M (2008) A gas-actuated anthropomorphic prosthesis for transhumeral amputees. IEEE Trans Rob 24(1):159–169

46. Weir R, Mitchell M, Clark S, Puchhammer G, Haslinger M, Grausenburger R, Kumar N, Hofbauer R, Kushnigg P, Cornelius V (2008) The intrinsic hand–a 22 degree-of-freedom artificial hand-wrist replacement. In: Proceedings of myoelectric controls/powered prosthetics symposium, New Brunswick, pp 233–237

47. Mitchell M, Weir R F (2008) Development of a clinically viable multifunctional hand prosthesis. In Proceedings of the 2008 myoelectric controls powered prosthetics symposium Fredericton, IEEE, New Brunswick, p 45–49

48. Dalley SA, Wiste TE, Withrow TJ, Goldfarb M (2009) Design of a multifunctional anthropomorphic prosthetic hand with extrinsic actuation. Mechatron, IEEE/ASME Trans 14(6):699–706

49. Kinea Design LLC. (2011). http://www.kineadesign.com. Accessed 08 Apr 2011

50. Liu H, Meusel P, Hirzinger G, Jin M, Liu Y, Xie Z (2008) The modular multisensory DLR-HIT-Hand: hardware and software architecture. IEEE/ASME Trans Mechatron 13(4):461–469

51. Liu H, Wu K, Meusel P, Seitz N, Hirzinger G, Jin M, Liu Y, Fan S, Lan T, Chen Z (2008) Multisensory five-finger dexterous hand: The DLR/HIT Hand II. In: IEEE/RSJ international conference on intelligent robots and systems IROS, IEEE, pp 3692–3697

52. Dechev N, Cleghorn W, Naumann S (2001) Multiple finger, passive adaptive grasp prosthetic hand. Mech Mach Theory 36(10):1157–1173

53. Laliberté T, Birglen L, Gosselin C (2002) Underactuation in robotic grasping hands. Machine Intell Robot Control 4(3):1–11

54. Yang D, Zhao J-d Gu, Y-k Wang X-q, Li N, Liu H, Jiang L, Huang H, D-w Zhao (2009) An anthropomorphic robot hand developed based on underactuated mechanism and controlled by emg signals. J Bionic Eng 6(3):255–263

55. Jiang L, Zeng B, Fan S, Sun K, Zhang T, Liu H A (2014) modular multisensory prosthetic hand. In: 2014 IEEE international conference on information and automation (ICIA), IEEE, pp 648–653

56. Zhang T, Liu H, Jiang L, Fan S, Yang J (2013) Development of a flexible 3-d tactile sensor system for anthropomorphic artificial hand. IEEE Sens J 2:510–518

57. Hioki M, Ebisawa S, Sakaeda H, Mouri T, Nakagawa S, Uchida Y, Kawasaki H (2011) Design and control of electromyogram prosthetic hand with high grasping force. In: 2011 IEEE international conference on robotics and biomimetics (ROBIO), IEEE, pp 1128–1133

58. Santos VJ, Valero-Cuevas FJ (2006) Reported anatomical variability naturally leads to multimodal distributions of Denavit-Hartenberg parameters for the human thumb. Biomed Eng, IEEE Trans 53(2):155–163

59. Rezzoug N, Gorce P (2008) Prediction of fingers posture using artificial neural networks. J Biomech 41(12):2743–2749

60. Hou M, Jiang L, Jin M, Liu H, Chen Z (2014) Analysis of the multi-finger dynamics for robot hand system based on EtherCAT. In: 2014 10th international conference on natural computation (ICNC), IEEE, pp 1061–1065

61. Hou M, Jiang L, Jin M, Liu H, Chen Z (2014) Strategies to optimize fingertip force for impedance control of robot hand based on EtherCAT.

In: Proceedings of the 2014 Asia-Pacific conference on computer science and applications (CSAC 2014) computer science and applications, CRC Press, Shanghai, p 245

62. Antfolk C, Balkenius C, Rosen B, Lundborg G, Sebelius F (2010) SmartHand tactile display: a new concept for providing sensory feedback in hand prostheses. Scand J Plast Reconstr Surg Hand Surg 44(1):50–53

63. Raspopovic S, Capogrosso M, Petrini FM, Bonizzato M, Rigosa J, Di Pino G, Carpaneto J, Controzzi M, Boretius T, Fernandez E (2014) Restoring natural sensory feedback in real-time bidirectional hand prostheses. Sci transl med 6(222):222ra219

64. Fang Y, Hettiarachchi N, Zhou D, Liu H (2015) Multi-modal sensing techniques for interfacing hand prostheses: a review. Sensors J IEEE 15(11):6065–6076

65. Castellini C, Artemiadis P, Wininger M, Ajoudani A, Alimusaj M, Bicchi A, Caputo B, Craelius W, Dosen S, Englehart K, Farina D, Gijsberts A, Godfrey SB, Hargrove L, Ison M, Kuiken T, Markovic M, Pilarski P, Rupp R, Scheme E (2014) Proceedings of the first workshop on peripheral machine interfaces: going beyond traditional surface electromyography. Front Neurorobotics 8 22:21. doi:10.3389/fnbot.2014.00022

66. Ning J, Dosen S, Muller KR, Farina D (2012) Myoelectric control of artificial limbs: is there a need to change focus? IEEE Signal Process Mag 29(5):148–152

67. Ravindra V, Castellini C (2014) A comparative analysis of three non-invasive human-machine interfaces for the disabled. Front neurorobotics 8:24. doi:10.3389/fnbot.2014.00024

68. Chen X, Zheng YP, Guo JY, Shi J (2010) Sonomyography (SMG) control for powered prosthetic hand: a study with normal subjects. Ultrasound Med Biol 36(7):1076–1088

69. Siesler HW, Ozaki Y, Kawata S, Heise HM (2008) Near-infrared spectroscopy: principles, instruments, applications. Wiley, New York

70. Fifer MS, Acharya S, Benz HL, Mollazadeh M, Crone NE, Thakor NV (2012) Towards electrocorticographic control of a dexterous upper limb prosthesis. IEEE PULSE 3(1):38–42

71. McMullen DP, Hotson G, Katyal KD, Wester BA, Fifer MS, McGee TG, Harris A, Johannes MS, Vogelstein RJ, Ravitz AD, Anderson WS, Thakor NV, Crone NE (2014) Demonstration of a semi-autonomous hybrid brain-machine interface using human intracranial eeg, eye tracking, and computer vision to control a robotic upper limb prosthetic. Neural Syst Rehabil Eng, IEEE Trans 22(4):784–796. doi:10.1109/TNSRE.2013.2294685

72. Markovic M, Dosen S, Cipriani C, Popovic D, Farina D (2014) Stereovision and augmented reality for closed-loop control of grasping in hand prostheses. J Neural Eng 11(4):046001

73. Markovic M, Dosen S, Popovic D, Graimann B, Farina D (2015) Sensor fusion and computer vision for context-aware control of a multi degree-of-freedom prosthesis. J Neural Eng 12(6):066022

74. Riso RR (1999) Strategies for providing upper extremity amputees with tactile and hand position feedback–moving closer to the bionic arm. Technol Health Care 7(6):401–409

75. Kyberd PJ, Mustapha N, Carnegie F, Chappell PH (1993) A clinical experience with a hierarchically controlled myoelectric hand prosthesis with vibro-tactile feedback. Prosthet Orthot Int 17(1):56–64

6

An image processing method for changing endoscope direction based on pupil movement

Yang Cao[1*], Satoshi Miura[1], Quanquan Liu[2], Yo Kobayashi[1], Kazuya Kawamura[3], Shigeki Sugano[1] and Masakatsu G. Fujie[1]

Abstract

Increased attention has been focused on laparoscopic surgery because of its minimal invasiveness and improved cosmetic properties. However, the procedure of laparoscopic surgery is considerably difficult for surgeons, thus paving the way for the introduction of robotic technology to reduce the surgeon's burden. Thus, we have developed a single-port surgery assistive robot with a master–slave structure that has two surgical manipulators and a sheath manipulator for the alteration of endoscope direction. During the development of the surgical robotic system, achieving intuitive operation is very important. In this paper, we propose a new laparoscope manipulator control system based on the movement of the pupils to enhance intuitive operability. We achieve this using a webcam and an image processing method. After the pupil movement data are obtained, the master computer transforms these data into an output signal, and then the slave computer receives and uses that signal to drive the robot. The details of the system and the pupil detection procedure are explained. The aim of the present experiment is to verify the effectiveness of the image processing method applied to the alteration of endoscope direction control system. For this purpose, we need to determine an appropriate pupil motion activation threshold to begin the sheath manipulator's movement. We used four kinds of activation threshold, measuring the time cost of a particular operation: to move the image of the endoscope to a specific target position. Moreover, we identified an appropriate activation threshold that can be used to determine whether the endoscope is moving.

Keywords: Non-rigid face tracking, Single-port endoscopic surgery, Master–slave structure, Image processing, Double-screw-drive mechanism, Activation threshold

Background

Laparoscopic surgery is a technique whereby a laparoscope and surgical instruments are inserted into the patient's body through an artificial or natural body cavity, followed by the surgeon operating the instruments based on the monitor image captured by the laparoscope [1]. Laparoscopic surgery has many advantages, such as shorter hospitalization times, lower physical burden on patients and cosmetic improvement [2] compared with open surgery. Although laparoscopic surgery has many advantages for patients as described above, the difficulty of performing the technique is so high that surgeons need to carry out long-term training at a special medical training center. Even if they have experienced professional training, they still suffer from high mental stress during operations, which may reduce their dexterity or judgment [3]. One of the causes of such problems is that laparoscopic surgery requires another operator to hold the laparoscope for the surgeon. Therefore, the coordination between operators has a significant effect on the process and result of laparoscopic surgery [4].

To tackle the problems mentioned above, robotic technology is an effective solution. Naviot [5] provides surgeons with the possibility of solo surgery, and requires the surgeon to use one hand to hold a controller when

*Correspondence: a15caoyang@fuji.waseda.jp
[1] Graduate School of Creative Science and Engineering, Waseda University, 3-4-1 Ohkubo, Shinjuku-ku, Tokyo 169-8555, Japan
Full list of author information is available at the end of the article

they need to alter the direction of the laparoscope. The Da Vinci Surgical System [6] can change the control mode between the surgical manipulator and the endoscopic manipulator for the operator using a foot pedal. However, such methods do not allow the surgeon to alter the operative field while simultaneously manipulating tissue [7].

Other systems, like the automatic endoscope optimal positioning system (AESOP) from Computer Motion Inc. [8] and ViKY EP [9] use a voice control system to allow the operator to control the robotic endoscopic holder. However, rather than being helped by voice control, surgical time is actually often increased because of its slow response [10].

Alongside voice control, eye-tracking is a further intuitive hands-free method that can be used as an input signal for manipulating the laparoscope holder. As a practical example, Ubeda et al. [11] used the electrooculography signal from electrodes attached around the user's eye to control a manipulator. In the video-oculography field, Noonan, et al. [12] used a stand-alone eye tracker to alter the laparoscope direction based on gaze.

Single port surgery (SPS) is one of form of laparoscopic surgery that requires only one incision port. SPS has attracted increasing attention from patients because of its cosmetic advantages [13]. Although robots enhance the performance of standardized laparoscopic techniques [14], current systems, including those mentioned in the previous paragraphs, are not suitable for SPS because they require multiple incisions. Therefore, our laboratory has developed two prototypes of an SPS assistive robot system. Prototype 1 [15] uses respective controllers for tool manipulators and laparoscope direction; in Prototype 2 [16], the control mode can be changed between tool manipulators and endoscope direction by pushing a foot pedal like in the Da Vinci Surgical System [6].

The purpose of this paper is to introduce a control method for the alteration of endoscope direction using pupil-tracking into the Prototype 2 [16] system, which is achieved via image processing and the use of a webcam. The system will translate the obtained image data into an output signal. We propose a threshold for pupil movement distance: the sheath manipulator, which is used for the alteration of endoscope direction, activates when the user's pupil movement distance exceeds a threshold value; the manipulator remains in a static state when the pupil movement distance is below the threshold. Therefore, an appropriate threshold for the output signal needs to be determined to judge whether the movement state of the sheath manipulator is dynamic or static. To determine this, we tested four threshold values one by one in a horizontal movement experiment. The experimental outcome variable was the completion time of moving the

field of view to a specific target, so that the most appropriate threshold could be judged from the minimum completion time. The first part of this paper will describe how we obtain the pupil movement data. Then, the general system framework, including how the system integration is achieved, will be discussed. The second half of this paper describes an experiment to verify the effectiveness of the system and obtain a proper activation threshold value, which is important for the operability of pupil tracking.

Methods
Acquisition method of pupil movement
In this study, a webcam was used to perform eye tracking. Webcams have an advantage compared with wearable gaze tracking systems, as they need not be mounted on the user's head. Such head-mounted displays or glasses-type head-wearable devices may induce additional physical and mental stress on the surgeon [17]. In contrast, the presence of a webcam has little or negligible burden on surgeons. During the development of this system, a near infrared (IR) LED webcam was eventually selected for one reason: the color of East Asian people's pupils in visible light is the same as that of the iris, but reflects white in near infrared light [18]. So the near infrared reflection features from the pupil were used in the image processing of the pupil movement data. Additionally, it is worth mentioning that near infrared light is harmless to the eye; this kind of light has been used in retina identification technology for a long time [19]. The infrared LED webcam (DC-NCR13U, Hanwha Q CELLS Japan Co., Ltd.) is used in this research, and its resolution is 1.3 million pixels. Figure 1a shows the control console of the prototype [20] and the set position of the webcam. Generally, the webcam is set on top of the display, but this may cause problems in capturing the whole eye because eyelashes or bangs may interfere. Thus, we set up the webcam in the position shown in Fig. 1b.

Applying non-rigid face tracking
An image processing method, non-rigid face tracking, was used to collect pupil movement data in this study. This method involves machine learning. The main merit of non-rigid face tracking, compared with traditional face tracking methods, is that it is non-rigid in that it can detect an individual face from the entire image and can also measure the relative distance between the feature positions, e.g., corners of the mouth, canthi and pupils. Thus, this method was used for pupil movement detection. The program used for image processing was OpenCV 2.4.4. The specific realization of non-rigid face tracking refers to [21]. An overview of the image processing is shown in Fig. 2. The complete procedure

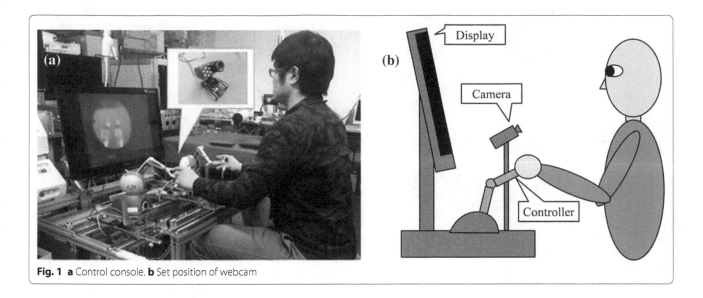

Fig. 1 **a** Control console. **b** Set position of webcam

consists of image capture, annotation, geometrical constraints, training detector, pupil tracking and setting the threshold.

Image capture

The first step was to use the IR LED webcam to capture as many of the operator's facial images as possible (Fig. 3). The more images the webcam captures, the more plentiful data the system has, which aids in training a robust face detector because the data include different lighting conditions, facial positions and eye directions.

Annotation

The second step was to annotate the tracking feature points on the facial image that could highlight the movement of the eyes. In computer vision, corner

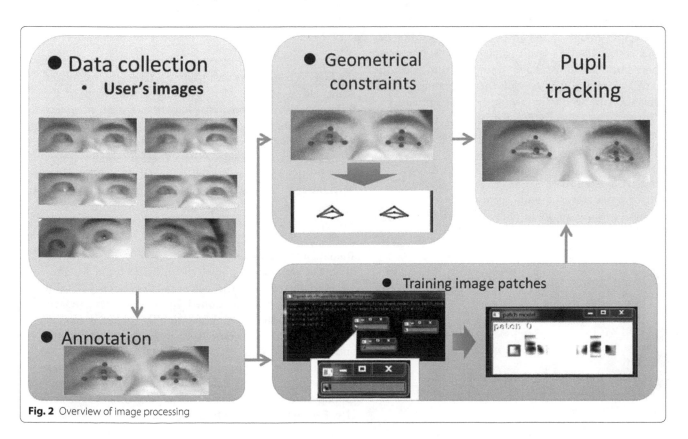

Fig. 2 Overview of image processing

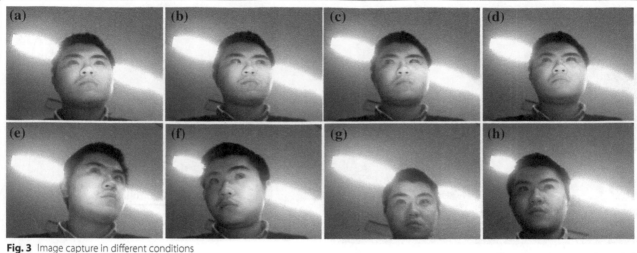

Fig. 3 Image capture in different conditions

points are easily detected, so we annotated the eye corners. Moreover, our system needs to be able to track pupil points. Hence, we also annotated the pupils as feature points (Fig. 4). After annotation, we indexed the symmetry points. For example, the two inner eye corners and two pupils can be regarded as symmetry points. Symmetry indices can be used to mirror the captured images, which can increase the training dataset. Then, we connected the annotated points (Fig. 5), which are used for visualizing the pupil-tracking effect. Finally, the remaining images need to be processed using the previous steps. The size and resolution of these images are the same. The annotation points of the previous image will remain and appear on the next image. There is a deviation between the annotation points and the facial feature points; therefore, we need to use a mouse pointer to drag the annotation points to the facial feature points (Fig. 6).

Geometrical constraints

The set of annotations should correspond to physically consistent locations on the human face. In face tracking,

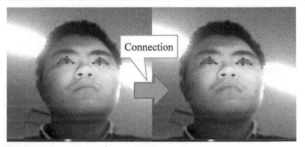

Fig. 5 Connecting the feature points

the face placement on an image will translate, rotate and scale. However, the corresponding relationship of the facial organs' positions, also known as facial geometry, will not change. Therefore, the corresponding relationship of the annotation points should also be constrained according to the facial geometry. These annotated points were combined as a geometrical model using Procrustes Analysis [22] to handle several variations on the pose of the user's face in the image such as translation, rotation and scale, and a shape model was established (Fig. 7).

Training image patches

To make annotation points to track the facial features, we need to train discriminative patch models [21]. The annotated images mentioned in the Annotation section comprise the training dataset. The image patch can be independently trained from every annotated point, and the trained patch is cross-correlated with an image region containing the annotated point. Then, the image patch will strongly respond to the part of the image containing the facial feature, and weakly respond everywhere else. Figure 8 shows the training process: the patches for

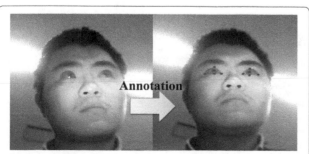

Fig. 4 Annotation of the feature points

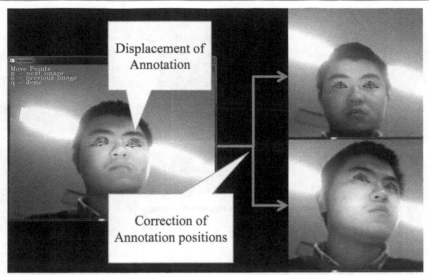

Fig. 6 Correcting the positions of annotation points

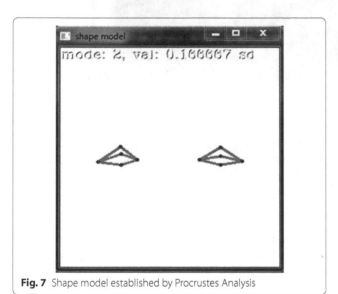

Fig. 7 Shape model established by Procrustes Analysis

detecting eye corners and pupils are being trained. Then, the result of the patch training is shown in Fig. 9.

Pupil tracking

Before processing pupil tracking, the last step is to combine the shape model and the image patches into one tracker model (Fig. 10). The reason for this step is that the image patch may respond strongly at an incorrect facial feature point. An accurate estimate of the facial feature position can be determined via the geometrical model, and then the facial features can be detected with the cross-correlation image patches, producing good robustness and efficiency. Therefore, we need the shape model to restrict the matching region of the image patches. Moreover, initialization of pupil tracking should result in the ability to place the tracker model on the user's face in the image. In this step we use OpenCV's built-in face detector to confirm the box region of the user's face in the captured images. Then, the tracker model can start the tracking operation from the box region of the user's face.

After completing the above steps, the face tracker was able to achieve pupil tracking because the positions of both intraocular angles and pupils could be obtained from the tracking trajectory. As shown in Fig. 11a, the midpoint of both pupils is labeled as the active point (red circle), while the midpoint of both inner canthus angles is labeled as the static point (blue circle). As shown in Fig. 11b, the variation in distance between the active point and the static point was taken as the output signal.

Besides up–down movement and left–right movement, human eye movement also includes some unconscious movements such as blinking, saccade and tremble [23]. The output signal caused by these movements is regarded as noise by the control system. Therefore, the moving average method was used to filter the noise signal. The average value was calculated from 10 samples and the window size of the filter was 400 (ms). As a result, the frequency of pupil tracking was 25 Hz.

Figure 12 shows the internal processing of pupil tracking for the control system. Two parameters, D (pixel) and ΔX (pixel), can be quantifiably obtained via image processing. D represents the distance between both inner canthus, and ΔX (pixel) represents the horizontal

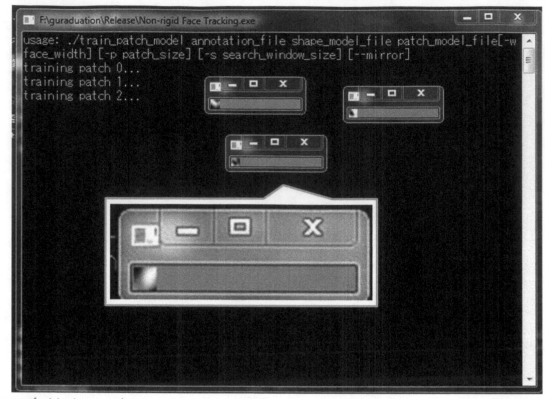

Fig. 8 Process of training image patches

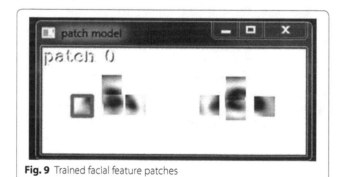

Fig. 9 Trained facial feature patches

distance between the static point and the active point. With the introduction of parameter t, D/t can be set as the changeable activation threshold. If the absolute value of ΔX exceeds the activation threshold value D/t, as shown in Eq. (1), the sheath manipulator will start to move.

$$|\Delta X| > \frac{D}{t} \tag{1}$$

Conversely, if the absolute value of ΔX does not exceed the activation threshold value D/t, as shown in Eq. (2), the sheath manipulator will remain static.

$$|\Delta X| \leq \frac{D}{t} \tag{2}$$

In this control system, Eq. (3) is used to control two servomotors, which are the actuators of the sheath manipulator.

$$V = \frac{TargetAngle\text{-}PresentAngle}{T_{cycle}} \tag{3}$$

In Eq. (3), V indicates the angular velocity of the servomotors, $TargetAngle$ indicates the target angle for the servomotors, $PresentAngle$ indicates the current angle of the servomotors, and T_{cycle} indicates the sampling period of the servomotors. ΔX can be thought of as the input signal to control the servomotors according to Eq. (4).

$$P \cdot (TargetAngle\text{-}PresentAngle) = \Delta X \tag{4}$$

P indicates the gain value and $P \cdot T_{cycle} = 800$.

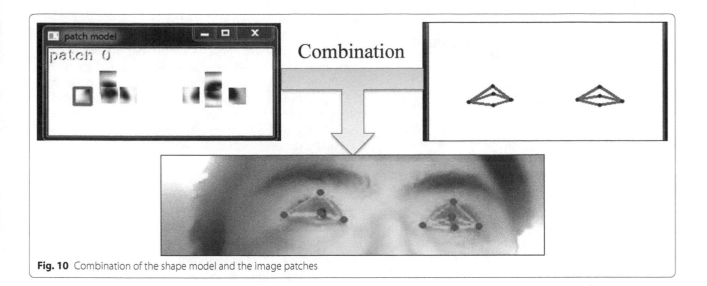

Fig. 10 Combination of the shape model and the image patches

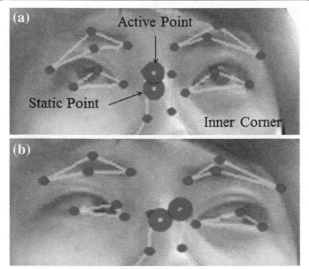

Fig. 11 a The static point is the middle position between the two canthi of the eyes, and the active point is the middle position between the pupils. **b** The position of the active point changed based on the movements of the pupils

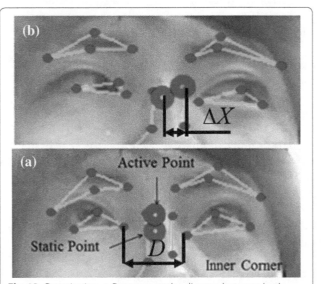

Fig. 12 Quantization: **a** D represents the distance between both inner canthus; **b** ΔX represents the horizontal distance between the static point and the active point

Through conversion from Eq. (1) to Eq. (4), Eq. (5) is obtained.

$$V = \begin{cases} \alpha \Delta X \left(|\Delta X| > \frac{D}{t} \right) \\ 0 \left(|\Delta X| \leq \frac{D}{t} \right) \cdot \left(\alpha = \frac{1}{P \cdot T_{cycle}} \right) \end{cases} \quad (5)$$

As shown in Eq. (5), ΔX is sent to the servomotors in a proportional manner, when the absolute value of ΔX exceeds the activation threshold value D/t. Conversely, no signal will be sent to the servomotors when the absolute value of ΔX does not exceed the activation threshold value D/t. Additionally, Eq. (5) can be graphed as in Fig. 13: *Limitation* indicates the limits of the pupil movement range.

Performance analysis

As mentioned in the previous research, the frequency of image processing was 25 (Hz), and therefore the time delay was 40 (ms). A frequency equal to or greater than 25 (Hz) is regarded as real-time [24, 25]. Moreover, the response of the manipulator was less than 100 (ms) [15]. Therefore, the overall delay was acceptable because it did

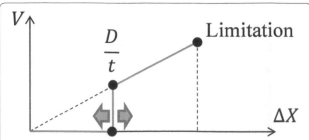

Fig. 13 Relationship between pupil tracking and output signal. *Horizontal axis* indicates the horizontal distance value. *Vertical axis* indicates the output signal value

not exceed 330 (ms), which was estimated as the maximum time delay compatible with the safe performance of surgical manipulations [26].

Before we design an experiment for determining the proper activation threshold value, we need to decide the range of D/t. If D/t is too big, the user has to rotate their eyes to move their gaze out of the screen; if D/t is too small, the sheath manipulator will activate even when the user keeps their eyes static. Therefore, three participants, engineering graduate students without glasses, tested the performance of the pupil-tracking method. During this test, we requested that the participants perform three kinds of eye motions: rotating the eyes to a maximum angle as much as possible, gazing at the edge of the monitor, and keeping the eyes static and gazing straight. Meanwhile, we recorded the variations in D and ΔX.

First of all, we made personal tracker models for each of the three participants as in the procedures mentioned in the previous sections. Then, we requested the participants to perform the three kinds of eye motions.

For rotating the eyes to a maximum angle as much as possible, we instructed each participant to move left and right twice. The purpose of this task was to obtain the maximum value of $\Delta X/D$. The result is shown as a graph (Fig. 14). On the vertical axis, with units of pixels, Nos. 1–4 represent the results for Participant 1, Nos. 5–8 represent the results for Participant 2, and Nos. 9–12 represent the results for Participant 3. The recorded values of D and ΔX are attached to the bottom of the graph. The average $\Delta X/D$ was calculated as 1/5.

Then, we requested participants to gaze at the edge of the monitor shown in Fig. 1a. The purpose of this task was to observe the maximum value of $\Delta X/D$ when the range of visibility was within the screen, because the user's gaze should not leave the monitor screen when controlling the endoscopic manipulator via pupil tracking. The distance between the screen and the participants' eyes remained at 600 [mm]. The participants had to gaze at the edge of the monitor for 3 s every trial, and

were instructed to gaze at the left and right sides of the edges twice. The results are shown as a graph (Fig. 15). On the horizontal axis, Nos. 1–4 represent the results of Participant 1, Nos. 5–8 represent the results of Participant 2, and Nos. 9–12 represent the results of Participant 3. The units of the vertical axis are pixels. The recorded values of D and ΔX are attached to the bottom of the graph. The D and ΔX for each trial are averaged over the 3 s. Finally, the average $\Delta X/D$ was calculated as 1/7.

Ideally, ΔX is equal to 0 when the eyes remain static and look straight. However, ΔX has some variation that needs to be observed. In this task, every participant was requested to keep their eyes static and look straight for 3 s. The results of the three participants are shown in Fig. 15. The average $\Delta X/D$ values shown in Fig. 16a, b are 1/40, and the average $\Delta X/D$ value shown in Fig. 16c is 1/25.

In the next section, we needed to identify an appropriate activation threshold that could be used to determine whether the sheath manipulator was moving or not. Using the above results, we confirmed that the activation threshold should be selected from a range between D/7 and D/25.

Experiment
Purpose of the experiment
The aim of the present experiment is to verify the effectiveness of the image processing method applied to the sheath manipulator control system. Moreover, to judge whether the sheath manipulator is dynamic or static, an appropriate output signal threshold needs to be obtained. In this experiment, the activation thresholds in four conditions were evaluated using operation time.

Sheath manipulator
In this experiment, we used the sheath manipulator of SPS robot prototype 2 [15] as an experimental platform. The sheath manipulator, designed for adjusting the direction of the endoscope, bends through double-screw-drive mechanisms (Fig. 17) to change the view orientation. The bending portion consists of three double-screw-drive mechanisms so that it can achieve several kinds of movement: up–down, left–right and diagonal turning. Furthermore, the screws between each double-screw-drive mechanism are connected by universal joints. Figure 18 shows that the upper two universal joints allow the endoscope to bend, while the one underneath is for support. If the upper two joints rotate at the same speed and in the same direction, the sheath manipulator will bend in the vertical direction. Conversely, if they rotate at the same speed but in opposite directions, the sheath manipulator will bend in the horizontal direction. In addition, if only one upper joint rotates, the sheath manipulator will bend in the diagonal direction.

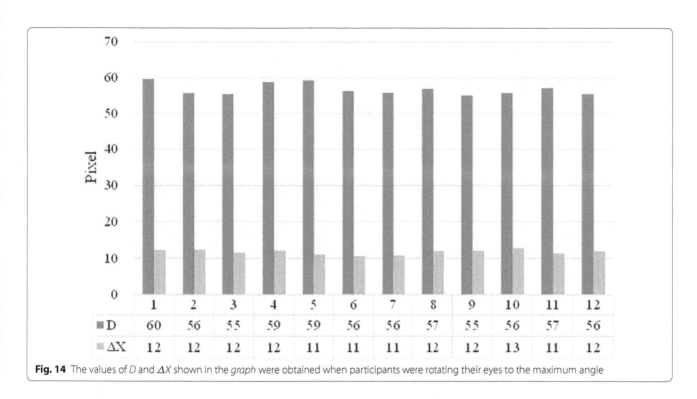

Fig. 14 The values of *D* and *ΔX* shown in the *graph* were obtained when participants were rotating their eyes to the maximum angle

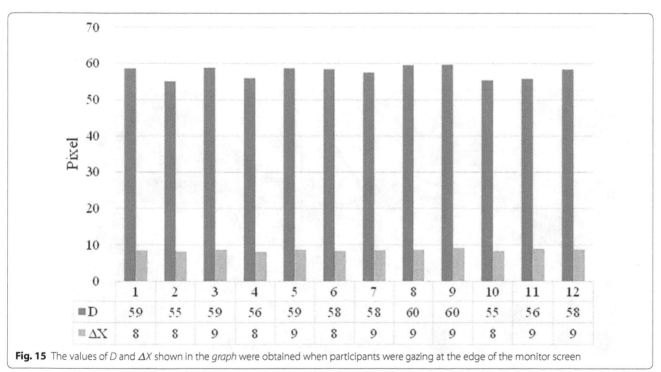

Fig. 15 The values of *D* and *ΔX* shown in the *graph* were obtained when participants were gazing at the edge of the monitor screen

Communication between master and slave

The SPS robot has a remote control function based on the master–slave structure. The master is a control console programmed using C/C ++ on a Windows PC. The slave is a dedicated computer that reads the encoders of the servomotors in real time and activates these servomotors according to the received signal from the master PC. The communication between master and slave is

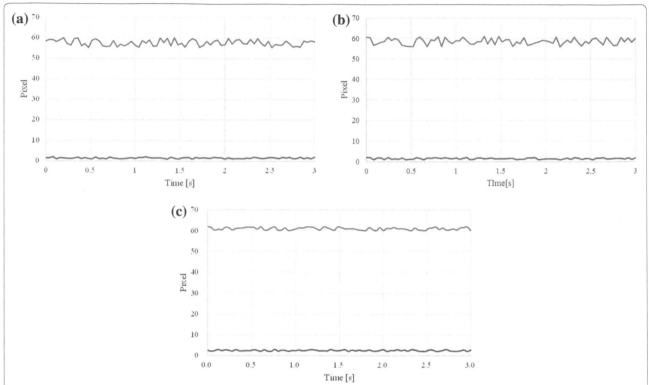

Fig. 16 The units of the *horizontal axis* are seconds; the units of the *vertical axis* are pixels. **a** indicates the results from Participant 1; **b** indicates the results from Participant 2; **c** Indicates the results from Participant 3

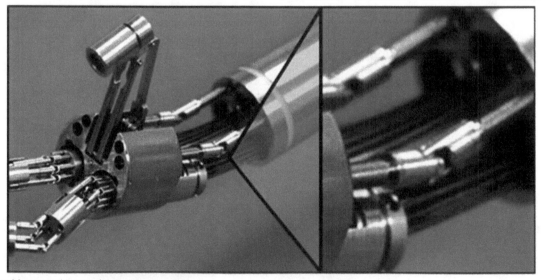

Fig. 17 Double-screw-drive mechanisms

realized using user datagram protocol (UDP) over Ethernet. Figure 19 shows the complete program architecture of this system.

The master computer was installed with Windows XP. The pupil-tracking program we proposed was merged into the operating system of prototype 2 and was

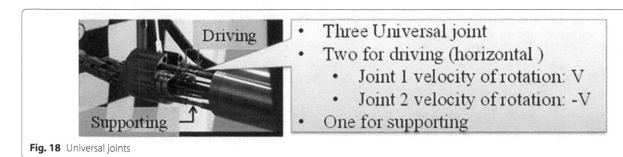

Fig. 18 Universal joints

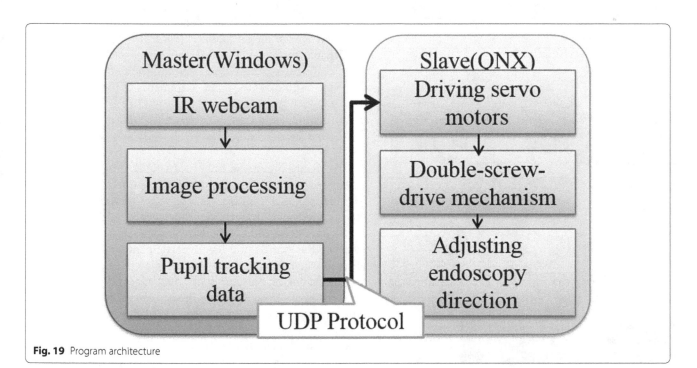

Fig. 19 Program architecture

compiled with Visual Studio 2008. As shown in Fig. 19, the Infrared LED webcam mounted on the master side initially captures the images. After this, the pupil movement data are extracted from the captured images by the pupil-tracking program and are then converted into an output signal for controlling the servomotor. Subsequently, the output signal is sent to the slave via UDP. After receiving the output signal, the slave computer activates the servomotors to drive the double-screw-drive mechanisms [15], thus adjusting the direction of the endoscope.

Experimental conditions

The participants in this experiment were five graduate students who had no experience of endoscopic surgery. Figure 20 shows the experimental setup. The distance between the display and the participant's face was set between 50 and 70 cm according to the most ergonomic view [27]. As shown in

Fig. 21, the operators use their pupils to operate the sheath manipulator, moving the image center point from "point 0" to "point 3" or "point 4", which were displayed on a chessboard with a 25×25 mm^2 lattice. The reason for using the chessboard was to facilitate the observation of the image changes. The image center of the endoscope matching "point 0" was set as the initial state. The activation threshold value D/t was determined in four conditions: when t was equal to 14, 18, 22, or 24. As mentioned in the Session Performance analysis, the four conditions were selected from the range between D/7 and D/25. Therefore, we selected the first condition as D/14, which was half the value of D/7, and obtained the value D/25 when participants kept their eyes static and looking straight. The manipulator could not be static if we had chosen a condition smaller than D/25. Therefore, we chose D/24 as the smallest condition. Beyond that, we selected two additional conditions.

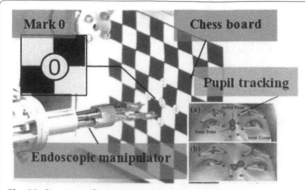

Fig. 20 Overview of experimental setup

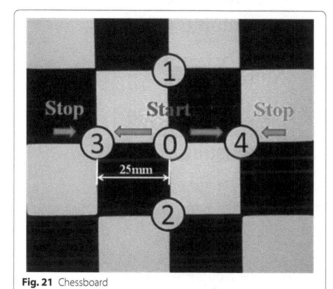

Fig. 21 Chessboard

Experimental procedure

Before starting the experimental procedure, the personal eye tracker model was established for each of the five participants as mentioned in the ("Methods") section. Then, the flow of the experiment proceeded as follows:

(a) The activation threshold was set for the control system.
(b) The system was tested as to whether it altered direction with the participant's pupil movement (Additional file 1).
(c) Initialization: the image center of the endoscope was changed to match "point 0" (Fig. 22).
(d) Each participant moved the image center of the endoscope from "point 0" to "point 3" four times, and from "point 0" to "point 4" four times. The time cost of every trial was recorded.
(e) The threshold value was changed and the above flow from step a to step c was repeated.

The change order was D/14, D/18, D/22 and D/24.

For the present experiment, it was important that the system reflected the operator's viewing intention so that the participants could confirm their location during the experiment. To judge whether the sheath manipulator was in the static state, each participant verbally confirmed when the operation was completed, i.e., when the image center of the endoscope had reached the target and stopped. Thus, the time measurement ended when the participant replied that the movement was complete.

Results and discussion

All participants could alter the direction of the sheath manipulator with their pupil movement. Figures 23 and

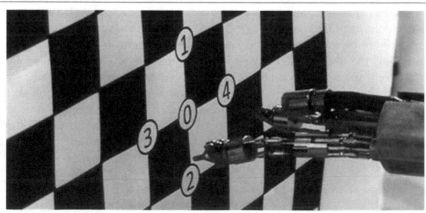

Fig. 22 Experimental initialization

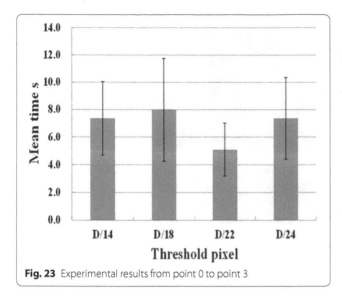

Fig. 23 Experimental results from point 0 to point 3

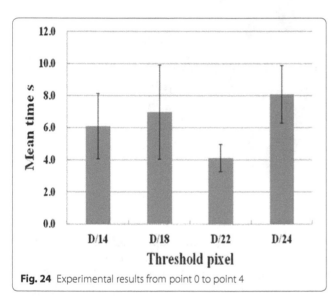

Fig. 24 Experimental results from point 0 to point 4

24 show the results of the experiment. In particular, Fig. 20 shows the results when moving the image center of the endoscope from "point 0" to "point 3" (Figs. 25), and 21 shows the results when moving the image center of the endoscope from "point 0" to "point 4" (Fig. 26). The horizontal axis represents the operation time and the vertical axis represents the activation threshold D/t (pixel). As shown in Fig. 23 and 24, the average operation time was shortest and the standard deviation of operation time was smallest when the activation threshold was set as D/22. In the D/22 condition, the average operation time was 5.2 s and the standard deviation was 1.9 s when the endoscope turned to the left (Fig. 25). In the other conditions, the average operation time was 4.2 s and the standard deviation was 0.8 s when the endoscope turned to the left (Fig. 26).

For the alteration of the endoscope direction operation system based on pupil position tracking, the shortest operation time was when the activation threshold value was equal to D/22. Similarly, the standard deviation values were smallest in both experiments when that threshold value was selected. Therefore, it is appropriate to use D/22 as an activation threshold for the operation system of the proposed SPS robot. In addition, we found that operating the sheath manipulator via pupil tracking can provide good stability and response when an appropriate threshold value is used. Also, the period between blinks was 6–8 s [28], which prevents the eye from fatiguing and from drying out. We observed an obvious difference for operation time and standard deviation in the D/22 condition in the two experiments. [29, 30] suggested that the center of monitor image should be aligned to the center of operative field. If the target points: "point 0", "point 3" or "point" is projected on the center of monitor image, ΔX will not exceed the activation threshold value. Therefore, the operator can stop the desired target position. The limitation of this experiment is that the numbers of participants and conditions were both rather small. Therefore, there is limited evidence to support the optimal threshold value. Moreover, the accurate relationship between the motion of the manipulator and the movement of the pupils was not confirmed. After causal analysis, we found that the difference arose from the intertwining flexible shafts that are set between the manipulators and servomotors; such a situation affects the stability and speed of rotation. As a solution to this problem, these flexible shafts need to be sheathed in pipes or fixed on a fixation device so as to avoid intertwining. Furthermore, we hope to achieve better pupil tracking in the vertical direction by using a higher resolution webcam.

To invite surgeons or medical trainees to participate the manipulation experiment in the future, we gave a presentation of the system and the experimental results to two endoscopic surgeons. After the presentation, the surgeons agreed with the usefulness of the proposal and stated that:

1. The proposed system is more intuitive than voice-control and pedals. The surgeons certainly need a hand-free strategy to manipulate the endoscope. Using webcam and image processing is a good approach because it does not require the surgeons to be attached to additional devices, which may increase their burden during an operation.
2. The proposed system is useful because the manipulator will stop when the center of the visual field aligns with the target. One of the fundamentals of manipulating the endoscope is aligning the center of the visual field with the center of operation.

Fig. 25 Sheath manipulator turning to the *left*

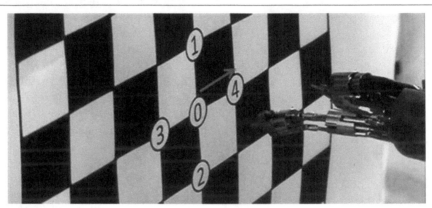

Fig. 26 Sheath manipulator turning to the *right*

3. The horizontal alternation of the visual field is greater than the vertical, but the vertical alternation is still indispensable.
4. A zoom function is indispensable for an endoscopic control system when the surgeon is performing a delicate operation. For example, the surgeon would ideally like to zoom in the visual field when they are peeling the tissue from around a vessel. Moreover, in a future system, it would be better to be able to adjust the rate of visual field alternation according to the magnification of the lens.
5. An emergency stop button is an indispensable part of the system, which is used to avoid a collision between the endoscope and tissue.

Conclusions

In this paper, a hands-free technique for controlling the alteration of endoscope direction using a pupil-tracking method via an image processing method was introduced. The novelty of the proposed method is its ability to achieve pupil tracking because the variation in distance

of both intraocular angles and pupils could be obtained from the tracking trajectory. In this method, an appropriate output signal threshold needs to be obtained for judging whether the sheath manipulator, which alters the endoscope direction, is dynamic or static. An experiment was performed to verify the effectiveness of the image processing method applied to the sheath manipulator control system, and the activation threshold of the control system had to be determined and used for the horizontal direction movement of the sheath manipulator. We found an activation threshold value that fulfils stability and response simultaneously. This time, we only verified the horizontal direction because of the limitations of our method. At present, it is quite difficult for the sheath manipulator to make vertical movements, because the vertical movement range of the eye is much less than its horizontal movement range. To realize vertical direction movement, we need a higher resolution webcam to detect the relatively small vertical movement of the eyes. As surgeons pointed out, a zoom function is indispensable for endoscope manipulation. Thus, using only the pupil position parameter as shown in this experiment is not sufficient to achieve the zoom function. In future work, we aim to develop an algorithm that includes a large number of operating conditions to judge the movement state and achieve more types of movements. Also, we will improve the current control system based on the surgeons' suggestions. Furthermore, we also plan to invite surgeons and medical trainees to be participants in the manipulation experiments.

Additional file

Additional file 1. Animation related to the verification of the effectiveness of the image processing method applied to the laparoscope manipulator control system.

Abbreviations
AESOP: automatic endoscope optimal positioning system; GUI: graphical user interface; UDP: user datagram protocol.

Authors' contributions
YC derived the basic concept of the overall system, technically constructed the system and drafted the manuscript. All authors read and approved the final manuscript.

Author details
[1] Graduate School of Creative Science and Engineering, Waseda University, 3-4-1 Ohkubo, Shinjuku-ku, Tokyo 169-8555, Japan. [2] Graduate School of Advanced Science and Engineering, Waseda University, 2-2 Wakamatsu-cho, Shinjuku-ku, Tokyo 162-8480, Japan. [3] Graduate School and Faculty of Engineering, Chiba University, 1-33 Yayoi-cho, Inage-ku, Chiba 263-8522, Japan.

Acknowledgements
The authors sincerely thank the volunteers for participating in our experiments. The work was supported in part by a research grant from JSPS Global COE Program: Global Robot Academia, JSPS Grant-in-Aid for Scientific Research (A) No. 20339716, JSPS Grant-in-Aid for Scientific Research (S) No.25220005, JSPS Grant-in-Aid for Exploratory Research No. 15K12606 and the Program for Leading Graduate Schools, "Graduate Program for Embodiment Informatics" of the Ministry of Education, Culture, Sports, Science and Technology.

Competing interests
The authors declare that they have no competing interests.

References
1. ASCRS, Laparoscopic surgery—what is it?, Available from: http://www.fascrs.org/patients/treatments_and_screenings/laparoscopic_surgery/
2. Horgan S, Vanuno D (2001) Robots in laparoscopic surgery. J Laparoendosc Adv Surg Tech A 11:415–419
3. Matern U, Koneczny S (2007) Safety, hazards and ergonomics in the operating room. Surg Endosc 21:1965–1969
4. Jihad K, George-Pascal H, Raj G, Mihir D, Monish A, Raymond R, Courtenay M, Inderbir G (2008) Single-Port laparoscopic surgery in urology: initial experience. Urology 71:3–6
5. Yasunaga T, Hashizume M, Kobayashi E, Tanoue K, Akahoshi T, Konishi K, Yamaguchi S, Kinjo N, Tomikawa M, Muragaki Y, Shimada M, Maehara Y, Dohi Y, Sakuma I, Miyamoto S (2003) Remote-controlled laparoscope manipulator system NaviotTM, for endoscopic surgery. Int Congress Ser 1256(2003):678–683
6. Intuitive Surgical Inc. [homepage on the Internet]. Available from: http://www.intuitivesurgical.com
7. Meyer A, Oleynikov D (2013) Surgical robotics, Mastery of Endoscopic and Laparoscopic Surgery, Vol 2, 4th edn. p 62–71
8. Kraft B, Jager C, Kraft K, Leibl B, Bittner R (2004) The AESOP robot system in laparoscopic surgery: increased risk or advantage for surgeon and patient? Surg Endosc 18:1216–1223
9. ViKY EP, available from: http://www.endocontrol-medical.com/viky_ep.php
10. Kolvenbach R, Schwiez E, Wasillijew S, Miloud A, Puerschel A, Pinter L (2004) Total laparoscopically and robotically assisted aortic aneurysm surgery: a critical evaluation. J Vasc Surg 39:771–776
11. Ubeda A, Ianez E, Azorin J (2011) Wireless and portable EOG-based interface for assisting disabled people. IEEE/ASME Transact Mechatron 16(5):870–873
12. Noonan D, Mylonas G, Shang J, Payne C, Darzi A, Yang G (2010) Gaze contingent control for an articulated mechatronic laparoscope. Proceedings of the 2010 3rd IEEE RAS and EMBS International Conference on Biomedical Robotics and Biomechatronics, The University of Tokyo, Tokyo, Japan, 26–29 September 2010
13. Autorino R, Cadeddu JA, Desai MM, Gettman M, Gill IS, Kavoussi LR, Lima E, Montorsi F, Richstone L, Stolzenburg JU, Kaouk JH (2010) Laparoendoscopic single-site and natural orifice transluminal endoscopic surgery in urology: a critical analysis of the literature. Eur Assoc Urol 59(1):26–45
14. Hubens G, Coveliers H, Balliu L, Ruppert M, Vaneerdeweg W (2003) A performance study comparing manual and robotically assisted laparoscopic surgery using the da Vinci system. Surg Endosc Other Intervent Tech 17(10):1595–1599
15. Kobayashi Y, Tomono Y, Sekiguchi Y, Watanabe H, Toyoda K, Konishi K, Tomikawa M, Ieiri S, Tanoue K, Hashizume M, Fujie GM (2010) The international journal of medical robotics and computer assisted surgery. Int J Med Robotics Comput Assist Surg 6:454–464
16. Kobayashi Y, Sekiguchi Y, Noguchi T, Liu Q, Oguri S, Toyoda K, Konishi K, Uemru M, Ieiri S, Tomikwa M, Ohdaira T, Hashizume M, Fujie GM (2015) The international journal of medical robotics and computer assisted surgery. Int J Med Robotics Comput Assist Surg 11:235–246
17. Rassweiler J, Gözen A, Frede T, Teber D (2011) Laparoscopy vs. robotics: ergonomics—does it matter? Robotics Genitourin Surg 2011:63–78
18. Ohno T, Mukawa N (2004) A free-head, simple calibration, gaze tracking system that enables gaze-based interaction. ETRA 2004, Eye Tracking Research and Applications Symposium, p 115–122
19. Hill B (1996) Retina identification, Biometrics, p 123–141
20. Liu Q, Kobayashi Y, Zhang B, Noguchi T, Takahashi Y, Nishio Y, Cao Y, Ieiri S, Toyoda K, Uemura M, Tomikawa M, Hashizume M, Fujie M G (2014) Development of a smart surgical robot with bended forceps for infant congenital esophageal atresia surgery. 2015 IEEE International Conference on Robotics and Automation, Hong Kong, p 2430–2435
21. Emami S, Ievgen K, Saragih J (2012) Non-rigid face tracking, mastering openCV with practical computer vision projects. Chapter 6:189–233
22. Gower J (1975) Generalized procrustes analysis. Phychometrika 40(1):33–51
23. Robinson D (1964) The mechanics of human saccadic eye movement. J Pysiol 1964(174):245–264
24. Marchand E, Chaumette F (2002) Virtual visual servoing: a framework for real-time augmented reality. Comput Graph Forum 21(3):289–297
25. Daniilidis K, Krauss C, Hanse M, Sommer G (1998) Real-time tracking of moving objects with an active camera. Real-Time Imaging 4(1):3–20
26. Marescaux J, Leroy J, Gagner M, Rubino F, Mutter D, Vix M, Butner SE, Smith MK (2001) Transatlantic robot-assisted telesurgery. Nature 413:379–380
27. Rempel D, Willms K, Anshel J, Jaschinski W, Sheedy J (2007) The effects of visual display distance on eye accommodation, head posture, and vision and neck symptoms. Hum Factors J Hum Factors Ergon Soc 49(5):830–838
28. Barbato G, Ficca G, Muscettola G, Fichele M, Beatrice M, Rinaldi F (2000) Diurnal variation in spontaneous eye-blink rate. Psychiatry Res 93(2):145–151
29. Hashizume M (2005) Fundamental training for safe endoscopic surgery, Innovative Medical Technology, Graduate School of Medical Science Kyushu University, p 49 (in Japaeses)
30. Donnez J (2007) Human assistance in laparoscopic surgery, Atlas of operative laparoscopy and hystereoscopy, 3rd edn, p 409

7

Development of a gas/liquid phase change actuator for high temperatures

Hiroki Matsuoka[1*], Koichi Suzumori[2] and Takefumi Kanda[1]

Abstract

Gas/liquid phase changes produce large volume changes in working fluids. These volume changes are used as the driving power sources in actuators such as micro-pumps and valves. Most of these actuators are utilized in ordinary temperature environments. However, the temperature range in which the phase change actuator can operate depends on the characteristics of the working fluid. We hypothesized that proper selection of the working fluid and the structure of the actuator can enable such actuators to be applied not only in ordinary environments but also in high temperature environments. Consequently, in this paper, we discuss the design and fabrication of a new gas/liquid phase change actuator for use in high temperature environments. Our proposed actuator consists of a bellow body, spring, heater, and working fluid. We used the Inconel super alloy, which is highly heat and corrosion resistant, for the bellow and moving parts of the actuator. For the working fluid, we prepared triethylene glycol, which has a boiling point of 287.3 °C and very low vapor pressure at ordinary temperature. As a result, our proposed actuator can be utilized in high temperature environments up to 300.0 °C. The results of several experiments conducted confirm that our proposed actuator generates 1.67 mm maximum displacement in a 300.0 °C atmospheric environment. In addition, we confirmed that the operation of the actuator is stable in that environment. Our results confirm that a gas/liquid phase change actuator can be used in high temperature environments.

Keywords: Actuator, Gas/liquid phase change, High temperature, Triethylene glycol, Bellow

Background

Phase changes in materials, resulting from temperature changes, produce huge volume changes, especially in liquid/gas phase changes. This attribute is utilized to provide a power source for micro-pumps and valves in combination with MEMS (micro-electromechanical system) heaters, micro-channels, diaphragms, and membranes [1–10]. Kato et al. used this phenomenon to provide a power source for actuators and robots. They made a metal bellow actuator to control cutting equipment and a pipe inspection robot [11, 12]. Phase change is used not only in actuators but also in some kinds of pressure sources. For example, Kitagawa et al. used the triple point of carbon dioxide as a mobile pressure source

[13], and Shibuya et al. developed a buoyancy control device for underwater robots using paraffin oil [14].

The actuators described above were developed for use in ordinary environments. In contrast, our aim is to utilize these phase change actuators in special environments. In particular, driving actuators in high temperature environments is a typical example of the special environments being considered. For example, in the hydrothermal synthesis method, which is one of the methods used to fabricate piezoelectric devices, the water solution inside the high temperature chamber needs to be agitated [15]. In one instance where this process was used, the water solution was agitated using an autoclave—an end-over-end shaker with heat. Fabrication of the (Pb, La)(Zr, Ti)O_3 (PLZT) film took 24 h. Not only the rotation condition but also the attitude of the sample will affect the quality of the fabrication. The actuators, which produce the inclination of the shaker, are predictably effective devices. Another example is the fabrication process for the ferric oxide crystal via the floating-zone melting method [16].

*Correspondence: matsuoka9@act.sys.okayama-u.ac.jp
[1] Okayama University, 3-1-1, Tsushima-naka, Kita-ku, Okayama 700-8530, Japan
Full list of author information is available at the end of the article

In this process, the partially melted sample is turned and pulled inside the chamber for growth. Both the processes require 8–24 h and a single directional drive in order to fabricate the tiny sample. We believe that an actuator that can realize linear motion in high temperature environments can rectify these quality problems.

In previous work, we targeted these environments for utilization of actuators and proposed gas/liquid phase change actuators. We subsequently fabricated an actuator driven by the gas/liquid phase change of water. This actuator consisted of a cylinder as a vessel and actuation device, an external heater to excite the phase change, and a spring that controlled the speed of motion. Our proposed actuator was driven in a 180 °C environment. Thus, we realized directional motion with gas/liquid phase changes in a high temperature environment using that actuator [17, 18].

In this paper, we discuss the development of another actuator for utilization in high temperature environments. Our proposed actuator is composed of an Inconel alloy bellow (which is highly heat and corrosion resistant), a spring, a heater with built-in temperature sensor, and working fluid. The bellow produces direct motion without friction when it is moving. This friction is significant in cylinder type actuators; it can cause leaking of the working fluid and disturb the motion of the actuator. Although it has the same bellow shape, our proposed actuator can be sealed very easily and has a more simplified structure.

Springs change the speed of an actuator; when a hard spring is attached, the actuator is driven at a slow speed. Conversely, a soft spring produces a high speed for the actuator. These characteristics were confirmed in our previous study.

In this paper, we explain how we chose the new working fluid to facilitate operation of the actuator in high temperature environments. The triethylene glycol (TEG) working fluid has a liquid phase and a boiling point of 287.3 °C under ordinary atmospheric pressure conditions [19]. These characteristics enable our proposed actuator to operate in a 300.0 °C environment.

In experiments conducted, in which the actuator was driven in a 300.0 °C atmospheric temperature environment by a heater assembled in the actuator, a maximum displacement of 1.67 mm with 2700 J of input power was produced. The results confirmed that our new gas/liquid phase change actuator can be driven in 300.0 °C high temperature environments with TEG working fluid.

Basic principle underlying the gas/liquid phase change actuator

Our proposed actuator is driven by the gas/liquid phase change of its working fluid. When the working fluid is heated, its vapor pressure increases and pushes out the chamber.

The fundamental structure of the proposed actuator is shown in Fig. 1. The actuator consists of a variable volume chamber, springs that produce the reactive force against vapor pressure, heating device, and working fluid.

A bellow is a variable volume vessel. In previous work, we used a cylinder, but this structure has a sliding member in its body. As a result, there was much friction when it was being driven and the working fluid also leaked. Considering the application of our actuator to high temperature environments, we surmised that these issues would be a bottleneck. Consequently, we designed the bellow to be driven without friction because the expansion and contraction characteristics of a bellow are produced by its distinctive structure. In addition, the chamber also includes the spring properties itself.

The springs produce a refractive force against the vapor pressure, in order to control the temperature sensitivities of the actuator. The fundamental theoretical characteristics can be calculated using the very simple equation

$$k(x_0 + x) = A \cdot P_{bellow} \tag{1}$$

where k is the spring constant of the actuator; x_0 and x are the initial and produced displacements, respectively; A is the pressurized area of the bellow; and P_{bellow} is the inner pressure of the bellow, which is the vapor pressure of the working fluid.

The general properties underlying the concept of the actuator are illustrated in Fig. 2. When a stiff spring is attached to the actuator, the actuator generates a small displacement with respect to the temperature increment. Conversely, when a soft spring is attached, the actuator generates a large displacement with respect to the temperature increment.

The characteristics of the actuator also depend on the properties of the working fluid. The inherent vapor

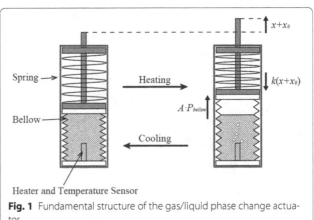

Fig. 1 Fundamental structure of the gas/liquid phase change actuator

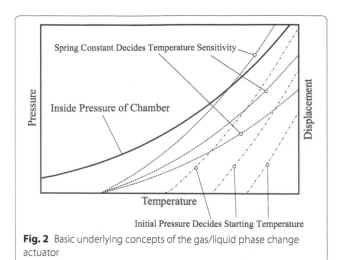

Fig. 2 Basic underlying concepts of the gas/liquid phase change actuator

pressure curve of the working fluid determines the driving temperature range of the actuator. Thus, by selecting an appropriate working fluid, we can set the operating temperature range of the actuator.

Various kinds of modes of heating exist; for example, fire, electrical heating, chemical reaction, induction heating, Peltier heating, and IR light.

In our previous work, we confirmed the adequacy of these fundamental characteristics of gas/liquid phase change actuators and designed an actuator that could be driven by the phase change of water in a 180.0 °C environment.

Design of the actuator

We designed a prototype actuator based on our proposed driving principle. As shown in Fig. 3, our actuator consisted of bellows, spring, chassis, and heater unit—which includes the temperature sensor.

The bellow, the most important part of the actuator, was fabricated using Inconel 600 super alloy—an austenite Nickel–Chromium based alloy. The target temperature, 300.0 °C, is the temperature used for low

temperature annealing of stainless steel. This heat treatment process causes stainless steel to lose its corrosion resistance and its mechanical properties to change. By contrast, Inconel alloy has better corrosion and heat resistance than stainless steel as a result of its stable passivating oxide layer. Hence, we chose Inconel super alloy for the actuator. This bellow had outer diameter 22.0 mm, inner diameter 15.2 mm, and length 65.0 mm—with a welded 40 mm diameter flange and a tiny cap, 0.12 mm double layer thickness wall, 19 replications, and a 13.5 N/mm spring constant. The working fluid was poured into this chamber.

The compression coil spring was set on the top of the bellow. This spring was made from oil tempered SiCr-alloyed valve spring wire (SWOSC-V). When the actuator is driven, the bellow extends, and the spring is compressed. Consequently, the spring constant of the actuator is calculated by summing those of the bellow and the spring.

Incidentally, the spring constants are also temperature dependent. In general, the spring constant of a coil spring is calculated using the equation

$$k_{spring} = \frac{Gd^4}{8N_a D^3} \qquad (2)$$

where k_{spring} is the spring constant, G is the modulus of rigidity, d is the diameter of the wire, N_a is the number of turns, and D is the average diameter of the coil. This equation reflects the fact that the spring constant of the coil spring depends on the modulus of rigidity of the metal. The modulus of rigidity of SWOSC-V can be calculated using the regression formula [20]:

$$G = 7308 + \left(-2.604 \times 10^7\right) \exp\left(\frac{-6800}{T + 273}\right) \qquad (3)$$

where T is temperature. Meanwhile, approximate equations for the spring constants of metal bellows k_{bellow} are given by [21]

$$k_{bellow} = \frac{2\pi m^2 E t^3}{3n\left(m^2 - 1\right)b^2} / \left\{c^2 - 1 - \frac{4c^2}{c^2 - 1}(\log c)^2\right\} \qquad (4)$$

$$m = \frac{1}{v} \qquad (5)$$

$$c = \frac{a}{b} \qquad (6)$$

where E is the modulus of longitudinal elasticity (Young's modulus), n is the number of replications, a and b are the radius of the outer and inner diameter, respectively, t is the thickness of the bellow's wall, and v is Poisson's ratio. From these equations we can estimate the temperature

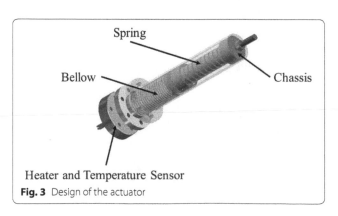

Fig. 3 Design of the actuator

dependence of the spring constant of the actuator. The spring constant of actuator $k_{actuator}$ can be calculated using the following equation:

$$k_{actuator} = k_{spring} + k_{bellow} \quad (7)$$

The results of calculations with the spring constant of the coil spring at 13.9 N/mm and of the bellow at 13.5 N mm under ordinary temperature are shown in Fig. 4. From these results, we determined that at 300.0 °C atmospheric temperature, the spring constant of the actuator declined by 5 %.

The static displacement of the actuator is affected not only by vapor pressure but also by thermal expansion of the working fluid. Thus, the theoretical static displacement of actuator x was calculated using the equation

$$x = \varepsilon L_0(T - T_0) + \frac{A \cdot P_{workingfluid}}{k_{actuator}} \quad (8)$$

where ε is the thermal expansion ratio of the working fluid, L_0 is the initial length of the bellow under ordinary temperature, T_0 and T are the ordinary temperature and the current temperature, respectively, and $P_{working\,fluid}$ is the vapor pressure of the working fluid.

As shown in Fig. 5, the heater is welded onto the flange. It has a rated output of 24 V at 15 W, and a K-type thermoelectric couple on the top of the heating rod, which is 8 mm in diameter and 24 mm in length. All of these parts were inserted into the chassis and the actuator finally assembled, as shown in Fig. 6. The actuator has a length of 220 mm, flange diameter 40 mm, and outer diameter 25 mm. The displacement produced can be observed from the slit in the middle of the chassis or the output shaft at the top of the actuator.

We chose TEG, which has melting point −14.3 °C and boiling point 287.3 °C at ordinary pressure, as the working fluid. This liquid is clear and colorless and has the chemical formula $C_6H_{14}O_4$. This working fluid has a very small vapor pressure at ordinary temperature [22]. This

Fig. 5 Flanged heater

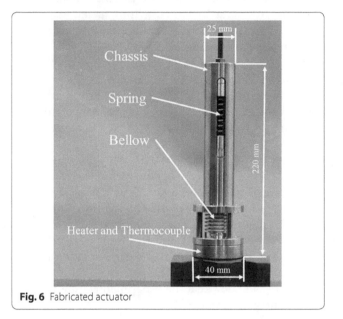

Fig. 6 Fabricated actuator

is because we chose a liquid that is suitable as a working fluid for driving the actuator in a 300.0 °C environment. The basic specifications of TEG are shown in Table 1.

Experiments and properties of the actuator
Static properties of the actuator
We conducted an experiment to evaluate the basic static properties of our prototype actuator. In this experiment, we determined the relationship between the atmospheric temperature and the displacement of the actuator, which is produced by atmospheric temperature. The temperature was controlled via a constant temperature reservoir and the displacement measured at 25.0, 100.0, 200.0, 275.0, and 300.0 °C, respectively. Prior to each measurement, the actuator was placed in the respective temperature environment for 1 h. This was carried out to enable the temperature of the actuator to accord with that of the constant temperature reservoir. The displacement was

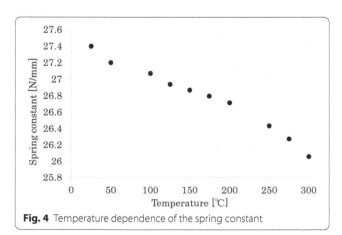

Fig. 4 Temperature dependence of the spring constant

Table 1 Basic characteristics of TEG

Chemical formula	$C_6H_{14}O_4$
Melting/boiling point	−14.3 °C/287.3 °C
Specific gravity	1.1255
Coefficient of expansion	0.00072

measured by image processing using a photograph taken from the outside of the constant temperature reservoir through the observation window. The experimental setup is shown in Fig. 7.

In this experiment, 15.9 g of degassed TEG was poured into the bellow and the actuator sealed while ensuring that no air was included. TEG has a thermal expansion ratio of 7.2×10^{-4}/°C at 55.0 °C [23]. Thus, an increase in temperature from room temperature to 300.0 °C results in a 22 % increase in the volume of TEG.

The SWOSC-V spring, with a spring constant of 13.9 N/mm, was inserted into the chassis. As discussed previously, the spring constant of an actuator has a small temperature dependence. In this case, a temperature increase from 25.0 to 300.0 °C resulted in a decrease in the spring constant from 27.4 to 26.1 N/mm.

The experimental results obtained are shown along with the theoretical displacement in Fig. 8. The actuator was displaced 13.1 mm at 300.0 °C as a result of the thermal expansion and vapor pressure of TEG. As shown in Fig. 8, the theoretical displacement calculated using Eq. (8) is consistent with the actual experimental value. The output force of the actuator that regards a spring as a load and the vapor pressure of TEG are shown in Figs. 9 and 10. The results show that the experimental values are identical to theoretical values.

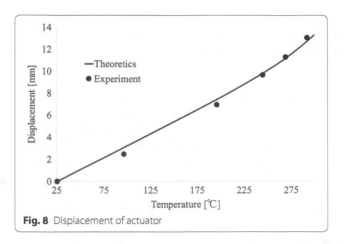

Fig. 8 Displacement of actuator

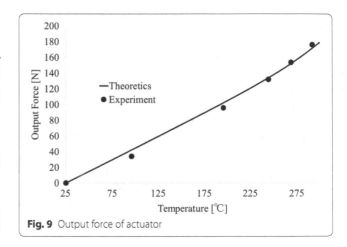

Fig. 9 Output force of actuator

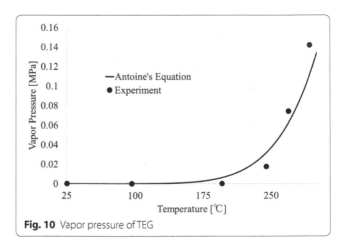

Fig. 10 Vapor pressure of TEG

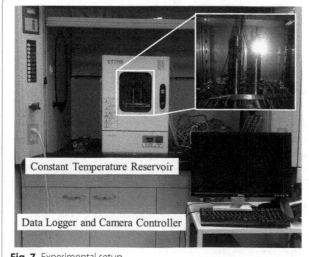

Fig. 7 Experimental setup

Dynamic properties of the actuator

We also evaluated the dynamic properties of the proposed actuator in order to verify that it can be utilized in a high temperature environment. In the experiment conducted, the atmospheric temperatures were set via a constant temperature reservoir to 25.0, 100.0, 200.0, 250.0, 275.0, and

300.0 °C, respectively. After leaving the actuator at each respective temperature for 30 min, 24 VDC was applied to the heater, which resulted in the working fluid being heated in 180 s. After being heated, the actuator was then cooled by heat radiation. The current applied to the heater was 0.6 A, and the applied energy was 2700 J. The experimental results obtained are shown in Fig. 11. In the experiment, the actuator produced 1.67 mm maximum displacement in the 300.0 °C environment with a 56.2 °C temperature increment. Further, in all the environments, the actuator ceased moving immediately as the heating stopped. In addition, the actuator did not revert to any of its initial states. In the initial states, the temperature sensor inside of the bellow indicated 25.5, 96.3, 196.1, 244.9, 269.1, and 292.3 °C, respectively. However, the K-type thermocouple has a 5 % measuring error.

We also applied the same energy five times in the 300.0 °C environment. The results are shown in Fig. 12. After the first heating, the actuator did not return to its initial state within 3000 s. We assumed that this phenomenon was because the initial temperature of the working fluid was lower than the atmospheric temperature. However, the other four experiments generated exactly the same displacements. In these experiments, the actuator took 40.8 N average force with 2700 J of input energy in 180 s. The results of these experiments confirm that our proposed actuator can produce stable movements in a 300.0 °C high temperature environment.

Conclusions

We proposed a new actuator that is driven by the gas/liquid phase of a working fluid for high temperature environments. Phase change actuators are driven by the heating of working fluids; thus, their characteristics depend on the characteristics of the working fluid and the springs that produce the reactive force against the vapor pressure.

We designed and fabricated an actuator that can be utilized in a 300.0 °C high temperature environment. The actuator consists of a bellow body, spring, heater with a built-in temperature sensor, and TEG working fluid. The bellow was fabricated from Inconel super alloy, which is highly heat and corrosion resistant, and can generate displacements without friction. The TEG working fluid has a very low vapor pressure at ordinary temperature and a boiling point of 287.3 °C. Consequently, the proposed actuator is controllable in 300.0 °C temperature environments.

The results of static evaluation of the characteristics confirmed that the theoretical properties are consistent with the experimental results. The actuator is heated by atmospheric temperature and produces a displacement using both vapor pressure and thermal expansion of the working fluid.

We operated this actuator using a heater assembled in the actuator in several temperature environments. The results showed that the actuator can be driven in all the environments tested. In the 300.0 °C environment, the actuator generated 1.67 mm maximum displacement using 2700 J of input power. In addition, we operated the actuator fitfully five times in a 300.0 °C environment in exactly the same condition. The results of this examination confirmed that the actuator is stably driven.

Thus, our proposed gas/liquid phase change actuator is suitable for operation in 300.0 °C environments. In order to put this actuator to practical use, the operating time needs to be reduced and a controlling system added. We believe that this actuator can be driven in even higher temperature environments in the future.

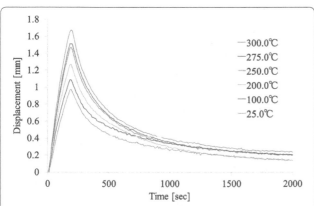

Fig. 11 Dynamic characteristics of the actuator at various temperatures

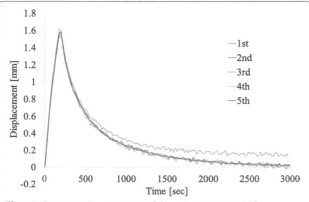

Fig. 12 Dynamic characteristics of the actuator at 300.0 °C

Authors' contributions
HM carried out design, experiment, evaluating of the actuator and drafted the manuscript. KS and TK conceived of the study, and participated in its design and coordination and helped to draft the manuscript. All authors read and approved the final manuscript.

Authors' information

HM was born in July 3, 1986. He received his B. Eng. and M. Eng. from Okayama University, Japan in 2010 and 2012, respectively. He has been a doctoral student at the graduate school of Okayama University since 2012. His current research interest is in the area of actuators for special environments.

KS was born in 1959. He received a Doctor's Degree from Yokohama National University in 1990. He worked for the Toshiba R&D Center from 1984 to 2001 and for the Micromachine Center, Tokyo, from 1999 to 2001. He was a professor at Okayama University, Japan since 2001–2014. He has been a professor at Tokyo Institute of Technology, Japan since 2014. He is a member of the Japan Society of Mechanical Engineers, the Robotics Society of Japan, IEEE and the Institute of Electrical Engineers of Japan.

TK was born in Fukuoka, Japan, on June 18, 1972. He received his B. Eng., M. Eng., and Dr. Eng. in precision machinery engineering from the University of Tokyo, Japan in 1997, 1999, and 2002, respectively. From 2002 to 2007, he was a research associate and lecturer at Okayama University, Japan. Since 2007, he has been an associate professor at Okayama University. His research interests are micro sensors, micro actuators, micro systems, and piezoelectric film. He is a member of the Japan Society for Precision Engineering, the Institute of Electrical Engineers of Japan, IEEE, the Japan Society of Mechanical Engineers, and the Robotics Society of Japan.

Author details

[1] Okayama University, 3-1-1, Tsushima-naka, Kita-ku, Okayama 700-8530, Japan. [2] Tokyo Institute of Technology, 2-12-1-I1-60, Ookayama, Meguro-ku, Tokyo 152-8550, Japan.

Acknowledgements

We sincerely appreciate the help received from Dr. Y. Sakata and Mr. Y. Yamada through valuable discussions held. This study was partly supported by the Fluid Power Technology Promotion Foundation.

Competing interests

The authors declare that they have no competing interests.

References

1. Henning AK (1998) Liquid and gas–liquid phase behavior in thermopneumatically actuated microvalves. In: SPIE3515, Micro Fluidic Devices and Systems, pp. 53–63, Sante Clara, CA
2. Takao H, Miyamura K, Ebi H, Ashiki M, Sawada K, Ishida M (2005) A MEMS microvalve with PDMS diaphragm and two-chamber configuration of thermo-pneumatic actuator for integrated blood test system on silicon. Sens Actuators A 119(2):468–475
3. Henning AK (2006) Comprehensive model for thermopneumatic actuators and microvalves. J Microelectromech Syst 15(5):1308–1318
4. Bardaweel HK, Anderson MJ, Weiss LW, Richards RF, Richards CD (2009) Characterization and modeling of the dynamic behavior of a liquid–vapor phase change actuator. Sens Actuators A 149(2):284–291
5. Song GE, Kim KH, Lee YP (2007) Simulation and experiments for a phase-change actuator with bistable membrane. Sens Actuators A 136(2):665–672
6. Ogden S, Jonsson J, Thornell G, Hjort K (2012) A latchable high-pressure thermohydraulic valve actuator. Sens Actuators A 188:292–297
7. Bardaweel HK, Bardaweel SK (2013) Dynamic simulation of thermopneumatic micropumps for biomedical applications. Microsyst Technol 19(12):2017–2024
8. Lee H, Richards CD, Richards RF (2013) Experimental and numerical study of microchannel heater/evaporators for thermal phase-change actuators. Sens Actuators A 195:7–20
9. Kong LX, Parate K, Abi-Samra K, Madou M (2015) Multifunctional wax valves for liquid handling and incubation on a microfluidic CD. Microfluid Nanofluid 18(5–6):1031–1037. doi:10.1007/s10404-014-1492-x
10. Lei KF, Chen KH, Chang YC (2014) Protein binding reaction enhanced by bi-directional flow driven by on-chip thermopneumatic actuator. Biomed Microdevices 16(2):325–332
11. Ono M, Izumi T, Kato S (2005) Proposal of a gas–liquid phase-change microactuator and its applications. In: Proceedings of the ASPE 2005 annual meeting, Norfolk, 2005, pp 138–141
12. Ono M, Kato S (2010) A study of an earthworm type inspection robot movable in long pipes. Int J Adv Rob Syst 7(1):85–90
13. Wu H, Kitagawa A, Tsukagoshi H (2005) Development of a portable pneumatic power source using phase transition at the triple point. In: Proceedings of the JFPS international symposium on fluid power, Tsukuba, 2005
14. Inoue T, Shibuya K, Nagano A (2010) Underwater robot with a buoyancy control system based on the spermaceti oil hypothesis development of the depth control system. In: Proceedings of 2010 IEEE/RSJ international conference on intelligent robots and systems (IROS) 2010, Taipei, 2010
15. Yamaguchi D, Tonokai A, Kanda T, Suzumori K (2013) Light-driven actuator using hydrothermally deposited PLZT film. IEEJ Trans Sensors Micromach 133(8):330–336
16. Ikeda N, Ohsumi H, Ohwada K, Ishii K, Inami T, Kakurai K, Murakami Y, Yoshii K, Mori S, Horibe Y, Kitô H (2005) Ferroelectricity from iron valence ordering in the charge-frustrated system $LuFe_2O_4$. Nature 436(7054):1136–1138
17. Suzumori K, Matsuoka H, Wakimoto S (2012) Novel actuator driven with phase transition of working fluid for uses in wide temperature range. In: Proceedings of 2012 IEEE/RSJ international conference on intelligent robots and systems, Algarve, 2012
18. Matsuoka H, Suzumori K (2014) Gas/liquid phase change actuator for use in extreme temperature environments. J Ref Int J Autom Technol 8(2):140–146
19. Stull DR (1947) Vapor pressure of pure substances. Organic and inorganic compounds. Ind Eng Chem 39(4):517–540
20. Committee for High Temperature Springs (1989) Long-term elevated temperature setting properties of oil tempered Cr–Si steel wire (SWOSC-V) and stainless steel wire (SUS304-WPB) springs. Trans Jpn Soc Spring Eng 1989(34):59–84 **(in Japanese)**
21. Takenaka T (1959) On the effective area and stiffness of bellows. Trans Jpn Soc Mech Eng 25(149):43–46 **(in Japanese)**
22. NIST WebBook (2015) National Institute of Standards and Technology, http://webbook.nist.gov/cgi/cbook.cgi?ID=C112276&Units=SI&Mask=4 #Thermo-Phase. Accessed 14 Jan 2016
23. Akram MS, Haider B, Afzal W (2010) Thermophysical behavior of some industrially important associating fluids: thermal expansion coefficients. Paper presented at the 3rd Symposium on Engineering Science, Lahore, Pakistan

8

A method for using one finger to press three separate keys on a three-dimensional keyboard designed to be mounted on a mouse

Tomohiro Suzuki[1*], Satoshi Miura[1], Yo Kobayashi[2] and Masakatsu G. Fujie[2]

Abstract

Here we propose and verify the feasibility of a new keyboard structure for keyboard-mounted mouse. Reducing the distance that fingers move while typing within a three-dimensional (3-D) key arrangement that fits along the fingers is effective for keyboard-mounted mouse, provided that each finger can separately press three different keys. Our objective was to design and test a method for separately being able to press three keys with a single finger, and in the process, reduce finger-moving distance. We analyzed 3-D finger motion while participants typed on a standard keyboard, and used this data to develop a 3-D layout for keys that are arranged along the fingers. Analysis of 3-D finger motion while using our new keyboard showed that the distance fingers traveled was 74 % less than that when using a standard keyboard ($p < 0.05$). Moreover, this did not result in typing mistakes caused by interference between the keys and finger movements. After typing 30 characters 20 times, the average input error rate in the 20th trial was 18 %, while the average error rate across all trials for the standard keyboard was 9.5 %. We conclude that our proposed 3-D keyboard can be used accurately with one finger while reducing the distance fingers must move. However, input mistakes were caused by finger linkage motion. In the future, we must devise a character-input algorithm that eliminates such erroneous input. We must also include the mouse function and evaluate the operability of the device in tasks that require keyboard and mouse use.

Keywords: Keyboards, Motion analysis, User interfaces, Human computer interaction, Ergonomics

Background

Social background

In the current information-technology society, people spend a great deal of time using computers at work and at home. Keyboards and mice are widely used as the de facto standard for operating personal computers, and almost all people are accustomed to using them [1]. Generally, keyboards are used for character input and mice are used to direct the cursor and make selections. However, because hands must be frequently moved between the two independent interfaces, typing efficiency and mouse operation are not as efficient as they could be. Thus, integrating the two interfaces, mounting mouse functions on the keyboard or mounting keyboard functions on the mouse, will be effective to reduce the movement between two interfaces and will improve the efficiency of typing and mouse operation. Developing an interface that combines keyboard and mouse function is therefore of keen interest.

Related works

TrackPoint is one of several interfaces that have integrated devices that mount mouse functions on the keyboard. The TrackPoint has a mouse trackpad that is placed in the center of the keyboard [2]. By using the TrackPoint, mouse-cursor movement can be controlled by using the fingertips. Another keyboard can sense hand movement with a built-in infrared sensor, and some mouse functions are possible via finger motion as if the keyboard is a touchpad mouse [3]. One keyboard can act as a touchpad mouse through touch sensors that

*Correspondence: suzuki-t03@toki.waseda.jp
[1] The Graduate School of Science and Engineering, Waseda University, Tokyo, Japan
Full list of author information is available at the end of the article

are placed on the keyboard surface [4, 5]. These studies considered mouse functions such as cursor movement and scrolling, but studies that examine clicking and dragging functions have not yet been conducted. In order to mount mouse function onto the keyboard, part of the keyboard needs to be made into something like a touchpad mouse. However, mouse function via touchpad has been reported to take more time than that using a conventional mouse [6]. Thus, improving efficiency of typing and mouse operation by integrating the devices, a more useful approach might be to mount the keyboard function onto a mouse that is operated by the hand and fingers.

Several attempts have been made to mount keyboard functions on a mouse. One is a multi-button mouse that can have several keys mounted on it [7]. With this type of mouse, keys such as 'Return' or 'Delete' can be input using buttons on the side of the mouse. CombiMouse, which has the entire right half of a standard keyboard connected to a mouse, can be used similarly as a standard mouse [8, 9]. Character input speed of CombiMouse has been reported to be 70 % that of a standard keyboard. To improve typing and mouse operation efficiency, ideally the keys can be pressed with small finger movements while a hand is placed on the mouse. In contrast, CombiMouse requires the hand to be released from the mouse portion of the device when using the keyboard portion, and a suitable key structure for mounting on a mouse has not yet been considered. DataHand is a mouse equipped with a keyboard with a special concave key structure around the fingertip [10]. Because the concave keys of DataHand can be pressed using small finger motions, the hand can stay on the mouse at all times, even when typing. However, character input speed with DataHand only reached 70 % that of the standard keyboard after practice for more than 13 h. Thus, learning the special key arrangement still presents some difficulties for character input. Based on these attempts of integrating a keyboard onto a mouse, an ideal interface should satisfy the following conditions: (1) the distance that fingers are required to move should be reduced and keys should be accessible while the hand is on the mouse and (2) learning the special key arrangement should be fairly easy. For this purpose, the best key arrangement is one that fits along the fingers and maintains the same direction of finger movement as is employed by the standard keyboard. This should eliminate difficulty learning a special key arrangement. Analyzing finger motion while typing on the standard keyboard is an effective way to determine the best arrangement based on these conditions.

Finger-motion analysis while typing on keyboards has been reported in several studies that measured finger-joint angle, joint-angle velocity, and joint-angle acceleration [11–13]. Although the displacement of finger-joint position trajectory is necessary to determine the key arrangement that requires the least finger movement, these data have rarely been reported.

Objective

We propose a keyboard-mounted mouse whose keys fit along the fingers, allowing minimal finger movement while the hand is placed on the mouse, and whose key arrangement is easy to learn. The conventional keyboard-mounted mouse is operationally satisfactory in terms of mouse function because the mouse is a simple interface that fits the hand nicely and whose only function is the control of cursor movement and selection via mouse click [8, 9]. However, the keyboard portion of the conventional keyboard-mounted mouse does not fit fingers effectively, and complex operations typically performed with a keyboard are difficult. Thus, slower character-input speed compared with the standard keyboard remains a problem for the conventional keyboard-mounted mouse [8–10]. One solution to this problem might be a keyboard design that fits along fingers, but studies of such a method have not been reported. If keys are to be fitted along fingers, multiple keys must be in close contact with a single finger. Because of this, a method for being able to separately press several different keys with a single finger must be part of the overall design. In this paper, we focus on designing the arrangement of keys that fit along fingers for a keyboard-mounted mouse. Our objective is to design and test a method for separately being able to press several keys with a single finger, and in the process, reduce finger-moving distance. To accomplish this, we tracked fingers while participants typed on a standard keyboard and used the data to derive a proposed key arrangement. We then developed a keyboard structure that incorporated this layout and tested our predictions. We used three-dimensional (3-D) finger-motion analysis to determine the distance that fingers traveled while performing a simple typing task, and recorded typing accuracy (Experiment 1). To determine how accuracy on the proposed keyboard compared with those of a standard keyboard, we computed error rates for participants who performed 20 trials of a typing exercise on each type of keyboard (Experiment 2).

Methods

Target keys on our keyboard-mounted mouse

We proposed a keyboard-mounted mouse that is used by the right hand because most people are right-handed and use a mouse with their dominant hands. Character keys are the most frequently used keys on a keyboard. When inputting character keys around the home position of the keyboard, people often "touch type" without actually looking at the keys. Normal typing divides the keyboard into a left and right side with the boundary being 'Y', 'H', and 'N' on the right and 'T', 'G', and 'B' on the left. However, 'Y', 'H', and 'N' could be typed with the left hand. Therefore, we focused on developing a keyboard for typing 12 characters around the home position on the right hand side: 'U', 'J', 'M', 'I', 'K', ',', 'O', 'L', ',', 'P', ';', '/'. The 12 keys are arranged in the usual four columns, with a different finger assigned to the three keys in each column. Mouse clicks and scrolling are done by assigning the click keys and scroll keys (up and down) to some of the 12 keys. The thumb, which is not used for typing character keys on the standard keyboard, is used for operating the switch that toggles the device's function between keyboard and mouse. The switch not only toggles between keyboard and mouse functions, but can also toggle the key map for the keys, similar to the 'shift' key on a standard keyboard. Thus, the 12 character keys can be made to represent the other keys on the right-hand side of a standard keyboard (Return, Delete, '-', '=', '[', ']', '\', '"', '←', '→', '↑', '↓') or the numeric keys (0–9).

Methods for using one finger to press three separate keys

A 3-D key arrangement that fits along the fingers may be effective for reducing the distance fingers move while typing, but because each finger joint touches different keys at the same time, typing accuracy might be reduced. However, with a careful key-arrangement design that considers the combinations and positions of finger-joint movements, a keyboard can be made such that three keys can be typed separately by each finger joint.

Fingers have three joints, the distal interphalangeal (DIP) joints, proximal interphalangeal (PIP) joints, and metacarpophalangeal (MCP) joints. PIP joints and MCP joints can move voluntarily, and DIP joints move subordinately to PIP joint movement [14]. Therefore, eight voluntary finger motion types are possible by combining two finger joints (MCP and PIP) and two directions of motion (flexion and extension) (Fig. 1a). Observing finger motion on a standard keyboard shows that as a finger moves from the middle key to the upper key, the MCP joint flexes and the PIP joint extends. Touching the middle key requires the MCP joint to flex, and moving to the lower key requires the PIP joint to flex and the MCP joint to extend (Fig. 1b).

Moving some fingers individually is difficult because the finger muscles and corticospinal neurons are linked between fingers [15]. Because finger extension causes more finger-linked movements than flexion [16, 17], using finger extension to separately type keys that are in close contact to fingers is difficult. Therefore, keys on our new keyboard are arranged such that they can be typed using flexion. By considering the finger motions depicted in Fig. 1b, we assumed that 3-D keys placed in the red areas depicted in Fig. 1c could be typed through flexion. We hypothesized that 3-D keys could be typed through flexion with limited finger movement and with as much accuracy as the standard keyboard if the keys in the red areas in Fig. 1c could sense initial finger motion and ignore subsequent interference by other joint movements. By analyzing finger motion, we should be able to determine the appropriate 3-D key positions for sensing the distinct and separate initial finger motions. We did so by determining which joint began moving first and how other joints subsequently moved during normal typing on a standard keyboard. Thus, we reasoned that placing the new keys in positions that sense the first movements of the joints, not interfering with tapping of the other keys, and matching the red areas in Fig. 1c, we would be able to develop a 3-D keyboard that can be used accurately through flexion but that does not require fingers to move as much as a standard keyboard.

Typing motion analysis

To minimize finger movement and to allow the three keys along each finger to be pressed separately, keys need to be arranged such that they are in close contact with the finger position that starts moving the earliest when typing, and such that extraneous joint movements do not interfere with pressing the target key. Thus, we analyzed typing motion on the standard keyboard. The objective of this experiment was to determine finger positions when typing on a standard keyboard, and then apply them to the 3-D arrangement.

Experimental protocol

We used the AURORA® magnetic 3-D position measurement system (Northern Digital Inc., Canada), a standard keyboard consistent with the ISO 2530:1975, and palm rest (Fig. 2).

Five healthy men (mean age 21.6 years; range 20–23 years; all right-handed) were asked to input character keys with the index finger, middle finger, ring finger, and little finger of the right hand. Informed consent was obtained from each subject before participation. The 12 characters were typed five times each, with each finger typing three different characters (60 character inputs for each participant). The keys 'J', 'K', 'L', and ';' were set as the home position, and participants were asked to return their fingers to these four keys after inputting each character. We measured the 3-D trajectory of fingertips, distal interphalangeal (DIP) joints, proximal interphalangeal (PIP) joints, and metacarpophalangeal (MCP) joints during typing. Task instructions were as follows:away from the other

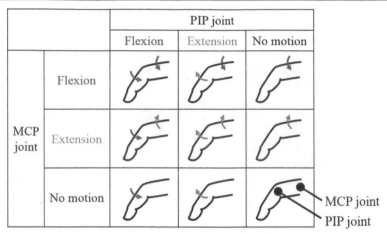

a   *Eight finger-motion types constructed by combining two finger joints and two directions of motion*

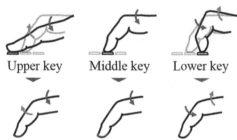

b   *Finger-motion types for pressing the upper, middle, and lower keys on a standard keyboard*

c   *Proposed 3-D key areas for pressing three keys with one finger through flexion*

Fig. 1 Proposed 3-D key areas for pressing three keys with one finger through flexion. The findings are based on finger motions constructed by combining two finger joints (MCP and PIP) and two directions of motion (flexion and extension). **a** Eight finger-motion types constructed by combining two finger joints and two directions of motion, **b** finger-motion types for pressing the upper, middle, and lower keys on a standard keyboard, **c** proposed 3-D key areas for pressing three keys with one finger through flexion

1. Place the palm on the palm rest and place the fingers naturally on the home-position keys.
2. Type the 'U', 'J', and 'M' keys five times each using the index finger.
3. Type the 'I', 'K', and ',' keys five times each using the middle finger.
4. Type the 'O', 'L', and '.' keys five times each using the ring finger.
5. Type the 'P' and ';', and '/' keys five times each using the little finger.

Results and discussion

The mean and standard deviation of finger displacement in the y- and z-axes are shown in Fig. 3 (300 character inputs, 60 inputs × 5 subjects). Initial finger position is the origin and has a displacement of zero. When we averaged the data, we normalized the time axis after approximating the data with a linear regression between the two points on the data plot.

Negative displacements in the y- and z-axes indicate that the joint was moved by bending another joint. By placing a key at the sites at which joints moved in a

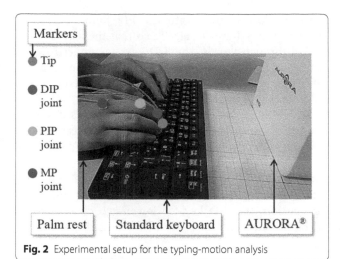

Fig. 2 Experimental setup for the typing-motion analysis

negative direction, the key can be pressed by bending the finger, which produces less finger-linked movements than does extension [16, 17]. The starting times for such movements are shown in Table 1 and Fig. 4. Because the push-in distance of keys on a standard keyboard is 3.0 mm [18], the displacement start time was set to the time when the displacement from the initial position reached 3.0 mm. In Table 1, a dash means that there were no negative displacements greater than 3.0 mm in a given axis.

As seen in Fig. 4a, when typing keys in the upper row, negative displacement in the z-axis is fastest in the PIP joint. The PIP joint starts moving significantly earlier than the DIP and MCP joints (df = 4, $p = 0.001$ vs. DIP joint z-axis, $p = 0.02$ vs. MCP joint z-axis). Thus, typing upper-row keys would be sensed earliest by arranging

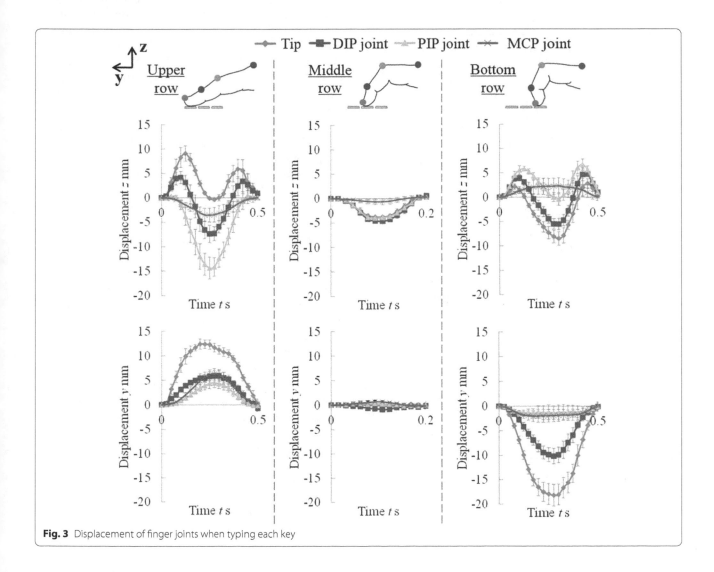

Fig. 3 Displacement of finger joints when typing each key

Table 1 Displacement start time of each finger part

	Tip	DIP joint	PIP joint	MCP joint
Upper key				
y-axis start time (s)	–	–	–	–
z-axis start time (s)	–	0.18	0.12	0.16
Middle key				
y-axis start time (s)	–	–	–	–
z-axis start time (s)	0.068	0.062	0.067	–
Lower key				
y-axis start time (s)	0.064	0.11	–	–
z-axis start time (s)	0.17	0.22	–	–

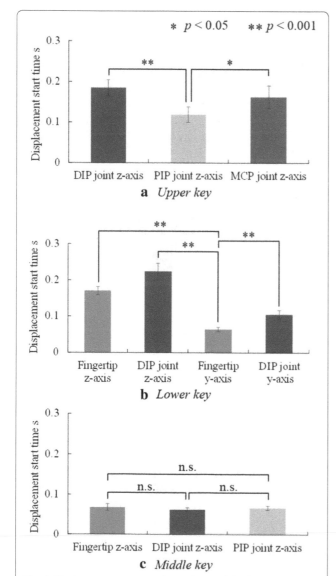

Fig. 4 Displacement start times for each part of the finger, **a** when typing upper keys, **b** when typing lower keys, **c** when typing middle keys

3-D keys in close contact with the PIP joint. Taking into account the red area in Fig. 1c so that the keys can be pressed through flexion, the 3-D keys for upper-row characters should be placed at the proximal phalanx (the part of the finger between the PIP and MCP joints) near the PIP joint on the negative z-axis (Fig. 5, left).

Data from Fig. 4b also show that when typing a character in the bottom (lower) row, negative displacement in the y-axis is significantly early in the fingertip (df = 8, $p < 0.001$ vs. DIP joint y-axis, $p < 0.001$ vs. fingertip z-axis, $p < 0.001$ vs. DIP joint z-axis). Taking into account the red area in Fig. 1c so that the keys can be pressed through flexion, the 3-D key for lower-row characters should be placed at the fingertip in the negative y-axis (Fig. 5, right).

Figure 4c also indicates that when typing characters from the middle row, onset of negative movement in the z-axis occurs almost at the same time for the fingertip, DIP joint, and PIP joint, and t-tests show no significant differences. Thus, typing middle-row characters can be sensed at the same time by making the 3-D keys in close contact with the fingertip, DIP joint, and PIP joint. Taking into account the red area in Fig. 1c so that the keys can be pressed through flexion, and considering the proposed 3-D key placements for upper and lower keys, the middle 3-D key should be placed either in the same position as on a standard keyboard or at the middle phalanx (the part of the finger between the DIP and PIP joints) (Fig. 5, center).

We next used the data to verify that the proposed 3-D key positions can be pressed separately with a single finger. PIP joint movement used to press an upper key cannot interfere with other proposed keys because when the PIP joint moves to press an upper key, the fingertip moves in the positive direction in the y- and z-axes (Fig. 3), which is away from the other proposed keys. Fingertip movement to press a lower key might interfere with a middle key placed at the middle phalanx because when the fingertip moves to press the lower key, the DIP joint moves in the negative direction in the y-axis (Fig. 3). In contrast, fingertip movement to press a lower key cannot interfere with other keys if the middle key is placed in the same position as on a standard keyboard because fingertip movement is in a positive direction in the z-axis, which avoids pressing a middle key, and the PIP joint moves in a positive direction in the z-axis, which avoids pressing the upper key (Fig. 3).

Fingertip movement to press a middle key placed in the same position as on a standard keyboard or at the middle phalanx might interfere with other keys because the PIP joint and fingertip move in the negative direction in the z-axis (Fig. 3). However, because pressing a middle key requires only a small movement, by modifying the

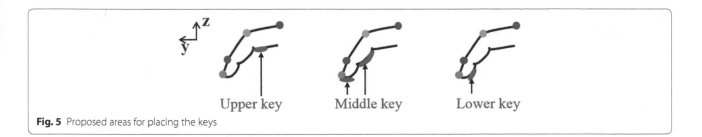

Fig. 5 Proposed areas for placing the keys

other key positions a little so that they cannot sense when the fingertip presses the middle key, fingertip movement should not interfere with the other keys.

In conclusion, the 3-D key arrangement, shown in Fig. 6 can sense the initial typing motion and pressing a target key does not cause other keys to be pressed, provided that upper and lower keys are set so that they do not sense the force applied when the user presses a middle key.

Upper key: Vertically below the PIP joint of the proximal phalanx.

Middle key: Vertically below the fingertip (in the same position as on a standard keyboard).

Lower key: At the fingertip when the PIP joint is flexed.

3-D keyboard design

Hardware

According to the typing-motion analysis, we designed a 3-D keyboard structure as depicted in Fig. 7, and developed it as shown in Fig. 8. A Flexi Force Sensor® (Tekscan Inc., USA) was built into each 3-D key, which could then sense the force of each finger. An Arduino Leonardo board (Arduino Srl, Italy) was used as a microcomputer to handle the A/D conversion of the force signal. The 3-D keys have adjustable mechanisms that allow the position and angle to be custom fit to a user's fingers. The optimal positions put the keys in close contact with the fingers so that the initial typing motion is sensed quickly. The upper and lower keys are set to a position such that the upper- and lower-key output from the force sensors become 0 N

when the user presses the middle key. The key pitch of the 3-D keyboard was set to 19 mm, which is the pitch of most keyboards. To reduce the distance fingers move, input is accomplished solely by touching, and pressing is not necessary. The 3-D keyboard corresponds to the character keys that would normally be pressed by the right hand on a standard keyboard (Fig. 9). Thus, the 3-D keyboard is used in conjunction with a standard keyboard. Figure 10 shows an overview of the proposed system.

Character input algorithm

The force sensor built into the 3-D keys senses the contact force applied by the fingers. When this force exceeds the threshold force, the character code of the key is sent to the computer. The flow chart of the character input algorithm is shown in Fig. 11.

Moving some fingers individually is difficult because some finger muscles and corticospinal neurons are linked between fingers, and the amount of finger linkage can vary from person to person [15]. Because the 3-D keys are in close contact with the fingers, the user may unintentionally press undesired keys because intentional movements of one finger are accompanied by unintentional movements of linked fingers. To prevent erroneous inputs resulting from finger linkage, the force threshold needs to be set individually for each person, depending on how much finger-linkage movement they experience. We therefore devised the following calibration method that can be used to set the threshold force for each individual before initial use:

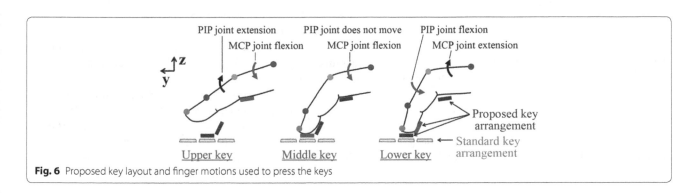

Fig. 6 Proposed key layout and finger motions used to press the keys

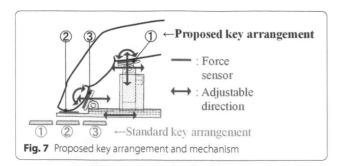

Fig. 7 Proposed key arrangement and mechanism

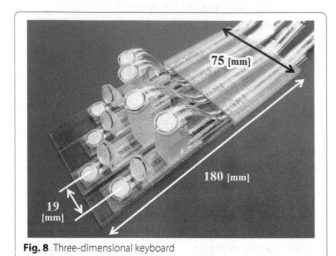

Fig. 8 Three-dimensional keyboard

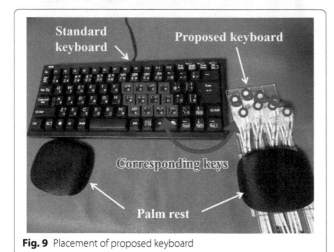

Fig. 9 Placement of proposed keyboard

1. Users place their hand on the keys.
2. Each key position is adjusted so as to be in close contact with the fingers.
3. Users press each 3-D key five times, and the output of the force sensor is recorded (Fig. 12).
4. The maximum contact force F_{max} N that was generated by unintentional touches is calculated.

5. Threshold is set to $F_{max} + 0.1$ N (0.1 N is the force sensor specification for the maximum value of repeatability).

With correct calibration, only character codes from keys that are touched intentionally will be sent to the computer.

Experiment 1: Evaluation of the distance moved while typing and typing accuracy

Experimental objective

The 3-D keyboard is designed to reduce the distance fingers must travel while typing, without reducing typing accuracy caused by inadvertently pressing undesired keys. The objective of this experiment was to evaluate whether the distance that fingers move when using the 3-D keyboard is less than that when using a standard keyboard, and whether the 3-D keyboard can be used as accurately. This was accomplished by measuring 3-D finger motion and the registered input when typing in a regular order.

Experimental protocol

We used the same equipment as was used in the analysis of typing motion. Participants were three healthy men (mean age 22.3 years; all right-handed). The experiment began after adjusting the keys to be in close contact with the fingers and performing the threshold calibration described above. The task was almost the same as in the typing-motion analysis experiment, except that each key was hit ten times. We measured the 3-D trajectory of the fingertips, DIP joints, PIP joints, and MCP joints during typing. The input characters were recorded in a text file. This experiment was carried out after obtaining informed consent from the subjects.

Results

The comparison of distance moved while typing each character key (upper, middle, and lower rows) is shown in Fig. 13. The distance for one character input was calculated from the 3-D position as the total distance moved by each finger joint in one finger when touching one key. Means and standard deviations were calculated from the ten trials for each key.

T-tests showed that when using the 3-D keyboard, the distance fingers traveled was reduced significantly in all keys compared with the standard keyboard (df = 4, $p = 0.03$ for upper-row keys, $p = 0.007$ for middle-row keys, $p = 0.02$ for lower-row keys). On average, distance was reduced 76 % for the upper-row keys, 58 % for the middle-row keys, and 88 % for the lower-row keys. When all keys were combined, the average reduction was

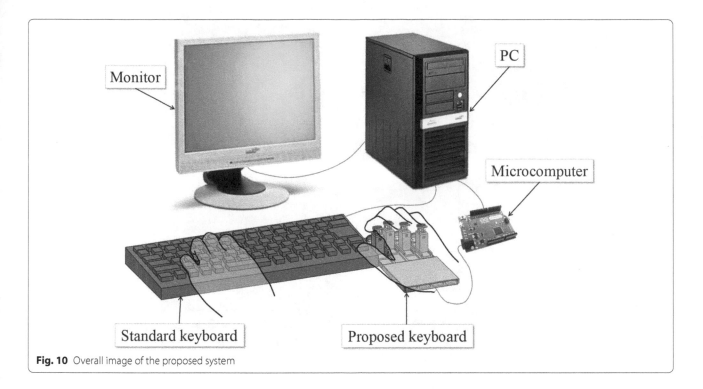

Fig. 10 Overall image of the proposed system

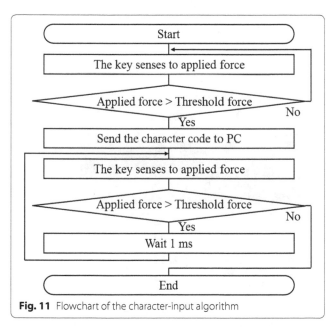

Fig. 11 Flowchart of the character-input algorithm

74 %. Participants did not make any typing errors in this experiment.

Experiment 2: Evaluation of typing accuracy when typing a series of characters
Experimental objective
In Experiment 1, we found that typing with 3-D keys is accurate when typing in a regular order. However, this is very simple and differs from actual keyboard use. In Experiment 2, our objective was to compare typing accuracy between the 3-D and standard keyboards when typing a series of characters.

Experimental protocol
We used a 3-D keyboard and a standard keyboard in conjunction with a Windows XP computer running the software "Stamina Typing Tutor 2.5 [19]", which can record error rate and typing speed.

Five healthy individuals (four men, one woman; mean age 29.2 years; range 20–54 years; four right-handed, one left-handed) who use standard keyboards daily participated in the experiment. None had experience using a 3-D keyboard. The experiment began after adjusting the key positions so that they were in close contact with the fingers and after performing the force-threshold calibration. Thirty pseudorandomly chosen characters were displayed on the monitor, and the task was to type them as fast as possible. To make trial difficulty consistent, the characters alternated between keys on the left and right sides of the keyboard (Fig. 14). Participants completed 40 sets, alternating between the standard and 3-D keyboards (20 sets each). A 15 s break was taken between each set. Each subject used the 3-D keyboard for about 20 min. The experiment was approved by Waseda University IRB (#2014-018) and was carried out after obtaining informed consent from the subjects.

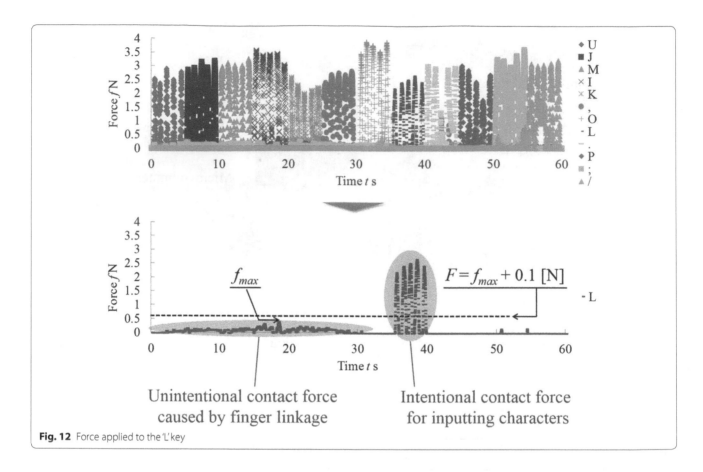

Unintentional contact force
caused by finger linkage

Intentional contact force
for inputting characters

Fig. 12 Force applied to the 'L' key

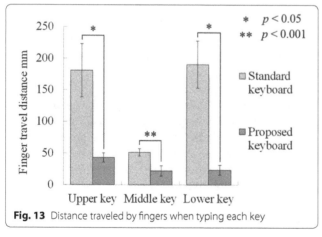

Fig. 13 Distance traveled by fingers when typing each key

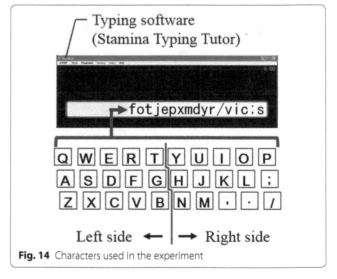

Fig. 14 Characters used in the experiment

Results

The total average error rate using the standard keyboard and error rates for each trial on the 3-D keyboard are shown in Fig. 15 for each participant. Comparing the 1st and 20th trial on the 3-D keyboard with t-tests showed that after about 20 min of practice, the error rate significantly decreased (df = 8; $p = 0.004$). The average error rate for the final trial when using the 3-D keyboard was 18 %, while that for the first trial was 49 %. In fact, subject B did not make any mistakes in the 10th trial and subject C had perfect performance in the 13th. The average error rate across all trials for the standard keyboard was 9.5 %.

Typing speed for each subject was calculated and is shown in Fig. 16. Comparing the 1st and 20th trial, typing speed significantly increased (df = 8; $p = 0.002$). By

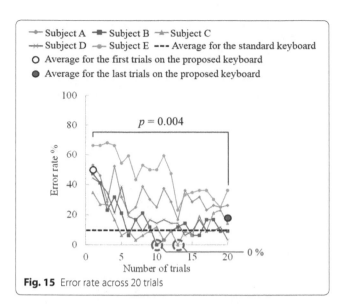

Fig. 15 Error rate across 20 trials

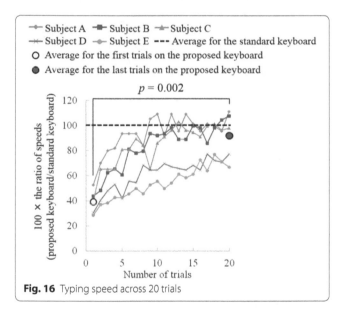

Fig. 16 Typing speed across 20 trials

the 20th trial, the average typing speed when using the 3-D keyboard reached 92 % of the speed when using the standard keyboard, an increase from 39 % observed at the first trial. For participants A, B, and C, the maximum typing speed with the 3-D keyboard was slightly higher than the average speed when using the standard keyboard.

Discussion

In Experiment 1, we found that our 3-D keyboard can be used accurately, and that joint movement does not interfere with pressing the target key. Thus, our proposed method satisfies the condition that three keys can be pressed separately by a single finger despite close contact with all three keys.

In Experiment 2, two subjects were able to type 30 characters without any mistakes once in the 20 trials. This result shows that the 3-D keyboard can be used accurately even when typing a series of characters. While the initial trials contained more errors because subjects did not yet recognize the arrangement of keys, this type of mistake decreased as subjects got used to layout. By the last trial, the average error rate was only 18 %. Thus, 20 min of using the proposed keyboard reduced the error rate from 49 to 18 %.

This error rate was almost double that for the standard keyboard. This can be attributed to a software error in the proposed keyboard. Some erroneous inputs were generated by contact forces that exceeded the input threshold during the experiment. These errors seem to have been caused by finger-linkage motion that was different from that when calibrating the force threshold. The degree of finger linkage might have changed because finger movements changed as subjects became habituated to the 3-D keyboard. In the future, we should devise a method for updating the input-force threshold according to any adaptations to the 3-D keyboard made by the participants. Because the erroneous inputs were caused by finger linkage, further consideration will have to be given to pattern recognition methods that use a combination of force sensors to differentiate intentional input from unintentional input.

In Experiment 2, we also calculated typing speed. Average typing speed during the last trial on the proposed keyboard reached 92 % of that on the standard keyboard. Some participants reached maximum typing speeds on the 3-D keyboard that were faster than the average typing speed on the standard keyboard. Therefore, about 20 min of practice is all that is required to become accustomed to the 3-D keyboard and reach a typing speed equivalent to that of a standard keyboard. Changes in the software could improve the final speed and rate of improvement within the same amount of practice time.

Although we found some significant differences, the sample size was quite small. Therefore, we should conduct further experiments with larger sample sizes.

Here, we only focused on the keyboard portion of the keyboard-mounted mouse. In the future, we must equip the mouse function on the 3-D keyboard, and we must make the mechanism and algorism for switching the function between keyboard and mouse, or between different sets of keys. We also need experiments that evaluate the operability of our new device in tasks using both the keyboard and mouse.

Conclusion

Our objective was to design and test a 3-D keyboard design that allows each finger to control three separate keys, and in the process, reduce finger-moving distance. We analyzed 3-D finger motion while subjects typed on a standard keyboard, and used these data to develop a keyboard that accomplishes this goal. According to the results, the 3-D keys can be pressed separately without interference from other joint movements. Additionally, the distance that fingers moved when using the proposed keyboard was 74 % less than that when using the standard keyboard. Moreover, after using the proposed keyboard for about 20 min, the average input error rate was reduced to 18 %. We conclude that our proposed 3-D keyboard reduces the distance fingers move while typing and allows individual fingers to control three separate keys. In the future, we will devise a better character input algorithm that eliminates erroneous input caused by finger linkage. Further, we must add mouse functionality and the switching mechanism that toggles between keyboard and mouse functions. Then we will evaluate the operability of our proposed keyboard-mounted mouse when using both keyboard and mouse functions.

Authors' contributions

TS conceived the study, developed the device, carried out all experiments, analyzed the data, and drafted the manuscript. SM, YK and MF participated in the research design and sequence alignment. All authors read and approved the final manuscript.

Author details

[1] The Graduate School of Science and Engineering, Waseda University, Tokyo, Japan. [2] The Faculty of Science and Engineering, Waseda University, Tokyo, Japan.

Acknowledgements

This research was supported in part by the Outstanding Graduate COE Support Subsidy "Global Robot Academia (GRA)", a Grant-in-Aid for Scientific Research (A) (26242061), and a Grant-in-Aid for Scientific Research (S) (25220005). All grants were issued by the Ministry of Education, Culture, Sports, Science and Technology in Japan.

Competing interests

The authors declare that they have no competing interests.

References

1. Atkinson S, Woods V, Haslam RA, Buckle P (2004) Using non-keyboard input devices: interviews with users in the workplace. Intern J Ind Erg 33:571–579
2. Armbrüster C, Sutter C, Ziefle M (2007) Notebook input devices put to the age test: the usability of trackpoint and touchpad for middleaged adults. Ergonomics 50(3):426–445
3. Keskin C, Hilliges O, Izadi S, Helmes J (2014) Type–Hover–Swipe in 96 bytes: a motion sensing mechanical keyboard. ACM CHI conference on human factors in computing systems, p 1695–1704
4. Tung YC, Cheng TY, Yu NH, Chen MY (2014) FlickBoard: enabling trackpad interaction with automatic mode switching on a capacitive-sensing keyboard. UIST'14 Adjunct proceedings of the adjunct publication of the 27th annual ACM symposium on user interface software and technology, p 107–108
5. Habib I, Berggren N, Rehn E, Josefsson G, Kunz A, Fjeld M (2009) DGTS: integrated typing and pointing. Human-computer interaction—INTERACT, Volume 5727 of the series Lecture Notes in Computer Science, p 232–235
6. Ulrich TA, Boring RL, Lew R (2015) Control board digital interface input devices—touchscreen, trackpad, or mouse. Resilience Week (RWS), p 1–6
7. Scarlett D (2005) Ergonomic mice: comparison of performance and perceived exertion. Usability news from Software Usability Research Laboratory at Wichita State University
8. Slocum J, Thompson S, Chaparro B, Bohan M (2013) Evaluating the CombiMouse: a new input device for personal computers. Usability news from Software Usability Research Laboratory at Wichita State University
9. Slocum J, Thompson S, Chaparro B, Bohan M (2004) Examining first-time usage of the CombiMouse™. In: Proceedings of the human factors and ergonomics society 48th annual meeting
10. Knight LW, Retter D (1989) DATAHAND: design, potential performance, and improvements in the computer keyboard and mouse. In: Proceedings of the human factors society 33rd annual meeting, p 450–454
11. Baker Nancy A, Cham Rakie, Hale Erin, Cook James, Redfern Mark S (2007) Digit kinematics during typing with standard and ergonomic keyboard configurations. Intern J Ind Erg 37:345–355
12. Baker Nancy A, Cham Rakie, Cidboy Erin Hale, Cook James, Redfern Mark S (2007) Kinematics of the fingers and hands during computer keyboard use. Clin Biomech 22:34–43
13. Soechting John F, Flanders Martha (1997) Flexibility and repeatability of finger movements during typing-analysis of multiple degrees of freedom. J Comput Neurosci 4:29–46
14. Leijnse JNAL, Quesada PM, Spoor CW (2010) Kinematic evaluation of the finger's interphalangeal joints coupling mechanism—variability, flexion–extension differences, triggers, locking swanneck deformities, anthropometric correlations. J Biomech 43:2381–2393
15. Aoki Tomoko, Francis Peter R, Kinoshita Hiroshi (2003) Differences in the abilities of individual fingers during the performance of fast repetitive tapping movements. Exp Brain Res 152(2):270–280
16. Yu WS, van Duinen H, Gandevia SC (2010) Limits to the control of the human thumb and fingers in flexion and extension. J Neurophysiol 103:278–289
17. Schieber MH (1991) Individuated finger movements of rhesus monkeys: a means of quantifying the independence of the digits. J Neurophysiol 65(6):1381–1391
18. Kim JH, Aulck L, Bartha MC, Harper CA, Johnson PW (2014) Differences in typing forces, muscle activity, comfort, and typing performance among virtual, notebook, and desktop keyboards. Appl Ergon 45(6):1–8
19. Stamina typing tutor by typingsoft. http://typingsoft.com/stamina.htm. Accessed 10 Jan 2015

9

Concept proposal and experimental verification of a sidewalk supporting system utilizing a smartphone

Shin-yo Muto[1*] ⓘ, Yukihiro Nakamura[1], Hideaki Iwamoto[1], Tatsuaki Ito[1] and Manabu Okamoto[2]

Abstract

In this report, we propose a system to be used as a basis for the efficient gathering of information on sidewalk usage and services for the purpose of building a system to support social infrastructures through which users can move safely and securely. We assume services for extracting and providing useful information to sidewalk users and local governments. Specifically, we propose a concept for a sidewalk supporting system that accumulates information from strollers, mobility scooters, bicycles, etc., through sidewalk users' smartphones, sensors, and other such devices to provide valid information. Next, we show a proposed basic system configuration utilizing data collection technologies that have been developed in robots and walking appearance analysis with regard to responses to the diversity of sensors and the characteristics of smartphones. Furthermore, we introduce techniques for estimating sidewalk conditions using the subjectivity of sidewalk passers-by with regard to the acquisition of training data to analyse data effectively. As a specific application case, we evaluated the estimation of the gradient and unevenness of sidewalks by machine learning from the sensor data during the passage of strollers, and verified the effectiveness of this method.

Keywords: Sidewalk supporting system, Personal mobility, Smartphone, Sensor

Background

Along with the spread of smartphones and cloud technology in recent years, an environment of collecting and accumulating records of individuals' activities or so-called life logs through networks, has been prepared. Life logs include internet operation histories, application portfolios in place, and GPS information, that occur during daily activities. As an example of its application, a system has been proposed to estimate individuals' preferences and life patterns from the accumulated data to provide a personalized information service [1]. In addition, wearable devices [2] have generally been spread in a rapid manner, and a wide variety of services using sensors and network terminals owned by users have been considered.

To date, robotic rooms [3, 4] that assist users by managing a wide variety of sensor information installed indoors and studies that assume robot services in the street or in multiple dwellings [5–7] have been proposed as a framework for managing and utilizing a plurality of sensor information. Authors have built a network robot platform for sharing information with robots that deals with many types of sensor information with a plurality of guide systems [8–10]. Here we introduce abstracted information from sensors and robots in the form of 4W (who/where/when/what) user information or robots based on the XML format. Furthermore, while expanding on this, we have built a system to perform operations such as the physical transportation of goods in addition to the guidance of only providing information, and we have linked the information from a wide variety of sensors and robots [11, 12]. An approach has been adopted that streamlines the system by handling only the information which has been abstracted to a limited form. Through these methods, it is possible to accumulate information from multiple sensors designed and arranged for the purpose of measuring the specific information in the particular experimental environment and to capture it as an actual

*Correspondence: muto.shinyo@lab.ntt.co.jp
[1] NTT Service Evolution Laboratories, 1-1 Hikarinooka, Yokosuka-Shi, Kanagawa 239-0847, Japan
Full list of author information is available at the end of the article

example of its application. It can also be said that these are examples where the data collector side built systems to collect data primarily with sensors whose specifications were defined in advance.

Among the wide variety of sensors and robots that are researched and developed, research and development of middleware and operating system from the viewpoint of software development include RT-Middleware [13] as a common platform that uses a network-distributed component technology. In RT-Middleware, devices such as servo motors, sensors, and cameras and the functional elements of robots such as various processing programs, are processed into RT components, which are then linked to develop a system. Moreover, in ROS [14], an environment is provided in which devices such as actuators and sensors and various processing programs are modularized or processed into components to develop software. They enhance the reusability of resources by processing various functional elements into software components to aim at efficient software development. Furthermore, in SensorML [15], where sensor information is managed by utilizing XML technology and focus is on the discovery of sensors and the acquisition of positional information, there are examples of utilization for the purpose of collecting information on the population and states of pollution by dividing the area into specific regions of widely distributed sensors.

In parallel with the evolution of demonstration experiments at laboratories or specific facilities and the preparation of system development environments as described above, services using sensors built in smartphones carried by users and information from connected sensors are thought to increase along with the rapid spread of smartphones in the future. If we look at such a point of view, in the demonstration of robotic room [3, 4] and multiple robot cooperation [5, 12], a particular device is determined in advance, it is also fixed function sharing the server and the client. Also the research and development of middleware [13–15] related development efficiency software is improved. But the overall system design is dependent on the developer. Here, the efficient and usable systems are considered to be important in the process from the early experiment stage to commercialization. It is considered to be important to change the flexible function sharing in consideration of the client's capabilities and performance in the experimental stage. In addition, responses to the diversity of smartphone models and changes in their performances as well as the standpoint of dealing with the subjective evaluation values of users, which cannot be automatically acquired by sensors, are also important.

The authors have conducted studies on understanding the health status in those who suffer from rheumatoid arthritis as an application in the medical field using smartphone sensors, based on a platform [8–10] for liaising information to the robot as described above. The joint destruction of limbs caused by rheumatoid arthritis lasts for many years, and has become a social problem leading to severe motility disturbance and dysfunction in daily life activities [16]. It is necessary to accurately grasp the daily health conditions of patients undergoing this treatment, and thus objective, quantitative, and temporal evaluation methods are sought. For these situations, the authors performed phased experiments to build a system for estimating the disease activity score 28 (DAS28) without using blood test values from the information on gait that was obtained from the acceleration information of patients' smartphones. This was done while patients walked and self-reported information that they entered in their smartphones, verifying its effectiveness [17–19].

In this study, we propose technology for efficient information gathering as to situations involving sidewalks based on sensor information gathered from the smartphones of sidewalk-users. A sidewalk supporting system is used as the basis of the service for properly managing the collected data and results and extracting useful information such as that on traffic availability. This is then provided to sidewalk users, local governments, and others with the goal of expansion into daily life, including the aforementioned application to the medical field.

As conventional studies for detecting the states of road surfaces, a method of detection has been proposed which uses a smartphone's acceleration sensor fixed to the vehicle against the differences in level that become roadway traffic obstacles [20]. With regard to sidewalks, there is a test example of detection of surface damage using mobile robot equipped with laser scanner [21]. And data specifications for organizing sidewalk information have been proposed [22], and appropriate information presentation has been demanded as a social need for people who find it difficult to pass along streets, such those who use wheelchairs or strollers. Technical points needed to build services systematically against such problems could include information gathering technologies through the carrier network from smartphone sensors with the aim of collecting sidewalk movement logs, which are less burdensome for users.

In this paper, we propose a concept for a system that accumulates information from strollers, mobility scooters, bicycles, etc. through sidewalk users' smartphones, sensors and other such technology to provide valid information for the purpose of building a system to support social infrastructure where users can move safely and securely. Then we show a proposed basic system configuration utilizing data collection technologies in robotics and gait analysis for responding to the diversity of sensors and the characteristics of smartphones. By using this system, it can be handled physical data of a stroller, a bicycle and a wheelchair, the user's movement as well as the user's subjectivity data and vital data. Furthermore, we will introduce a method for estimating the conditions of sidewalks utilizing the subjectivity of sidewalks passers-by with regard to the acquisition of training data to analyze data effectively. As a specific application case, we evaluate the estimation of the gradient and the unevenness of sidewalks through machine learning from the sensor information and users' subjective information during the passage of strollers to show the effectiveness of this technique.

Methods

Basic concept

The basic concept of the sidewalk supporting system is intended to store sensor information from strollers, mobility scooters, and bicycles that utilize sidewalks as well as the personal mobility that has been developed in recent years to provide effective information to moving people and those that manage infrastructures. The basics of the system and service configuration are first assumed to collect information about sidewalk situations through sidewalk users' smartphones, utilize the collected data and results of their analysis, and provide sidewalk users

and local governments with information whether it's possible and suitable to pass (Fig. 1).

We assume a method of efficient data collection utilizing users' terminals by utilizing a mechanism for transmitting information on the sensors built into the smartphones (such as acceleration sensors, GPS, and attached sensors as the sensor information from bicycles, strollers, wheelchair, etc. that use sidewalks) into the cloud server system through the smartphones possessed by users. Based on the acceleration and GPS information, etc., it becomes possible to register information on whether users could move smoothly on the sidewalks they passed through or if they faced any difficulty in moving in situations with relatively significant acceleration changes.

What should be the focus of this report is the type of sensor and server connection function to move with the user mobility; and a system configuration with greater coverage is considered possible as a form to supplement this by linking to a system in which sensors are fixed, etc., as shown in Fig. 1(f). For example, it is desirable to collect information in sufficient temporal spatial density with a fixed sensor when measuring the sunshine and temperature on the sidewalk, etc., but it can be assumed to be difficult to install a device having the capability of network connection to the server in the sensor function

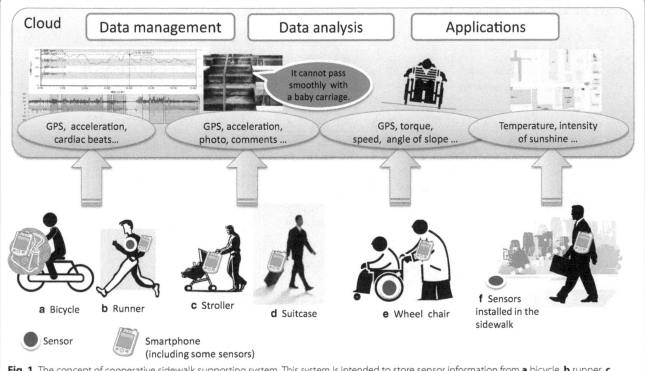

Fig. 1 The concept of cooperative sidewalk supporting system. This system is intended to store sensor information from **a** bicycle, **b** runner, **c** stroller, **d** suitcase, **e** weel chair and **f** sensors installed in the sidewalk

at a sufficient density in terms of its costs. In this situation, the construction of a more efficient information collection system is considered by introducing the smartphone connecting to the server and the sensor having a radio function such as nearby WiFi and Bluetooth.

Basic configuration of the data collection system

In this chapter, the authors propose a sidewalk supporting system that utilizes the XML technology based on the network robot platform [8–10], and on the gait measurement system that is conducted from the acceleration information of a smartphone [17].

Basic framework

Figure 2 shows the basic configuration of the information system. Here we provide the strata of the client, the local server, and the main server. We assume a smartphone carried by the user as the client. Here, in order to respond to the characteristics of clients that are diverse and quick technical innovations, we carry out the generalization of the information defined by the XML technology. Specifically, we introduce a protocol that enables the description of the definition of information and actual data in one XML file and consider a function that allows the management of information required for services without any changes in the database for the data in accordance with this description. By doing so, it becomes possible to address not only the format of the sensor information or sensor performance differences but also the abstracted information such as the user's subjectivity.

As a local server, we assume a computer having a network function disposed within a certain extent of equipment. The specific processing contents of the local server are considered to include processing that handles information for several clients, the execution of processing of information that the client cannot or preferably should not process due to the CPU power or batteries, and the execution of processing in cases where the network connection has not been made with the main server or where the environment of the network connection with the main server is bad. In addition, the execution of processing information, including the destruction of information that does not have to be left on the main server on a long-term basis, the destruction or anonymity of information that users do not desire to leave in the long term, and the calculation of statistical data. If there is no need for the processing described above, it is possible to configure the system with only the client and the main server.

As the main server, we assume a server system that is built in a cloud environment. Based on the capability of accumulating and managing data if they are those described with a predetermined XML technology, it facilitates their continued accumulation even if the client's equipment type is changed. The construction of an application that is able to handle a wide range of data is anticipated through the installation of a management function that is available to an application user with the specified authority.

We consider the deployment of common functions for each stratum of the client, the local server, and the main server as follows:

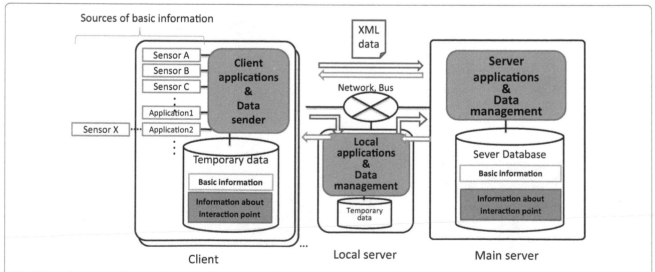

Fig. 2 Basic framework of information system. The sensory information is saved temporarily in user's smartphone. And this information is transmitted to a server in the state where the network is connected appropriately

1. Client: a common function that turns the data transmission and reception functions with the server into the library and an application function requiring a high level of responsiveness such as real-time sensor feedback within the client.
2. Local server: an application function that requires several clients' information.
3. Main server: a common function that accumulates and manages smartphones, sensors and application functions using such information.

The database in each stratum handles the basic information of sensor values themselves and the information on road surfaces, enabling the description of an application in accordance with the response speed and the computer power in each stratum. When considering the construction of several applications in the system configuration described above, it becomes possible to share the data management unit of the main server and make responses efficiently only by building application portions of each stratum. Furthermore, we consider that this system can make responses flexibly when there is a variation in measurement characteristics such as sampling time due to a change in terminal types, when the function as the application is improved by adding new information, or when the function deployment is modified among the client, the local server, and the main server.

Information description with XML
Information transmitted from a smartphone to a server, etc., is described in FDML (Field Data Mark-up Language) format [9]. This description format is the XML format that has been developed to uniformly describe a variety of sensor information, which is a configuration allowing flexible responses even when an environment is generated where new sensor information can be acquired or when there is a need for transmitting the information that the users input into their smartphones. On the server side, the management unit for executing information management stores the information transmitted from the smartphone into the appropriate database with the assumption that it is information in the FDML format and manages the information in the database to be made available in response to requests by the applications.

By generating sensor data and information associated therewith in the smartphone as described above, it would be possible to handle the information from sensors built into smartphones and devices or sensors connected to WiFi and Bluetooth (e.g. control information for electric wheelchairs) in a unified manner. The XML format introduced here has past records of performances in handling data from the manufacturing equipment and sensors in factories, the information from sensors in an

environment utilized by the guidance service robot, and the information from various types of smartphone sensors from the earlier days to date; and thus it is considered able to respond flexibly to the future development of these devices and the sophistication of the personal mobility.

Because it can handle users' input values to the smartphone applications similarly, subjective information, such as users' comments on how difficult their movement is and the information that users can understand at a glance when they see it, can be collected efficiently, even if it cannot be automatically determined with the sensor. It is difficult to impart all of such users' subjective data to the sensor data. But the advancement of the automated analysis can be expected with the techniques such as machine learning by collecting a set of users' subjective data and sensor data to be a certain amount of training data. It can be handled physical data of a stroller, a bicycle and a wheelchair, the user's movement as well as the user's subjectivity data and vital data synthetically by introducing FDML.

Examples of sensor data and applications
Figure 3 is an example of data from the smartphone acceleration sensor attached to the stroller, representing the difference between data from a gravel sidewalk with severe unevenness and a maintained sidewalk. The left side of the gravel sidewalk is in a situation where a considerable amount of force is required to push the stroller, and since a significant amount of acceleration was measured in such a situation, there is the possibility of the data measurement reflecting the state of the road surface even with the sensor equipped in a smartphone.

In addition, Fig. 4 represents an assumed application when the data as described above are stored, indicating a map showing the easiness of passage on the street sidewalks along the route from the station to the park on the basis of the acceleration information during the passage of the stroller and the GPS information. Besides such an example, we consider that the creation of new services linked with services regarding the movement information of cars and new services linked with users' biometric sensor information can be expected in the future.

Estimation experiment of road surface conditions of the sidewalk using the training data of users' subjective view.

Here, we discuss the method of estimating sidewalk conditions using the sidewalk users' subjectivity and results of the verification experiment as the basic experimental study of the information collection system for sidewalk road conditions. We assume that it is intended to improve the convenience for sidewalk passers by analyzing the information that can be collected easily through the passage on the sidewalk without conducting

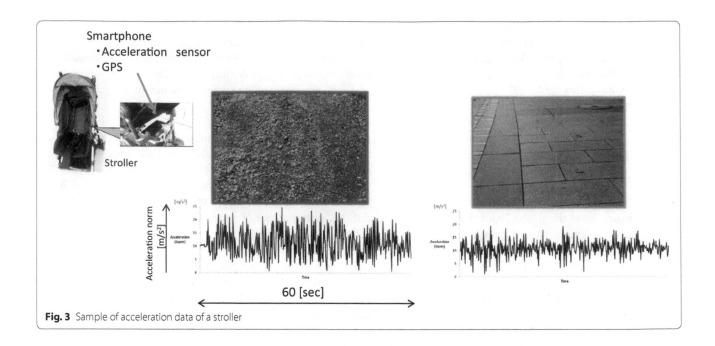

Fig. 3 Sample of acceleration data of a stroller

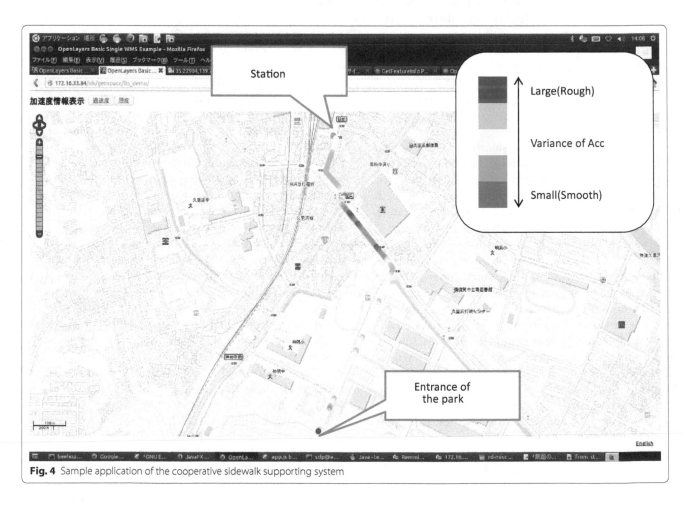

Fig. 4 Sample application of the cooperative sidewalk supporting system

special research activities on each sidewalk and providing information on hindrances. As a specific example, we show the results of evaluating the estimation by machine learning for the gradient and unevenness of the sidewalk from the sensor data during the passage of a stroller.

Experimental method

In the experiment, a smartphone was fixed to the stroller to acquire sensor data. There are various types of sidewalk utilization, including walking, bicycles, wheelchairs, and strollers. The utilization of strollers is assumed as the first phase of the study here in light of the situations where the child-rearing generation has growing needs for comfortable outings and the size of strollers is increased [23]. In addition, according to the study by Kozuka et al., level differences and tilting are attributed to disorders in using strollers [24]. With reference to its findings, the verification experiment was determined to aim at collecting sensor data during the actual passage of strollers as useful information on sidewalks so as to estimate the following labels.

[Shape of road surface] flatness/slope/composite gradient.
[Property of road surface] smoothness/unevenness.

The gradient is comprised of the upstream/downstream gradient (longitudinal gradient) and the left/right gradient (cross grade) with respect to the traveling direction. The composite gradient refers to the shape of the road surface that includes the cross grade addition to the longitudinal gradient. There is a report that such a gradient becomes a major factor in increasing stroller users' body burdens and senses of danger [25]. The composite gradient was determined to be the label of the road surface shape in addition to that of the flatness and slope, and considered to be an example of a place recognized subjectively to be difficult to pass through, as it is difficult to be measured by objective criteria. Furthermore, it was expected that problems to the label estimation model would be clarified by adding the composite gradient.

In the data collection, sensor data were acquired by passing through the sidewalks with a gradient or unevenness located around the NTT Yokosuka Research and Development Center, while a smartphone was fixed to the stroller. In order to collect information on what impedes passage through sidewalks, such as gradients and unevenness, data were acquired and recorded from built-in sensors in regard to GPS, angles (between the gravity direction and the smartphone), and angular velocity at a regular interval during the passage through the sidewalk. It was assumed that the angle represents the slope and that the angular velocity reflects the road surface properties such as unevenness. The data acquisition interval was determined to be

approximately 1 s for GPS and 20 ms for others. Additionally, the duration of sensor data acquisition is approximately 45 min. We used the "Caldia Auto 4-Cass Egg Shock HB" manufactured by Combi Corporation for the stroller and ARROWS NX F-01F (OS Android 4.2) by NTT DoCoMo for smartphones. The smartphone was fixed to the pipe-shaped frame so as to lay to the left and right rear wheels of the stroller using fixing equipment manufactured by National Products, Inc.: RAM MOUNTS X-Grip Holder.

The following labeling was manually applied to the acquired sensor data:

[Road surface shape] flatness/slope/composite gradient.
[Road surface property] smoothness/unevenness.

In reality, all of the sensor data along all the routes travelled by the stroller were not labelled manually, but typical places where the traffic was felt to be subjectively difficult were selected for manual labelling. Data at 50 points corresponding to 1 s, acquired at an interval of 20 ms, were treated as one set of sensor data, whereas labelled data were acquired at 150 points for each of the flatness/slope/composite gradients in the road surface shapes and at 500 points for each of the smoothness/unevenness in the road surface properties. Figure 5 is a photo

a Sample of composite gradient

b Sample of unevenness surface

Fig. 5 Examples of labeled point

showing an example of the road surface conditions with the composite gradient and unevenness actually labelled. Data was acquired by the condition that a stroller moves by approximately one constant speed along the road this time.

Results and discussion

Cross-validation of the obtained data was conducted by machine learning. The SVM was used for machine learning, and the kernlab library ksvm function of the open source free software in the R language was used for statistical analysis. The feature amounts used for learning were the mean and variance of the pitch angle of the angle sensor and the variance of the roll angle with respect to the road surface shapes. The mean and variance of the pitch angle of the angular velocity and the mean and variance of the roll angular velocity with respect to the road surface properties were used. Figures 6 and 7 indicate the distribution of feature values of labeled points. Figure 6 denotes the pitch and roll angle with respect to the road surface shapes. The variance of pitch and roll angle of composite gradient point tends to become large. And the mean of pitch angle of slope point trends to become large. Figure 7 denotes the pitch and roll angular velocity with respect to the road surface properties. The variance of pitch and roll angular velocity of unevenness

point tends to become large. And the mean of pitch angular velocity of unevenness point tends to become small. Table 1 shows the conditions of experiments. The frequency of cross-validation is 450 times for the road surface shapes and 1000 times for the road surface properties. In other words, it is considered that one sensor datum is not labelled for road surface shapes, and the estimation of road surface conditions was repeated 450 times as the total number in the estimation model which was constructed from the sensor data corresponding to 449 labels. One set of data was estimated in the model which was constructed from 999 sets of data for road surface properties as well. Results of estimation in the cross-validation of road surface shapes and road surface properties are shown in Tables 2 and 3 respectively.

The figure in each column in Tables 2, 3 represents the number of data whose respective labels were estimated for the sensor data considered to be unlabeled. Italics numbers represent the number of data whose estimated results are correct. In addition, the recall ratio represents the percentage of the data correctly estimated for labels. The precision factor represents the ratio of correct answers for the estimated results.

Looking at the results, angle sensors and angular velocity sensors were effective in determining road surface shapes and properties, respectively. However, the

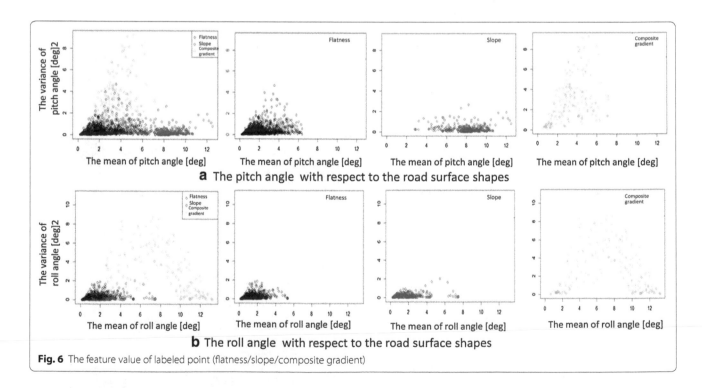

a The pitch angle with respect to the road surface shapes

b The roll angle with respect to the road surface shapes

Fig. 6 The feature value of labeled point (flatness/slope/composite gradient)

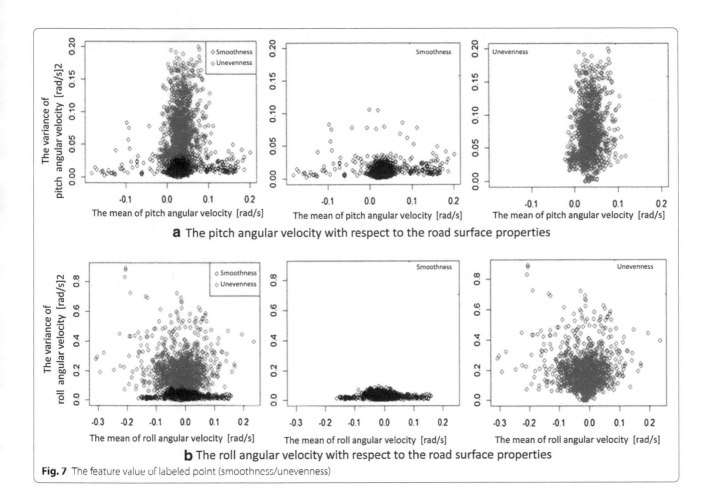

a The pitch angular velocity with respect to the road surface properties

b The roll angular velocity with respect to the road surface properties

Fig. 7 The feature value of labeled point (smoothness/unevenness)

Table 1 Conditions of experiments

Experiment	Label	Number of labeled data	Feature value	Data acquisition interval
Road surface shape	Flatness	150	Mean and variance of pitch angle (1 s = 50 points) mean($|\theta pi|$), var($|\theta pi|$),	GPS: 1 s Others: 20 ms
	Slope	150		
	Composite gradient	150	Mean and variance of roll angle (1 s = 50 points) mean($|\theta ri|$), var($|\theta ri|$),	
Road surface property	Smooth	500	Mean and variance of pitch angular velocity (1 s = 50 points) mean(ωpi), var(ωpi) Mean and variance of roll angular velocity (1 s = 50 points) mean(ωri), var(ωri)	
	Unevenness	500		

relevance factor of the flatness for the composite gradient: 83.4 %, and the recall ratio of the composite gradient: 88 %, are lower values as compared to others in the estimated results for the road surface shapes. The estimation of composite gradients is likely to be less sufficient

with only angle sensors. In addition, composite gradient points involve the difficulty or reluctance to press for the travelling of strollers as uneven points. In order to improve estimation accuracy, we plan to proceed with the efficient collection of sensor data and user-subjective

Table 2 Evaluation result for road surface shape

	Result			Total	Recall (%)
	Flatness	Slope	Composite gradient		
Originaldata					
Flatness	*141*	3	6	150	94.0
Slope	7	*143*	0	150	95.3
Composite gradient	21	0	*129*	150	86.0
Total	169	146	135	450	91.8
Precision (%)	83.4	97.9	95.6	91.8	91.8

Table 3 Evaluation result for road surface property

	Result		Total	Recall (%)
	Smoothness	Unevenness		
Original data				
Smoothness	*495*	5	500	99.0
Unevenness	45	*455*	500	95.3
Total	540	460	1000	95.0
Precision (%)	91.7	98.9	95.0	95.0

training data based on the proposed system, considering that if such subjective experiences in the difficulties in passage affect the acceleration and the angular velocity, there is a possibility to improve the estimation accuracy of the composite gradients using their feature amounts.

Figure 8 denotes pictures of the surface judged to be uneven at a part besides the data used for learning. The shape of the road surface changed and these places were a part where a pedestrian feels the resistance more than usual to push a stroller. These results confirm the effect

of the machine learning. Measuring error of GPS is several meters. Therefore the uneven spot strictly can not be specified, but we think that the outline of the shape of the sidewalk can be useful information. It was validity confirmation by the limited condition by an experiment of this first step. As the future works, we consider expansion of input sensor information, introduction of vital information, a variation of strollers and a variation of movement. And we examine the scalability of the proposed system.

Conclusions

In this report, we proposed a system to be the basis for efficient information gathering on sidewalk situations and services for extracting useful information and providing it to sidewalk users and local governments for the purpose of building a system to support social infrastructures through which users can move safety and securely. Specifically, we proposed a concept for a sidewalk supporting system that accumulates information from strollers, mobility scooters, bicycles, etc. through sidewalk users' smartphones, sensors, and other such devices to provide valid information. Next, we showed a proposed basic system configuration utilizing data collection technologies that have been developed in robots and walking appearance analysis with regard to responses to the diversity of sensors and the characteristics of smartphones. Furthermore, we introduced techniques for estimating sidewalk conditions using sidewalk passers' subjectivity with regard to the acquisition of training data to analyse data effectively. As a specific case of application, we evaluated the estimation of the gradient and unevenness of sidewalks by machine learning from the sensor data during the passage of strollers, verifying the effectiveness of this technique.

Fig. 8 The samples of estimated unevenness surface

Abbreviations
DAS28: Disease Activity Score 28; FDML: Field Data Mark-up Language; ROS: Robot Operating System; SVM: Support Vector Machine; XML: eXtensible Markup Language.

Authors' contributions
SM proposed the concept of the sidewalk supporting system, designed the information system, analyzed experimental results and drafted the manuscript. YN helped to design the information system, built an experimental system and did the verification experiments. TI and HI built an experimental system and did the verification experiments. MO analyzed experimental results and helped to draft the manuscript. All authors read and approved the final manuscript.

Author details
[1] NTT Service Evolution Laboratories, 1-1 Hikarinooka, Yokosuka-Shi, Kanagawa 239-0847, Japan. [2] NTT Media Intelligence Laboratories, 1-1 Hikarinooka, Yokosuka-Shi, Kanagawa 239-0847, Japan.

Competing interests
The authors declare that they have no competing interests.

References
1. Tezuka H, Ito K, Murayama T, Seko S, Nishino M, Muto S, Abe M (2011) Restaurant recommendation service using lifelogs. NTT Tech Rev 9(1):1–6
2. Mann S (1997) Wearable computing: a first step toward personal imaging. Computer 30(2):25–32
3. Sato T, Harada T, Mori T (2004) Environment-type robot system "robotic room" featured by behavior media, behavior contents and behavior adaptation. IEEE/ASME Trans Mechatron 9(3):529–534. doi:10.1109/tmech.2004.834650
4. Noguchi H, Mori T, Sato T (2006) RDF sensor description for heterogeneous sensors in human behavior monitoring environment. In: Proc. of the 2006 IEEE Int. Conf. on robotics and automation, pp 4369–4371, 15–19 May 2006
5. Sanfeliu C (2009) URUS Project: Communication systems. In: Proc. of workshop on network robot systems at 2009 IEEE/RSJ Int. Conf. on Intelligence robots and systems (IROS09-WS), 12:31, 10–15 Oct 2009
6. Kim H, Cho Y.J, Oh S.R (2007) Implementation and application of URC server framework. In: Proc of workshop on Network Robot System at 2007 IEEE Int. Conf. on Robotics and Automation (ICRA07-WS), No.SF-2-5, 23:27, 10–14 April 2007
7. Dustbot (2006) Dustbot - Networked and cooperating robots for urban hygiene http://www.dustbot.org/. Accessed 08 May 2015
8. Tezuka H, Katafuchi N, Nakamura Y, Machino T, Nanjo Y, Iwaki S, Shimokura K (2006) Robot platform architecture for information sharing and collaboration among multiple networked robots. J Robot Mechatron 18(3):325–332
9. Nakamura Y, Machino T, Motegi M, Iwata Y, Miyamoto T, Iwaki S, Muto S, Shimokura K (2008) Framework and service allocation for network robot platform and execution of dependent services. Robotics and Autonomous Systems Journal 56(831):842
10. Muto S, Shimokura K, Nakamura Y, Tezuka H, Abe M (2010) Feasibility study of platform-based network robot systems through field experiments. IEICE Trans Inform Syst D J93-D(10), 2240:2256 **(in Japanese)**
11. Nakamura Y, Muto S, Motegi M (2012) Mizukawa M (2012) Proposal of active information supplement platform based on 4w information. IEEE/SICE Int. Symp Sys Integr (SII) 951(956):16–18. doi:10.1109/SII.2012.6427284
12. Nakamura Y, Muto S, Maeda Y, Motegi M, Takashima Y (2014) Proposal of framework based on 4W1H and properties of robots and objects for development of physical service system. J Robot Mechatron 26(6):758–771
13. Ando N, Suehiro T, Kitagaki K, Kotoku T, Woo-Keun Y (2005) RT-middleware: distributed component middleware for RT (robot technology). In: Proc. of IEEE/RSJ Int. Conf. on intelligent robots and systems (IROS2005), pp 3933–3938, 2–6 Aug. 2005
14. Quigley M, Gerkey B, Conley K, Faust J, Foote T, Leibs J, Berger E, Wheeler R, Ng A (2009) ROS: an open-source robot operating system, In: Proc. of open-source software workshop of the Int. Conf. on robotics and automation (ICRA), 12–17 May 2009
15. Open Geospatial Consortium (2007) SensorML. http://www.ogcnetwork.net/SensorML. Accessed 08 May 2015
16. Murphy D (1996) Lycra working splint for the rheumatoid arthritic hand with MCP ulnar deviation. Aust J Rural Health 4(4):217–220
17. Nishiguchi S, Yamada M, Nagai K, Mori S, Kajiwara Y, Sonoda T, Yoshimura K, Yoshitomi H, Ito H, Okamoto K, Ito T, Muto S, Ishihara T, Aoyama T (2012) Reliability and validity of gait analysis by android-based smartphone. Telemed e-Health 18(4):292–296
18. Yamada M, Aoyama T, Mori S, Nishiguchi S, Okamoto K, Ito T, Muto S, Ishihara T, Yoshitomi H, Ito H (2012) Objective assessment of abnormal gait in patients with rheumatoid arthritis using a smartphone. Rheumatol Int 32(12):3869–3874
19. Muto S, Nakamura Y, Ito T, Ishihara T, Shinohara A, Nishiguchi S, Yamada M, Aoyama T, Okamoto K, Yoshitomi H, Furu M, Ito H (2014) Examinations for the extensibility of gait outline monitoring applications that use information gathering system for sensors in mobile terminal. Transactions of the JSME 80(819):324. doi:10.1299/transjsme.2014dr0324 **(in Japanese)**
20. Perttunen M, Mazhelis O, Cong F, Kauppila M, Leppänen T, Kantola J, Collin J, Pirttikangas S, Haverinen J, Ristaniemi T (2011) Riekki J (2011) Distributed road surface condition monitoring using mobile phones. Proc Ubiquitous Intell Comput 64(78):2–4
21. Yamada T, Ito T, Ohya A (2013) Detection of road surface damage using mobile robot equipped with 2d laser scanner. In: Proc. of the 2013 IEEE/SICE International Symposium on System Integration, 364:369, 15–17 Dec. 2013, doi: 10.1109/SII.2013.6776679
22. Ministry of Land, Infrastructure, Transport and Tourism (2010) Walking space network data maintenance specification plan. http://www.mlit.go.jp/common/000124059.pdf. Accessed 08 May 2015 **(in Japanese)**
23. Ministry of Land, Infrastructure, Transport and Tourism (2010) Research report about the state of the environmental improvement that everyone can rear children surely. http://www.mlit.go.jp/common/000116746.pdf. Accessed 08 May 2015 **(in Japanese)**
24. Kodoka K, Sinzaki A Hatoko M (2003) Extracting problems from existing barrier-free transportation policy in terms of parents with babies and infants. In: Proc. of infrastructure planning and management, JSCE, 28, VIII(287), 6–8 Jun. 2003 **(in Japanese)**
25. Ogami H, Saito S, Okura M, Muraki S (2005) Investigation of Physical Strain and Danger in Sidewalk in Stroller Users. In: Proc. of the 3rd annual meeting of the Japanese Society for Wellbeing Science and Assistive Technology, 1P2-C4, 8–9 Dec 2005 **(in Japanese)**

10

Interaction force estimation on a built-in position sensor for an electrostatic visuo-haptic display

Taku Nakamura[*] and Akio Yamamoto

Abstract

This paper discusses a force sensing method using a built-in position sensing system for an electrostatic visuo-haptic display. The display provides passive haptic feedback on a flat panel visual monitor, such as LCD, using electrostatic friction modulation via multiple contact pads arranged on a surface-insulated transparent electrode. The display demonstrated in previous studies measured the positions of the pads in a similar manner to surface-capacitive touch-screens. This paper extends the sensor such that the system can monitor the electrostatic interaction force provided to the contact pads. The extension is realized by estimating the capacitance between the contact pads and the display surface. The paper investigates its basic characteristics to show that the force estimation is possible, regardless of the pad positions and the pushing force exerted by users. The force estimation capability is used for feedback control of interaction force, which improves the accuracy of the interaction force. The paper further extends the method such that the system can detect the moving direction of contact pads. By dividing the electrode of a contact pad and comparing their capacitances, the system can detect in which direction the user is trying to move the pad. Such capability is effective for solving the sticky-wall problem, which is known to be a common problem in passive haptic systems. A pilot experiment shows that the proposed system can considerably reduce the sticky-wall effect.

Keywords: Haptic display, Passive haptics, Surface haptics, Tabletop interaction, Multi-touch, Electrostatic adhesion, Built-in sensor

Introduction

Haptic interaction is imperative for intuitive operation of computer systems. In modern computer devices, the touch-screen technology has realized intuitive operations that allow us to directly touch graphical icons on the screen. However, users are sometimes frustrated due to the lack of appropriate feedback; implementing haptic feedback is expected to enrich the user experience. On flat screens, providing force feedback in vertical direction is considerably difficult, and thus many studies have tried to implement lateral force feedback, in which the feedback force is provided within lateral two dimensions. Such lateral force feedback has been regarded as effective enough, since surface geometry recognition of humans is closely related with lateral force as reported in some studies [1, 2]. For example, lateral force can create an illusion of corrugated shapes on flat surfaces [2].

In recent studies, lateral haptic feedback on a flat visual displays, such as liquid crystal displays (LCD), has often been realized utilizing ultrasonic vibration [3, 4] or electrovibration [5, 6]. These technologies modulate surface friction for haptic rendering, by inducing a squeeze film effect or an electrostatic adhesion force between a fingertip and a flat surface. The change of friction, however, is not large enough for kinesthetic force rendering; they can only provide up to 0.1 N of friction change [7], which can only realize surface texture rendering, such as rendering of surface roughness; it is not possible to render, for example, virtual walls. Some studies have mentioned that electrostatic adhesion can realize large force almost up to 2 N between a fingertip and an electrode by using Johnsen-Rahbeck effect [8]. However, the effect requires

*Correspondence: taku_nkat@aml.t.u-tokyo.ac.jp
Department of Presicion Engineering, The University of Tokyo, Hongo 7-3-1, Tokyo 113-8656, Japan

special electrode material, which is not transparent, and thus it has not been applied to haptic feedback on an LCD. In addition, the technology needs further investigation on the time response, as it seems the effect becomes dominant only when DC voltage is applied.

To extend the possibility of haptic feedback on an LCD, the authors have proposed an indirect electrostatic visuo-haptic rendering system that can provide large haptic feedback force to multiple users or fingers [9–12], without using Johnsen-Rahbeck effect. The proposed system utilizes "contact pads" that have electrodes on their bottom surfaces. They are placed on an LCD surface to exert large enough electrostatic adhesion force. The user obtains haptic feedback via the contact pads, and therefore, the rendering method is called "indirect". One of the prototypes, which is used in [10–12], is shown in Fig. 1. The system is implemented on an off-the-shelf 40-inch LCD monitor without a touch-sensing function. The surface of the LCD monitor is covered with a transparent electrode, which is electrically grounded. Multiple contact pads placed on the transparent electrode are connected to different voltage sources, such that each pad can generate different electrostatic adhesion force. When the pad is moved by a user, the adhesion force is converted into friction force, which is perceived as haptic feedback; the resulting haptic feedback is passive due to the nature of friction force. The system is capable of providing more than 1 N of friction change to multiple fingers [9], which is large enough for kinesthetic haptic rendering including virtual wall rendering.

To provide haptic feedback in accordance with the visual information displayed on the LCD, it is imperative to detect the pad positions. The prototype systems detected the positions by superposing sensing signal on the haptic voltage; the positions were estimated in a similar manner as surface-capacitive-type touch-screens. Using the position sensing capability, the previous studies demonstrated a hockey game with haptic feedback, in which users can feel impact when hitting a puck or can feel walls of the hockey arena [10, 12].

A next challenge for the visuo-haptic system is to improve the quality of the rendering force. In the previous implementations, the system controlled the force through the applied voltage in an open-loop manner, assuming that the electrostatic adhesion force is proportional to the square of the applied voltage. Such simple rendering was effective enough for typical applications, such as games as demonstrated in [10]. However, for some applications that require higher accuracy of haptic rendering, such open-loop rendering was not precise enough due to the following two reasons.

The first is fluctuation and limited time-response of the electrostatic adhesion force. When voltage is applied between the pad electrode and the screen electrode, the electrodes deform due to the electrostatic adhesion force. The resulting gap variation between the electrodes can fluctuate the electrostatic force. Especially, when a soft conductive rubber is utilized as the pad electrode, the electrostatic adhesion force suddenly increases as the voltage is gradually increased, probably due to a sudden change of the gap between the electrodes [12]. The deformation of the electrodes also affects the time-response of the electrostatic force. As a result, on the actual device, the force does not simply follow the square-law.

The second is about the sticky wall problem, which is well-known problem in passive haptic systems [13, 14]. Due to passive nature of the friction force, virtual walls rendered by passive haptic systems tend to be sticky; a user feels resistive force during retrieving from the wall. To solve the problem, the system needs to detect the direction of the operation force given by the user.

These two problems require monitoring of interaction forces, such as the haptic feedback force and the operation force from users in tangential directions. To achieve such monitoring, this paper proposes an interaction force estimation method on the built-in position sensing system of the electrostatic visuo-haptic display. The basic concept, which was simply introduced in [11], is to measure the total amount of electric current flowing in the sensing system. The current amount would correspond to the capacitance between the pad and the screen electrode, which could be utilized for the force estimation. Based on the concept proposed in the previous work [11], this paper investigates detailed characteristics of the force estimation and applies it to closed-loop force control and to elimination of the sticky wall problem.

The structure of this paper is as follows. "Built-in force estimating system" introduces the concept of the proposed method and shows basic performance of the capacitance measurement on a prototype device. "Haptic force estimation" discusses monitoring of the haptic force, where the measured capacitance is used for estimation of the electrostatic adhesion force. Then, the

Fig. 1 Appearance of prototyped visuo-haptic display. (Reprinted from [11])

estimation is utilized for closed-loop control of haptic force in "Closed-loop control of haptic force". "Wall rendering with lateral-force-direction sensing" proposes another application of the sensing, which estimates direction of the operation force from the difference of capacitances of divided electrodes, to solve the sticky wall problem.

Built-in force estimating system
Concept
Figure 2a shows the built-in sensing system for the electrostatic visuo-haptic display, which was briefly introduced in [11]. The system shares most of its components with haptic feedback system. The shared components include a transparent surface-insulated indium-tin-oxide (ITO) electrode that covers the whole screen surface, and multiple contact pads with electrodes on their bottom surfaces. Current sensors are inserted between the ITO electrode and the ground at six peripheral points, as depicted in the figure.

Haptic rendering is realized by applying high voltages to the contact pads. The sensing system superposes high-frequency (much higher than that of the haptic voltage) signal to the haptic voltage using transformers, which results in corresponding high-frequency signals flowing through the current sensors. The signals are used for simultaneous estimation of position and force, whose

concept is depicted in Fig. 2b. Position and force estimation for multiple pads are realized by time division; high-frequency sensing signal is alternately applied to each pad, one by one, within a short switching period.

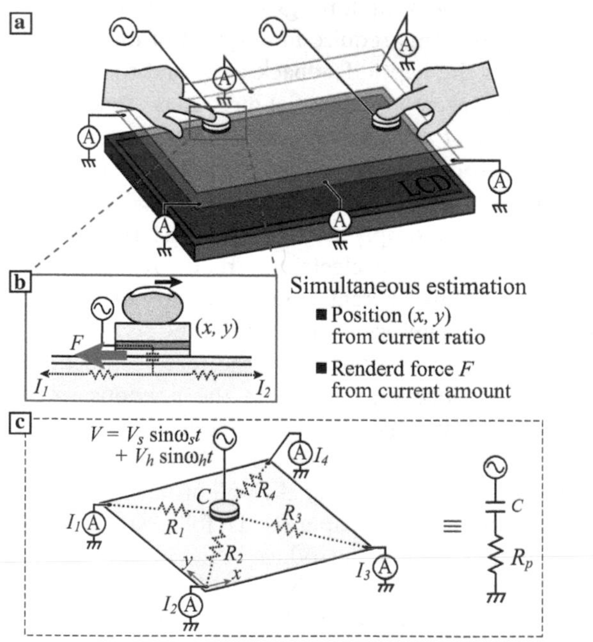

Before discussing the force estimation, the position estimation method, whose details are addressed in [12], is reviewed using the circuit model shown in Fig. 2c. When the sensing signal, $V_s \sin(\omega_s t)$, and the haptic voltage, $V_h \sin(\omega_h t)$ (note that $\omega_h \ll \omega_s$), are applied on a pad, corresponding current running through each terminal becomes

$$I_i = \frac{1}{R_i}\left(\frac{j\omega_s C R_p}{1 + j\omega_s C R_p}V_s + \frac{j\omega_h C R_p}{1 + j\omega_h C R_p}V_h\right), \quad (1)$$

where V_s and V_h are the sensing signal and the haptic voltage in complex representation respectively, R_i is the resistance of the ITO electrode from the contact point to the i-th terminal, C is the capacitance between the pad and the ITO, and R_p is the combined resistance of ITO resistances R_i. The resistances R_i are assumed to correspond to the distance between the terminals and the pad. Due to the resistances, the currents running toward the terminals depend on the pad position. Therefore, from the ratios of the current amplitudes, which are calculated as,

$$r_x = \frac{-I_1 - I_2 + I_3 + I_4}{I_1 + I_2 + I_3 + I_4} = \frac{-\frac{1}{R_1} - \frac{1}{R_2} + \frac{1}{R_3} + \frac{1}{R_4}}{\frac{1}{R_1} + \frac{1}{R_2} + \frac{1}{R_3} + \frac{1}{R_4}},$$

$$(2)$$

$$r_y = \frac{I_1 - I_2 - I_3 + I_4}{I_1 + I_2 + I_3 + I_4} = \frac{\frac{1}{R_1} - \frac{1}{R_2} - \frac{1}{R_3} + \frac{1}{R_4}}{\frac{1}{R_1} + \frac{1}{R_2} + \frac{1}{R_3} + \frac{1}{R_4}}, \quad (3)$$

where I_i is amplitude of $\boldsymbol{I}_i$, the system can estimate the position of the pad in x–y coordinate. In the actual setup, current amplitudes corresponding to the sensing signal are distilled from the total currents.

The force estimation method discussed in this paper, on the other hand, utilizes the sum of the current amplitudes measured at all the peripheral points. From the sum, the system can estimate capacitance between a pad and the ITO electrode. Since the capacitance should correspond to the normal force applied on the pad, including the electrostatic adhesion force, the system can estimate the force from the capacitance.

In the sensor system, the position and force sensing are independent from each other. Our previous work revealed the position sensing is not affected by the change of capacitance or interaction force [12]. The following calculation reveals the opposite: the force (or capacitance) sensing is not affected by the position change of the pads.

Fig. 2 Schematic (**a**), concept (**b**), and circuit model (**c**) of the built-in sensing system for the electrostatic visuo-haptic display. (Modified from [11])

The total current running through a pad can be calculated as

$$I = \frac{j\omega_s C}{1 + j\omega_s C R_p} V_s + \frac{j\omega_h C}{1 + j\omega_h C R_p} V_h, \tag{4}$$

in complex representation. For the force sensing, amplitude of the current corresponding to the sensing signal is focused, since the current amplitude, unlike the current phase, becomes independent from the pad position in a certain condition as discussed in the following. The amplitude corresponding to the sensing signal, which can be detected by using synchronous detection, becomes

$$A = \frac{\omega_s C}{\sqrt{1 + (\omega_s C R_p)^2}} V_s. \tag{5}$$

The amplitude includes not only the capacitance, C, but also the combined resistance, R_p, which depends on the position of the pad. However, if the combined resistance is small enough to satisfy $R_p^2 \ll \left(\frac{1}{\omega_s C}\right)^2$, the detected amplitude becomes

$$A \approx \omega_s C V_s, \tag{6}$$

which is independent from the pad position.

Capacitance measurement on prototype

The capacitance measurement is fundamental for the force estimation. To investigate the capacitance measurement performance in the actual setup, currents were measured in the setup used in [12]. The ITO screen electrode in the setup has approximately 860 mm × 500 mm in its size and 150 Ω/sq. of sheet resistance. The pad electrode is ϕ30-mm round-shaped conductive rubber sheet. The surfaces of the two electrodes are insulated with PET film whose thickness is 8 μm. As the sensing signal, a 100-kHz sine wave was used. In the setup, the combined resistance, which was measured at several points on the ITO electrode using a circuit tester, is approximately 500 Ω in maximum, while the capacitive impedance between the two electrodes, which was measured using an impedance analyzer, is approximately 5 kΩ. This large difference in impedance satisfies the condition for Eq. (6). Therefore, on this setup, the capacitance estimation, and in turn the force estimation, should be independent from the position of the pad.

Figure 3 shows the sum of the outputs of the current sensors, which corresponds to A in Eqs. (5) and (6), against the position of the pad in several load conditions. The load, representing user's pushing force, was applied by putting a weight on the pad. As expected, these results are almost flat regardless of the pad position, which verifies the independence of the capacitance sensing from the pad position.

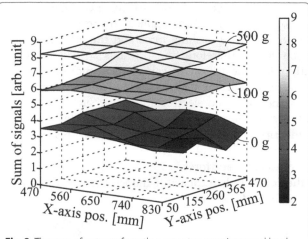

Fig. 3 The sum of outputs from the current sensors in several load and position conditions

Figure 4 shows the relation between the load and the sum of the signals in several haptic voltage conditions, measured at a fixed pad position. The error bar indicates the maximum and the minimum values within three-time measurements. It is clearly observed that two factors affect the total current: electrostatic adhesion force represented by the haptic voltage and the user's pushing force. The next section will focus on the effect of electrostatic adhesion force on the rendering force estimation. The later section, "Wall rendering with lateral-force-direction sensing", on the other hand, will utilize the effect of pushing force, to realize lateral-force-direction sensing.

Haptic force estimation

This section investigates the estimation of haptic force caused by the electrostatic adhesion. The previous work

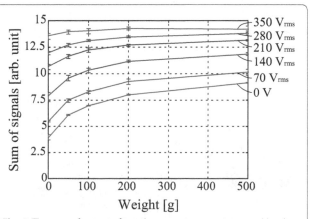

Fig. 4 The sum of outputs from the current sensors in several load and voltage conditions

[11] simply assumed that the capacitance change is caused by the normal force applied to the pad, which is the sum of user's pushing force and the electrostatic adhesion force. Therefore, the previous work tried to directly relate the measured capacitance (more exactly, the sum of the current sensor outputs) with the sum of the pushing force and electrostatic force. Such assumption, however, showed large errors probably due to the non-linearity of pad deformation. Therefore, this paper tries to estimate the haptic force alone using a theoretical model of electrostatic force.

In the parallel-plate capacitor model, the electrostatic force can be expressed as

$$F_e = \frac{(CV_h)^2}{2\varepsilon S}, \qquad (7)$$

where C is the capacitance between the pad and the ITO electrode, V_h is the haptic voltage applied between the electrodes, ε is the dielectric constant, and S is the area of the interface. Thus, the haptic force (increment of friction force), which is proportional to the electrostatic adhesion force, can be estimated as

$$F_h = k(C'V_h)^2, \qquad (8)$$

where k is a coefficient of the estimation and $C' = \beta C$ is the output of the sensing system with β being a proportionality coefficient.

To verify the haptic force estimation, an experiment on a 1-DOF setup was conducted. Figure 5 shows a schematic of the experimental setup, which consists of a simplified electrostatic haptic system on a motorized stage and a load-cell. The simplified system includes a current sensor, which consists of an I/V converter and a synchronous detection circuit to detect the current amplitude in the sensing signal frequency. By pushing the pad toward the load-cell using the motorized stage, the setup measures friction force between the pad and the ITO. In the experiment, a weight was put on the pad to represent pushing force from a user's finger. The voltage on the pad (which is the voltage after the transformer for superposing sensing signal) was measured as the haptic voltage.

First, a calibration for the estimation was conducted, whose results are summarized in Fig. 6. Figure 6a shows the change of the measured friction in several haptic voltage conditions. Square and round markers represent two weight conditions, 100 g and 20 g respectively. In the plot, the two series of the measured friction force have offsets due to the weight. Since the objective is to estimate the haptic force, which is the increment of the friction from the no-voltage condition, the increment is plotted in Fig. 6b using blue markers. A quadratic curve is also plotted using a black line for reference, which shows that the haptic force does not follow the square-law. This behavior of the haptic force is probably due to the change of the gap between the pad electrode and the ITO screen electrode. Since a rise of the haptic voltage would decrease the gap through increase of electrostatic attraction force, the electrostatic force would be enhanced deviating from the square-law. This assumption is supported by the sensor outputs measured at the same time, which are plotted in red color with reference to the right vertical axis. The sensor output, which is assumed to represent the

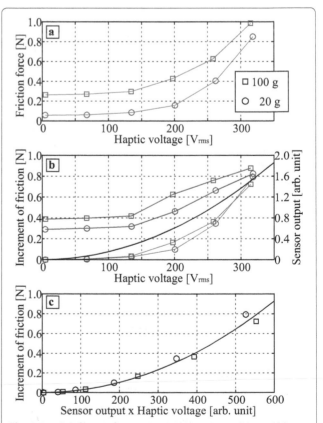

Fig. 6 Results of haptic force estimation in static conditions. **a** Measured friction in several haptic voltage conditions. **b** Increment of the friction and sensor output. **c** Relation between the increment of the friction and the product of the sensor output and the haptic voltage

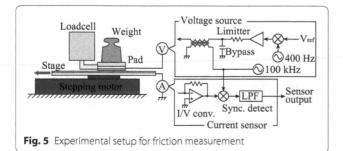

Fig. 5 Experimental setup for friction measurement

capacitance, suddenly increases above 150 V_{rms}, which means that the gap decreased above that voltage.

As Eq. (8) indicates, the haptic voltage should be estimated from the product of the sensor output and the haptic voltage. Figure 6c plots the relation between the product and the increment of friction, with a quadratic fitting curve. The plot clearly shows that Eq. (8) provides a good estimation of the haptic force, regardless of the pushing force. The maximum error was found less than 0.1 N for these conditions.

By using the calibration result, the step response of the friction force was estimated and compared with the output of the load-cell. Figure 7 shows the result when a weight of 50 g was applied and haptic voltage of 350 V_{rms} was commanded. It should be noted that the commanded haptic voltage is different from the actual haptic voltage measured at the pad, due to the voltage drop at the transformer and the current-limiting resistor. The upper plot indicates the temporal change of the applied voltage, and the lower plot shows the response of the friction force measured by the load-cell (blue line) and the estimated force (green line).

When the haptic voltage was raised, the friction force gradually increased. The measured voltage on the upper plot shows gradual decrease, which is due to the change of the capacitance between the pad and the ITO. As the normal force, and in turn friction, increases, the capacitance and the corresponding current also increase. The increased current leads to larger voltage drops at the current limiting resistor and the transformer that were arranged between the voltage source and the pad. The sudden rises of the friction at around 1.0 s and 2.8 s would correspond to "pull-in", in which the gap suddenly decreases when the balance between the linear elastic

force (from the insulating material) and non-linear electrostatic force is broken. The estimation successfully described the change of the friction force, including the sudden change due to pull-in. In addition, in contrast to the previous work [11], there was no residual output in the estimated value after the haptic voltage turned off. The error for this dynamic measurement was within the maximum error of 0.1 N that was found for the static measurement.

Closed-loop control of haptic force

Our previous system [9, 12] has controlled the haptic force in an open-loop manner, based on the assumption that the haptic force is proportional to the square of the haptic voltage. As shown in the previous section, however, the assumption is not true due to the capacitance change; the haptic force fluctuates even when a constant haptic voltage is applied. Such behavior can be compensated by closed-loop control using the haptic force estimation, which is demonstrated in this section.

Haptic forces with open-loop and closed-loop control were compared on the same setup as in Fig. 5. Figure 8 shows the two controllers used in the experiment. One is the open-loop feedforward controller used in the previous system. Supposing that the haptic force is proportional to the haptic voltage squared, the command voltage is controlled as

$$V_{ref} = K\sqrt{F_t}, \qquad (9)$$

where F_t is the target haptic force and K is a coefficient and was empirically set to 600 $V/N^{-\frac{1}{2}}$ in this particular work. The other is a feedback controller combined with the previous feedforward controller. The feedback controller is a simple proportional one, whose proportional gain, K_p, was empirically chosen to 1000 V/N. The coefficient for the estimation [k in Eq. (8)] was manually adjusted.

Figure 9 shows the results of the two controllers for several conditions of target force: square wave and single-sided sine wave in different amplitudes. Left-side figures are the results of the previous controller, and the right-side figures are the results of the proposed controller. For

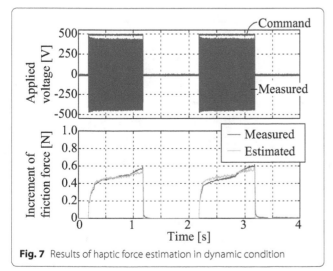

Fig. 7 Results of haptic force estimation in dynamic condition

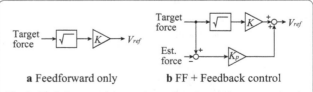

Fig. 8 *Block diagram* of the two controllers: **a** open-loop control and **b** closed-loop control

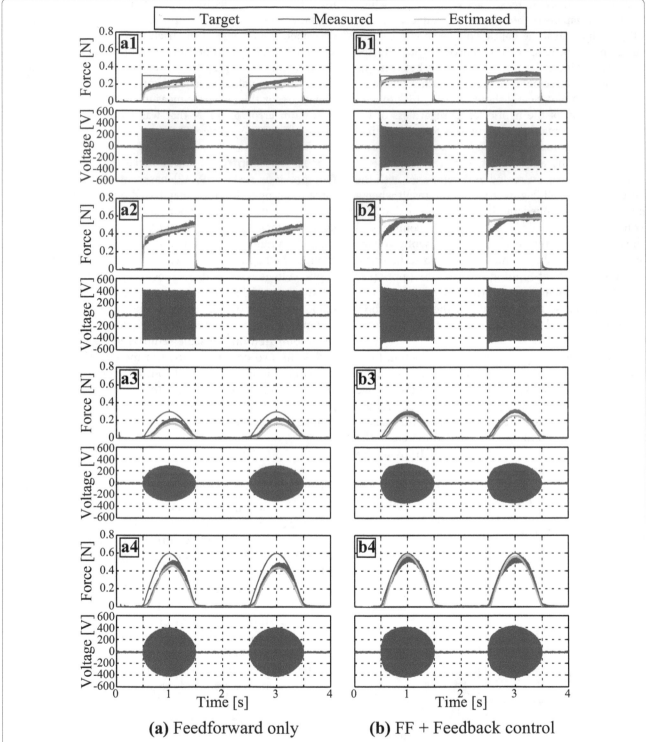

(a) Feedforward only   **(b)** FF + Feedback control

Fig. 9 Results of open-loop (**a**) and closed-loop (**b**) control for several target-force conditions: (1) 0.3 N square, (2) 0.6 N square, (3) 0.3 N single-sided sine, and (4) 0.6 N single-sided sine

each condition, the upper plot and the lower plot indicate the haptic force and the applied haptic voltage over time, respectively. Red, blue, and green lines represent target force, the measured force, and estimated force, respectively. Comparison of the two controllers in the square-wave case shows that the closed-loop controller improves the rising of the haptic force by increasing the applied voltage. In sine-wave case, whereas the previous controller lost the shape of the force from the target, the closed-loop controller achieved the force closer to the target. These results suggest that the closed-loop control with haptic force estimation can render the haptic feedback more precisely by improving the instability of the electrostatic adhesion.

Wall rendering with lateral-force-direction sensing

The system provides haptic feedback passively by using friction. Such a passive haptic system suffers from the sticky wall problem, which can be solved by considering direction of the force/movement of user's finger [13, 14]. Figure 10 explains the problem, which occurs when the reaction force from a virtual wall is rendered using passive haptic force, such as brake force and friction. When the pad enters the virtual wall, the brake or friction force acts as the reaction force from the wall (Fig. 10a). When the pad is retrieving from the wall, the brake force is still exerted until the pad completely getting off from the wall (Fig. 10b). This extra brake force is perceived as stickiness of the wall. To solve the problem, wall rendering should consider the direction of the force or the movement given from the user; the force rendering should be cut when the operating direction is the retrieving one. The direction of the movement, which can be obtained from the position sensor for the pad, however, has a limited effect in solving the sticky-wall problem, since the movement occurs after the operating force overcomes the maximum static friction under the haptic voltage application. If we can directly monitor the direction of the operating force, the haptic voltage can be cut before the operating force reaches the maximum static friction; the resulting sticky force would be much smaller. Toward the monitoring of the operating force direction, this section modifies the force sensing described in the previous sections.

Lateral-force-direction sensing

Figure 11 shows the concept of the lateral-force-direction sensing. The concept is to divide the pad electrode into a few parts and to detect the eccentric amount of the pressure as the lateral-force direction. Each part of the pad electrode, connected to an independent voltage source, detects its 2-D position, x_i (i is the index for each part), and normal force, F_i. The center position of the whole pad can be calculated by the mean position of all the electrode parts as $x_c = \frac{\sum_{i=1}^{n} x_i}{n}$. Here, n is the number of the divided parts. Then, the center of the pressure (CoP) can be calculated as $x_p = \frac{\sum_{i=1}^{n} F_i x_i}{\sum_{i=1}^{n} F_i}$. Finally, the subtraction of

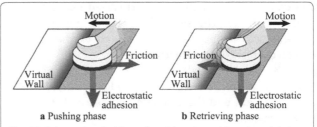

Fig. 10 Schematics of sticky wall problem. In pushing phase (**a**), the friction properly acts as reaction force from the virtual wall. In retrieving phase (**b**), however, the friction acts as sticking force

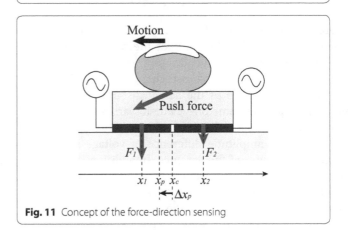

Fig. 11 Concept of the force-direction sensing

the CoP from the center position, $\Delta x_p = x_p - x_c$, would correspond to the lateral force direction.

To validate the proposed method, an experiment on a 1-DOF prototype was conducted. The experimental setup, whose schematic is shown in Fig. 12, consists of a 1-DOF prototype pad, a sensing table, and an indenter. The 1-DOF pad has two sectioned electrodes connected to independent voltage sources, and sensing signal is alternately applied to each pad for multiple sensing. On the sensing table, an ITO electrode sheet covering an acrylic base plate are connected to current sensors at the two ends to detect the force and 1-D position of each sectioned electrode of the pad. In the experiment, the actual eccentric amount and the sensor output were compared in several haptic voltage conditions. This particular experiment directly uses the signal ratios between the two current sensors, $\tilde{x}_i = (-I_1 + I_2)/(I_1 + I_2)$ [1-DOF

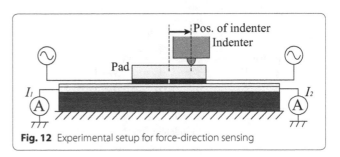

Fig. 12 Experimental setup for force-direction sensing

version of Eq. (2), $i = 1, 2$], as a normalized position of each pad electrode. The sum of the signal amplitudes, $a_i = I_1 + I_2$, was supposed to correspond to the force applied on each sectioned electrode.

Figure 13 shows the relation between the actual eccentric distance (displacement of the indenter from the center of the pad) and the estimated eccentric distance, which was calculated as

$$\Delta \tilde{x}_p = \tilde{x}_p - \tilde{x}_c = \frac{\sum_{i=1}^{2} a_i \tilde{x}_i}{\sum_{i=1}^{2} a_i} - \frac{\sum_{i=1}^{2} \tilde{x}_i}{2}, \quad (10)$$

where $\tilde{x}_p$ and $\tilde{x}_c$ represent the normalized CoP and center position respectively. The color of the plots represents the haptic voltage as the legend shows. For all conditions, the estimated eccentric distance changes approximately linearly to the change of the actual distance. The estimated distance has some offset due to variance of sensitivity of each electrode. Fabrication error of the sectioned electrodes and amplitude difference of voltage sources would be the cause of the variance. In addition, as same as the pushing force measurement, the sensitivity of force direction measurement decreases as the haptic voltage increases. The dependence of the offset and the degrading sensitivity due to the haptic voltage complicates the sensing of the lateral force, and should be improved in the future work. However, for solving the sticky wall problem, the important thing is not detecting the magnitude of the lateral force but just direction of the force. The prototype has enough performance in detecting the force direction.

Experiment for wall rendering

To demonstrate the capability of the force-direction sensing, a wall rendering experiment was conducted on the modified setup of the previous subsection. The appearance of the setup is shown in Fig. 14. The setup added a force sensor for evaluation, which was attached on a newly introduced linear guide for supporting the 1-DOF

Fig. 14 Appearance of experimental setup for wall rendering

movement of the pad. In the experiment, the rendered forces with/without considering force direction were compared when the pad was moved manually through the force sensor to interact with a virtual wall. The basic virtual wall (without consideration of force direction) was rendered by controlling the haptic voltage as

$$V_h = K_w \tilde{x}_c \quad (\tilde{x}_c < 0). \quad (11)$$

The wall rendering with considering the force direction was as follows:

$$V_h = K_w \tilde{x}_c \quad (\tilde{x}_c < 0 \cap \tilde{x}_c \cdot \Delta \tilde{x}_p \leq 0). \quad (12)$$

Figure 15 shows the results of the wall rendering. From above, these plots indicate time-variable plot of the position, the lateral force (eccentric distance) estimated by the built-in sensor, haptic voltage, and haptic force measured by the evaluation force sensor. Left-side plots show the result without considering force direction, and the right-side plots for considering force direction. The negative position means that the pad is within the virtual wall. In the left-side plots, the sticking force was exerted until the pad position returns to positive. On the other hand, sticking force was rapidly switched off in the right-side plots. This clearly shows that the force-direction sensing can reduce the stickiness of the virtual wall.

However, this simple rendering does not completely solve the problem. The sensor sometimes mis-detected the force direction when the pad is going to enter the virtual wall, which vanished the wall. In addition, during the retrieving phase, chattering vibration was sometimes observed. The reason of these behavior was low accuracy/sensitivity of the prototyped force-direction sensing, which might be solved by adding some signal processing filters. Combination use of motion direction might be useful to compensate the sensing accuracy. These solutions would be tackled in our future work.

Conclusion

This paper discussed interaction force sensing using a built-in sensing system for an electrostatic visuo-haptic display. The sensing system can estimate the capacitance

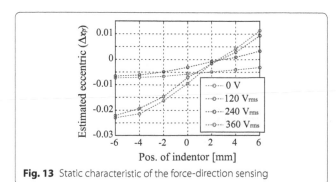

Fig. 13 Static characteristic of the force-direction sensing

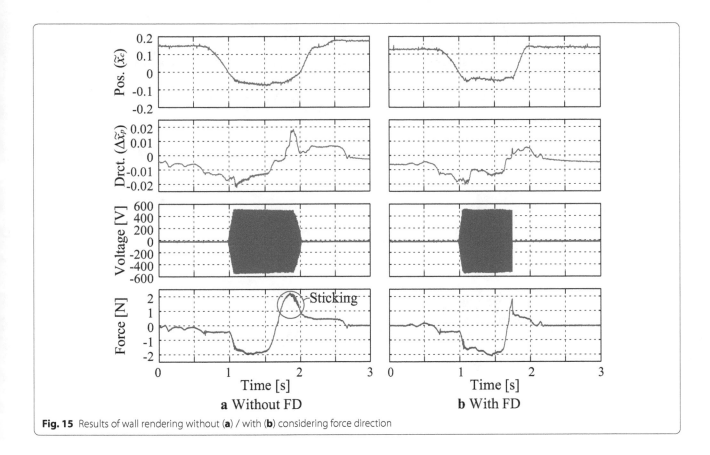

Fig. 15 Results of wall rendering without (**a**) / with (**b**) considering force direction

between the contact pad and the ITO electrode, in addition to, and independent from, the position sensing. Based on the estimated capacitance, the sensing system is able to estimate the haptic feedback force. The estimation enables precise control of haptic feedback in a closed-loop manner. The control reduces the error from the target force and improves the response speed.

By modifying the structure of the pad, the sensing method can be extended to detect the direction of the operating force given by the user. The direction sensing successfully reduced the stickiness of the virtual wall. These interaction force sensing has just opened the door to high-fidelity haptic feedback on flat visual displays. The rendering method would be further investigated in our future work.

Authors' contributions
TN contributed to the conception, design of experiments, acquisition of data, and drafting of the manuscript. AY took part in the conception, interpretation of data, and revising of the manuscript. Both authors read and approved the final manuscript.

Acknowledgements
This work was supported in part by Grant-in-Aid for JSPS Fellows (No. 26 9272) and Grand-in-Aid for Scientific Research (B) (No. 26280069) from JSPS, Japan.

Competing interests
The authors declare that they have no competing interests.

References
1. Christou C, Wing A (2001) Friction and curvature judgement. In: Proc. Eurohaptics
2. Robles-De-La-Torre G, Hayward V (2001) Force can overcome object geometry in the perception of shape through active touch. Nature 412(6845):445–448
3. Takasaki M, Kotani H, Mizuno T, Nara T (2005) Transparent surface acoustic wave tactile display. In: 2005 IEEE/RSJ international conference on intelligent robots and systems (IROS 2005), pp 3354–3359
4. Lévesque V, Oram L, MacLean K, Cockburn A, Marchuk ND, Johnson D, Colgate JE, Peshkin MA (2011) Enhancing physicality in touch interaction with programmable friction. In: Proceedings of the SIGCHI conference on human factors in computing systems. CHI '11, pp 2481–2490
5. Bau O, Poupyrev I, Israr A, Harrison C (2010) Teslatouch: electrovibration for touch surfaces. In: Proceedings of the 23nd annual ACM symposium on user interface software and technology. UIST '10, pp 283–292
6. Linjama J. Mäkinen V (2009) E-sense screen: novel haptic display with capacitive electrosensory interface. Presented at HAID 09, 4th Workshop for haptic and audio interaction design, Dresden, Germany
7. Meyer DJ, Peshkin MA, Colgate JE (2013) Fingertip friction modulation due to electrostatic attraction. In: 2013 world haptics conference (WHC), pp 43–48
8. Shultz CD, Peshkin MA, Colgate JE (2015) Surface haptics via electroadhesion: expanding electrovibration with johnsen and rahbek. In: World haptics conference (WHC), IEEE, pp 57–62
9. Nakamura T, Yamamoto A (2013) Multi-finger electrostatic passive haptic feedback on a visual display. In: World haptics conference (WHC), IEEE, pp 37–42

10. Nakamura T, Yamamoto A (2014) Built-in capacitive position sensing for multi-user electrostatic visuo-haptic display. In: Asia haptics 2014

11. Nakamura T, Yamamoto A (2015) Simultaneous measurement of position and interaction force on a multi-user electrostatic visuo-haptic display. In: 2015 IEEE world haptics conference

12. Nakamura T, Yamamoto A (2015) A multi-user surface visuo-haptic display using electrostatic frcition modulation and capacitive-type position sensing. IEEE Transactions on Haptics, no. 1, pp. 1, PrePrints. doi:10.1109/TOH.2016.2556660

13. Colgate JE, Peshkin MA, Wannasuphorasit W (1996) Nonholonomic haptic display. In: Proceedings of 1996 IEEE international conference on robotics and automation, vol. 1, pp 539–544

14. Furusho J, Sakaguchi M, Takesue N, Koyanagi K (2002) Development of er brake and its application to passive force display. J Intell Mater Syst Struct 13(7–8):425–429

11

Tracking control of piezoelectric actuator using adaptive model

Tran Vu Minh[1*†], Nguyen Manh Linh[2†] and Xinkai Chen[3]

Abstract

Piezoelectric actuators (PEAs) have been widely used in micro- and nanopositioning applications due to their fine resolution, rapid responses, and large actuating forces. However, a major deficiency of PEAs is that their accuracy is seriously limited by hysteresis. This paper presents adaptive model predictive control technique for reducing hysteresis in PEAs based on autoregressive exogenous model. Experimental results show the effectiveness of the proposed method.

Keywords: Piezoelectric actuator, Hysteresis, Adaptive model predictive control

Background

The use of piezoelectric actuator (PEA) has become very popular recently for a wide range of applications, including atomic force microscopes [1–3], adaptive optics [4], computer components [5], machine tools [6], aviation [7], internal combustion engines [8], micromanipulators [9] due to their subnanometer resolution, large actuating force, and rapid response. However, PEA exhibits hysteresis behavior in their response to an applied electrical energy. This leads to problems of inaccuracy, instability, and restricted system performance.

The control of PEA has been extensively studied recently. Ge and Jouaneh [10] discuss a comparison between a feedforward control, a regular PID control, and a PID feedback control with Preisach hysteresis. In this research, the nonlinear dynamics of piezoelectric actuator is first linearized and then reformulated the problem into a disturbance decoupling problem. In [11], an explicit inversion of Prandtl–Ishlinskii model is used to control a piezoelectric actuator. Webb et al. [12] proposed an adaptive hysteresis inverse cascade with the system, so that the system becomes a linear structure with uncertainties. Another adaptive control approach is fused with the Prandtl–Ishlinskii model without constructing a hysteresis inverse, since the inverse is usually difficult to be obtained [13]. In this concept, the implicit inversion of Prandtl–Ishlinskii model is developed and is associated with an adaptive control scheme. A new perfect inverse function of the hysteresis (which is described by Bouc–Wen model) is constructed and used to cancel the hysteresis effects in adaptive backstepping control design [14].

In this paper, the dynamics of the piezoelectric actuator is identified as a linear model with unknown parameters. These parameters will be updated online by using least square method. Then, a model predictive controller using estimated parameters is designed to achieve the desired control behavior. The experimental results show the effectiveness of the proposed method.

This paper is organized as follows. In "Modeling method" section, the adaptive model of PEA is given. In "Controlling method" section, the model predictive control design is presented. The experimental results are shown in "Result" section. "Discussion" section will conclude this paper.

Modeling method

In this section, the dynamics of piezoelectric actuator can be identified as a linear model as follows

$$m\ddot{y}(t) + k\dot{y}(t) + cy(t) = u(t) \tag{1}$$

*Correspondence: minh.tranvu@hust.edu.vn
†Tran Vu Minh and Nguyen Manh Linh contributed equally to this work
[1] School of Mechanical Engineering, Hanoi University of Science and Technology, Hanoi, Vietnam
Full list of author information is available at the end of the article

where $y(t)$ denotes the position of piezoelectric actuator, $u(t)$ is the force generated by PEA, m is the mass coefficient, k is the viscous friction coefficient of the PM, and c is the stiffness factor.

Now, express (1) as

$$\frac{\mathrm{d}}{\mathrm{d}t}\begin{bmatrix} y(t) \\ \dot{y}(t) \end{bmatrix} = \begin{bmatrix} 0 & 1 \\ -\frac{c}{m} & -\frac{k}{m} \end{bmatrix}\begin{bmatrix} y(t) \\ \dot{y}(t) \end{bmatrix} + \begin{bmatrix} 0 \\ \frac{1}{m} \end{bmatrix} u(t). \quad (2)$$

Let T be the sampling period and suppose $y(t)$ is constant during the sampling instant. By discretizing system (2), the input–output discrete time expression of system (1) can be given by

$$y(k) = a\left(q^{-1}\right)y(k-1) + b\left(q^{-1}\right)u(k) \quad (3)$$

where q^{-1} is the delay operator and $a(q^{-1})$ and $b(q^{-1})$ are polynomials defined by

$$a\left(q^{-1}\right) = -a_1 - a_2 q^{-1}$$
$$b\left(q^{-1}\right) = b_1 + b_2 q^{-1} \quad (4)$$

The parameters a_1, a_2, b_1, b_2 are unknown.
Let θ be the vector of unknown system parameters

$$\theta = [a_1,\ a_2, b_1,\ b_2]^T$$

Equation (2) can be written as

$$y(k) = \phi^T(k-1)\theta \quad (5)$$

where $\phi^T(k-1) = [y(k-1), y(k-2), u(k-1), u(k-2)]$.

Let $\hat{\theta}(k) = \begin{bmatrix} \hat{\theta}_1(k) & \hat{\theta}_2(k) & \hat{\theta}_3(k) & \hat{\theta}_4(k) \end{bmatrix}$ be the estimated of θ. Applying the least square method [15], the estimated parameters vector will be updated as follows

$$\hat{\theta}(k) = \hat{\theta}(k-1) + \frac{P(k-1)\phi(k)}{1 + \phi(k)^T P(k-1)\phi(k)}$$
$$\left(y(k) - \phi(k-1)^T \hat{\theta}(k-1)\right) \quad (6)$$

$$P(k-1) = P(k-2) - \frac{P(k-2)\phi(k-1)\phi(k-1)^T P(k-2)}{1 + \phi(k-1)^T P(k-2)\phi(k-1)} \quad (7)$$

where $P(k)$ is the covariance matrix with $P(-1)$ is any positive define matrix P_0. Usually, P_0 is chosen as $P_0 = \lambda I$, where λ is a positive constant, I is the identity matrix.

Controlling method

Using the estimated parameters, Eq. (3) can be rewritten as

$$y(k) = -\hat{\theta}_1(k-1)y(k-1) - \hat{\theta}_2(k-1)y(k-1)$$
$$+ \hat{\theta}_3(k-1)u(k-1) + \hat{\theta}_4(k-1)u(k-2) \quad (8)$$

Defining $x_1(k+1) = x_2(k) = y(k)$, it gives

$$\begin{cases} x_1(k+1) = x_2(k) \\ x_2(k+1) = -\hat{\theta}_1(k)x_2(k) - \hat{\theta}_2(k)x_1(k) \\ \qquad + \hat{\theta}_3(k)u(k) + \hat{\theta}_4(k)u(k-1) \end{cases} \quad (9)$$

Introducing new state variable $u(k) = u(k-1) + \Delta u(k)$, Eq. (9) becomes

$$\begin{bmatrix} x_1(k+1) \\ x_2(k+1) \\ u(k) \end{bmatrix} = \begin{bmatrix} 0 & 1 & 0 \\ -\hat{\theta}_2(k) & -\hat{\theta}_1(k) & \hat{\theta}_3(k)+\hat{\theta}_4(k) \\ 0 & 0 & 1 \end{bmatrix}$$
$$\begin{bmatrix} x_1(k) \\ x_2(k) \\ u(k-1) \end{bmatrix} + \begin{bmatrix} 0 \\ \hat{\theta}_3(k) \\ 1 \end{bmatrix}\Delta u(k)$$
$$y(k) = \begin{bmatrix} 0 & 1 & 0 \end{bmatrix}\begin{bmatrix} x_1(k) \\ x_2(k) \\ u(k-1) \end{bmatrix} \quad (10)$$

For simplicity, denote

$$M = \begin{bmatrix} 0 & 1 & 0 \\ -\hat{\theta}_2(k) & -\hat{\theta}_1(k) & \hat{\theta}_3(k)+\hat{\theta}_4(k) \\ 0 & 0 & 1 \end{bmatrix},$$
$$N = \begin{bmatrix} 0 \\ \hat{\theta}_3(k) \\ 1 \end{bmatrix} \text{ and } Q = \begin{bmatrix} 0 & 1 & 0 \end{bmatrix}.$$

Introducing the cost function

$$P = \sum_{i=1}^{N_p} \omega(i)\left(\hat{y}(k+i|k) - y_d(k+i|k)\right)^2$$
$$+ \sum_{i=1}^{N_p} \rho(i)\left(\Delta\hat{u}(k+i|k)\right)^2 \quad (11)$$

where $\hat{y}(k+i|k)$ is the ith step predicted output from time k, $y_d(k+i|k)$ is the ith step reference signal from time k, $\Delta\hat{u}(k+i|k)$ is the difference between ith step predicted input from time k and control input at time k, N_p is the number of predicted steps, and ω and ρ are weighting coefficients.

In order to minimize the cost function (11), output predictions over the horizon must be computed. Predictive outputs can be obtained by using (10) recursively, resulting in:

$$\hat{y}(k+j) = QM^j\hat{x}(k) + \sum_{i=0}^{j-1} QM^{j-i-1}N\Delta u(t+i) \quad (12)$$

Now, the predictions along the horizon are given by

$$\hat{y}(k) = \begin{bmatrix} \hat{y}(k+1|k) \\ \hat{y}(k+2|k) \\ \vdots \\ \hat{y}(k+N_p|k) \end{bmatrix}$$

$$= \begin{bmatrix} QM\hat{x}(k) + QN\Delta u(k) \\ QM^2\hat{x}(k) + \sum_{i=0}^{1} QM^{1-i}N\Delta u(k+i) \\ \vdots \\ QM^{N_p}\hat{x}(k) + \sum_{i=0}^{N_p-1} QM^{N_p-1-i}N\Delta u(k+i) \end{bmatrix} \quad (13)$$

For simplicity, define

$$\hat{Y} = F\hat{x}(k) + H\Delta U \quad (14)$$

where $\hat{Y} = \begin{bmatrix} \hat{y}(k+1|k) & \hat{y}(k+2|k) \dots \hat{y}(k+N_p|k) \end{bmatrix}^T$ is the predicted future output, $\Delta U = \begin{bmatrix} \Delta u(k) & \Delta u(k+1) \dots \Delta u(k+N_p-1) \end{bmatrix}^T$ is the vector of future control increments, the matrix H defined

$$\text{as } H = \begin{bmatrix} QN & 0 & \cdots & \cdots & 0 \\ QMN & QN & \ddots & \ddots & \vdots \\ \vdots & & \ddots & \ddots & \vdots \\ QM^{N_p-2}N & \ddots & & \ddots & QN & 0 \\ QM^{N_p-1}N & QM^{N_p-2}N & \cdots & QMN & QN \end{bmatrix}, \text{ and}$$

matrix F is defined as $F = \begin{bmatrix} QM & QM^2 & \dots & QM^{N_p} \end{bmatrix}^T$.

Consider the case where $\omega(i) = 1$ and $\rho(i) = \rho$. The control sequence Δu is calculated minimizing the cost function (10) that can be written as:

$$P = \left(H\Delta U + F\hat{x}(k) - Y_d \right)^T \left(H\Delta U + F\hat{x}(k) - Y_d \right)$$
$$+ \rho(\Delta U)^T(\Delta U) \quad (15)$$

An analytical solution exists that can be calculated as follows

$$\Delta U = \left(H^T H + \rho I \right)^{-1} H^T \left(y_d - F\hat{x}(k) \right) \quad (16)$$

It should be noted that only $\Delta u(k)$ is sent to the plant and all the computation is repeated at the next sampling time.

Result

The experimental setup on piezoelectric actuator is shown in Fig. 1. Figure 2 shows the experimental scheme. The PEA is PFT-1110 (Nihon Ceratec Corporation). The specification of PFT 1110 is shown in. The displacement is measured by the noncontact capacitive displacement

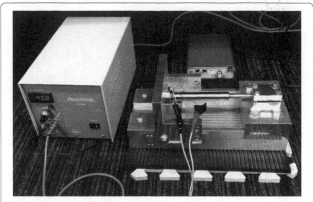

Fig. 1 Experimental setup

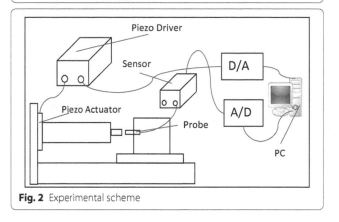

Fig. 2 Experimental scheme

sensor (PS-1A Nanotex Corporation) which has 2-nm resolution. The experiments are conducted with 2 desired output s $y_{d1}(k) = 10 \sin(2\pi \times k \times \Delta t)$ μm and $y_{d2}(k) = 7 \sin(2\pi \times 5 \times k \times \Delta t) + 3 \cos(2\pi \times 0.5 \times (1.5^{-k \times \Delta t}) \times k \times \Delta t)$ μm, where Δt is sampling period and be chosen as 0.5 ms. The experiment results of proposed method are compared with those getting from PID controller.

Table 1 shows the experimental setting parameters.

Figure 3 shows the control input for the experiment with $y_{d1}(k)$. The estimated parameters are shown in Fig. 4.

Figure 5 shows the tracking result. The tracking error is shown in Fig. 6. It can be seen that the maximum error at steady state is about 0.4 %.

Figure 7 shows the control input for the experiment with $y_{d2}(k)$. The estimated parameters are shown in Fig. 8.

Table 1 Experimental setting parameters

	N_p	$\omega(i)$	$\rho(i)$	λ	$\hat{\theta}(0)$	Offset (V)	Δt (ms)
$y_{d1}(k)$	3	1	0.1	0.1	0.2	30	0.5
$y_{d2}(k)$	3	1	0.1	0.1	0.2	30	0.5

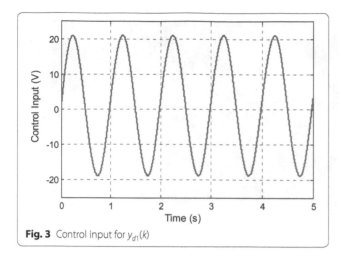

Fig. 3 Control input for $y_{d1}(k)$

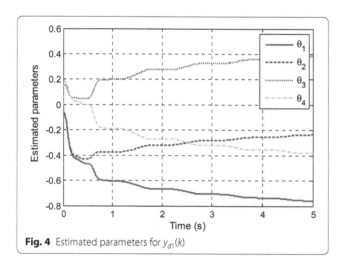

Fig. 4 Estimated parameters for $y_{d1}(k)$

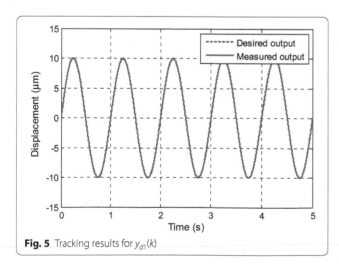

Fig. 5 Tracking results for $y_{d1}(k)$

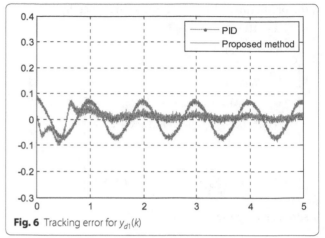

Fig. 6 Tracking error for $y_{d1}(k)$

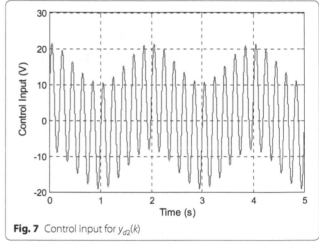

Fig. 7 Control input for $y_{d2}(k)$

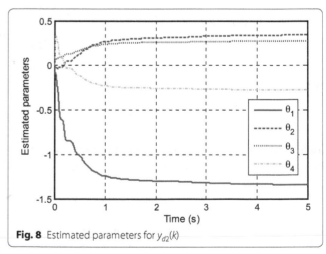

Fig. 8 Estimated parameters for $y_{d2}(k)$

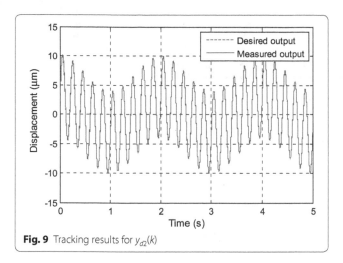

Fig. 9 Tracking results for $y_{d2}(k)$

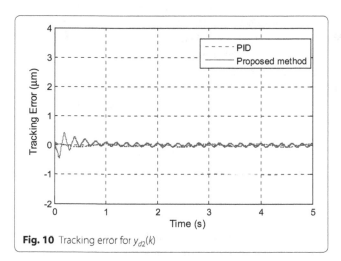

Fig. 10 Tracking error for $y_{d2}(k)$

Authors' contributions
The contribution of this paper is that the positioning performance of PEA with nonlinearity hysteresis phenomenon can be achieved by fusing model predictive control with adaptive linear model. All authors read and approved the final manuscript.

Author details
[1] School of Mechanical Engineering, Hanoi University of Science and Technology, Hanoi, Vietnam. [2] Graduate School of Engineering and Science, Shibaura Institute of Technology, Saitama 337-8570, Japan. [3] Department of Electronic and Information Systems, Shibaura Institute of Technology, Saitama 337-8570, Japan.

Competing interests
The authors declare that they have no competing interests.

References
1. Croft D, Shed G, Devasia S. Creep, hysteresis, and vibration compensation for piezoactuators: atomic force microscopy application. J Dyn Syst Meas Control. 2001;123(1):35–43.
2. Zou Q, Leang KK, Sadoun E, Reed MJ, Devasia S. Control issues in high-speed AFM for biological applications: collagen imaging example. Asian J Control. 2004;6(2):164–78.
3. Kassies R, Van der Werf KO, Lenferink A, Hunter CN, Olsen JD, Subramanian V, Otto C. Combined AFM and confocal fluorescence microscope for application in bio-nanotechnology. J Microsc. 2005;217(1):109–16.
4. Song H, Vdovin G, Fraanje R, Schitter G, Verhaegen M. Extracting hysteresis from nonlinear measurement of wavefront-sensorless adaptive optics system. Opt Lett. 2009;34(1):61–3.
5. Yang W, Lee S-Y, You B-J. A piezoelectric actuator with a motion-decoupling amplifier for optical disk drives. Smart Mater Struct. 2010;19(6):1–10.
6. Stöppler G, Douglas S. Adaptronic gantry machine tool with piezoelectric actuator for active error compensation of structural oscillations at the tool centre point. Mechatronics. 2008;18(8):426–33.
7. Viswamurthy SR, Rao AK, Ganguli R. Dynamic hysteresis of piezoceramic stack actuators used in helicopter vibration control: experiments and simulations. Smart Mater Struct. 2009;20(4):387–99.
8. Senousy MS, Li FX, Mumford D, Gadala M, Rajapakse RKND. Thermo-electro-mechanical performance of piezoelectric stack actuators for fuel injector applications. J Intell Mater Syst Struct. 2007;20(4):1109–19.
9. Wei J-J, Qiu Z-C, Han J-D, Wang Y-C. Experimental comparison research on active vibration control for flexible piezoelectric manipulator using fuzzy controller. J Intell Rob Syst. 2010;59(1):31–56.
10. Ge P, Jouaneh M. Tracking control of a piezoceramic actuator. IEEE Transaction on Control Systems Technology. 1996;4(3):209–16.
11. Krejci P, Kuhnen K. Inverse control of systems with hysteresis and creep. IEEE Proc Control Theory Appl. 2001;148(3):185–92.
12. Webb GV, Lagoudas DC, Kurdila AJ. Hysteresis modeling of SMA actuators for control application. J Intell Mater Syst Struct. 1998;9(6):432–48.
13. Chen X, Hisayama T, Su C-Y. Adaptive control for uncertain continuous-time systems using implicit inversion of Prandtl–Ishlinskii hysteresis representation. IEEE Trans Autom Control. 2010;55(10):2357–63.
14. Zhou J, Wen C, Li T. Adaptive output feedback control of uncertain nonlinear systems with hysteresis nonlinearity. IEEE Trans Autom Control. 2012;57(10):2627–33.
15. Goodwin GC, Sin KS. Adaptive filtering prediction and control. New York: Dover; 2009.

Figure 9 shows the tracking result. The tracking error is shown in Fig. 10. It can be seen that the maximum error at steady state is about 1 %.

Discussion

This paper has discussed the adaptive model predictive control for piezoelectric actuators, where the model of PEA is regarded as linear model. The unknown parameters in the model are estimated online. The proposed method shows its effectiveness in tracking performance. Moreover, it is simple and easy to be implemented. In the future, we will try to employ the proposed method to control piezo-actuated systems with load.

12

Tongue interface based on surface EMG signals of suprahyoid muscles

Makoto Sasaki[1*], Kohei Onishi[1], Dimitar Stefanov[2], Katsuhiro Kamata[3], Atsushi Nakayama[4], Masahiro Yoshikawa[5] and Goro Obinata[6]

Abstract

The research described herein was undertaken to develop and test a novel tongue interface based on classification of tongue motions from the surface electromyography (EMG) signals of the suprahyoid muscles detected at the underside of the jaw. The EMG signals are measured via 22 active surface electrodes mounted onto a special flexible boomerang-shaped base. Because of the sensor's shape and flexibility, it can adapt to the underjaw skin contour. Tongue motion classification was achieved using a support vector machine (SVM) algorithm for pattern recognition where the root mean square (RMS) features and cepstrum coefficients (CC) features of the EMG signals were analyzed. The effectiveness of the approach was verified with a test for the classification of six tongue motions conducted with a group of five healthy adult volunteer subjects who had normal motor tongue functions. Results showed that the system classified all six tongue motions with high accuracy of $95.1 \pm 1.9\%$. The proposed method for control of assistive devices was evaluated using a test in which a computer simulation model of an electric wheelchair was controlled using six tongue motions. This interface system, which weighs only 13.6 g and which has a simple appearance, requires no installation of any sensor into the mouth cavity. Therefore, it does not hinder user activities such as swallowing, chewing, or talking. The number of tongue motions is sufficient for the control of most assistive devices.

Keywords: Tongue interface, Motion classification, Support vector machine (SVM), Surface electromyography (EMG), Suprahyoid muscles

Background

A tongue is an intra-oral locomotorium that can be moved quickly and precisely according to one's own will. Anyone can set their own tongue position precisely and can smoothly change the magnitude of the force imposed on the teeth or palate. In fact, tongue motor functions are usually preserved even in people with cervical spinal cord damage. Various studies have demonstrated that people with a high level of movement paralysis can use tongue motions to control home appliances such as a PCs and electric wheelchairs [1, 2].

An interface system based on a small joystick operated by the tongue has been presented in the literature [3]. The joystick is fixed in a suitable position via a special arm mount. The application of such an interface is limited to people with sufficient head motion that is able to reach the joystick. The same design might hinder conversation, eating, and drinking because a part of the joystick is located in the mouth cavity during use. In addition, such a solution might allow flow of excess saliva out of the mouth.

Numerous studies have examined control interfaces containing an artificial palate with buttons activated by the tongue tip [4–6]. In other solutions, a few pairs of light emitting diodes and photodiodes are mounted on an artificial palate to detect the tongue position [7, 8]. By changing the tongue position and thereby activating different sensors, the user sets control commands. A benefit of such solutions is that the tongue interface device remains hidden to others (intra-oral interface). In addition, the number of sensors and their location on the artificial palate can be customized easily. However, because the artificial palate must remain in the mouth cavity for prolonged periods of time, such a design might require

*Correspondence: makotosa@iwate-u.ac.jp
[1] Graduate School of Engineering, Iwate University, Morioka, Iwate, Japan
Full list of author information is available at the end of the article

additional efforts for maintaining oral hygiene and might entail various difficulties related to talking and eating.

Some recent studies present solutions that consist of a small magnet or a piece of ferromagnetic material attached to the tongue tip via gluing or piercing, and a sensor array that detects the tongue tip position [9–13]. Sensor systems that include a small permanent magnet fixed to the tongue tip and an array of magnetic sensors have been presented in the literature [9–11]. Some earlier reports introduced a tongue interface that includes an air-cored coil with inductance changed by a small ferromagnetic stud attached to the tongue [12, 13]. Although such an approach is a simple interface solutions, the (ferro) magnetic stud on the tongue tip might cause some inconvenience to users.

The electromyography (EMG) signals created by skeletal muscles have been used for many years in human movement studies and for control of prostheses [14–26]. Our earlier study specifically addressed the potential of EMG signals and explored the viability of a tongue interface based on surface EMG signals detected at the underside of the jaw [27]. The initial interface system consisted of nine single-surface electrodes attached on the underside of the jaw and connected via multiple lead wires. The proposed tongue interface, which is based entirely on analysis of extra-oral EMG signals, requires no insertion of a palatal plate or a joystick in the mouth, attachment of a magnet or ferromagnetic studs to the tongue, or physical contact of the tongue with any sensor. An artificial neural network (ANN) with three layers of neurons (input, hidden and output) was used as the motion classifier. During the ANN training stage, three thin-film force sensors were installed on an upper jaw mouthpiece to deliver training data for voluntary tongue motions of three types: right, left, and forward. After this initial experiment, a new experiment was conducted for classification of the same voluntary motions without using signals from force sensors [28]. These initial experiments demonstrated that the tongue motions are classifiable from the EMG signals of the suprahyoid muscles. They have some potential for use in control interfaces. However, the initial interface system classified only a small number of tongue motions. Additionally, ANN-based classifiers are well known to have a few important shortcomings: long learning time, local optimal solution depending on the initial value of parameters, and complicated procedures for selection of the number of neurons in the hidden layer. Furthermore, the initial sensor module consists of single electrodes and wires, which can pose severe difficulties. For practical application, the initial interface system required further improvement in few main directions: increased number of classified voluntary tongue motions, improvement of the classification accuracy, and redesign of the electrode module.

This paper proposes a novel tongue interface based on classification of the tongue motions from surface EMG signals of the suprahyoid muscles detectable at the underside of the jaw. The interface allows classification of six tongue motions, which are sufficient for the control of PCs and electric wheelchairs. The new system was evaluated using a computer simulation experiment to assess control of an electric wheelchair.

EMG-based tongue interface

EMG measurement approach

Tongue motions are produced by the coordinated actions of intrinsic muscles, which control tongue posture and tongue tip position, and extrinsic muscles, which control tongue protrusion and retraction [29, 30]. The EMG activity of the lingual muscles has been studied using tungsten microelectrodes and hook-wire electrodes [31] and surface electrodes [32] placed within the oral cavity. However, intra-oral electrodes are unsuitable for the practical control of assistive devices.

The EMG signals of the suprahyoid muscles are detectable via electrodes placed on the skin of the underside of the jaw [33–35]. The suprahyoid muscles comprise several muscle groups such as digastric muscles, stylohyoid muscles, mylohyoid muscles, and geniohyoid muscles, as presented in Fig. 1 [29, 30]. The suprahyoid muscles control the position of the hyoid according to the direction, position, and force of the tongue tip. Therefore, they contain sufficient information about the performed tongue motions. However, the suprahyoid muscles also contribute to motions that are unrelated to the tongue position. Such motions produce EMG signals when such motions are performed. For example, suprahyoid muscles help jaw-opening by pulling the mandible down when the hyoid position is fixed by the infrahyoid muscles. They also pull the hyoid up to assist swallowing when the mandible position is fixed to the muscles used for mastication. A great challenge to the design of a reliable tongue interface is the identification and suppression of EMG signals that do not originate from voluntary tongue motions. That difficulty cannot be resolved merely by electrode positioning because measured EMG signals are always composed of several signals from different muscles around the electrode.

In this study, the EMG signals of the suprahyoid muscles are measured at multiple points of the skin using a multi-electrode array. The multi-electrode approach makes the interface system less sensitive to eventual positioning errors of the electrode unit. Moreover, it enables people with little experience or little knowledge of EMG measurement to apply the sensor. Current research was based on initial experiments conducted for the classification of tongue motions from the EMG signals patterns [27, 28].

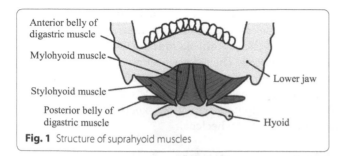

Fig. 1 Structure of suprahyoid muscles

Sensor module and signal pre-processing

The electrode module was designed as a thin flexible boomerang-shaped patch attached to the underside of the jaw (Fig. 2). The prototype sensor dimensions were decided by considering the average size of the lower jaw and curvature near the lower jaw and neck of the subjects in the tests (see "Experiments and data acquisition" section below). The sensor was designed to cover the entire jaw. The number of the electrodes was determined experimentally. The electrodes were positioned on equal inter-electrode distances. The interface assembly, which consisted of 22 active electrodes shaped as $\phi 2 \times 2.5$ mm

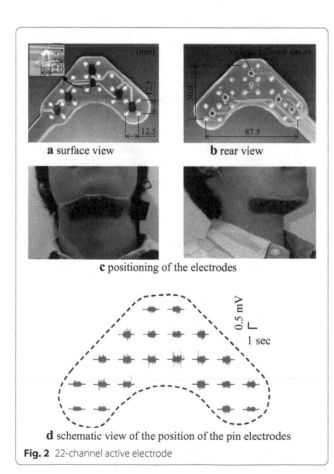

a surface view   **b** rear view

c positioning of the electrodes

d schematic view of the position of the pin electrodes

Fig. 2 22-channel active electrode

pure silver rods, was positioned at the inter-electrode distance of 12.5 mm on a polyimide film. The interface unit was 50.0-mm-long and 87.5-mm-wide. The thickness of the entire substrate including the reinforcement film was 0.3 mm. The electrode tips were shaped as hemispheres to facilitate the skin contact. Voltage follower circuits were incorporated into the same interface mount to reduce the output impedance. The electrode base thickness was 1.7 mm. For electric insulation of the electronic parts, both sides of the substrate were covered with a layer of silicon. The interface system was only 13.6 g. For the experiments, the interface module was adhered to the underside of the jaw of the subject. A ground electrode and an active common electrode were connected respectively to the left and right earlobes via ear clips (Fig. 2c). The electric potential between each electrode and the active common electrode was amplified using a separate differential amplifier. The gain of the differential amplifiers was set to 2052. A band-pass filter with a passband from 16 to 440 Hz bandwidth was used to remove the direct current component and high-frequency noise superimposed on the EMG signals. The EMG signals of all 22 EMG channels were digitized by a 16 bit analog-to-digital converter (USB-6218; National Instruments Corp.). In general, the EMG signal frequency range is 0–1000 Hz. Its usable energy is limited to 0–500 Hz [36]. Therefore, the sampling rate was set to 2000 Hz in compliance with the Nyquist theorem.

Classification of tongue motions

Figure 3 portrays the tongue motion classification procedure. It comprises the EMG measurement, feature extraction, and motion classification.

Feature extraction

The feature extraction process was based on the overlapped windowing technique proposed by Englehart et al. [37]. It allows faster system response. The EMG signals measured from all 22 channels were segmented for feature extraction into windows consisting of 256 samples, as portrayed in Fig. 4. The length of each window was 128 ms. The next sampling segment slides over the current segment with an increment time of 16 ms. For composition of the feature vector, the root mean square (RMS) and the cepstrum coefficients (CC) of the EMG signals were calculated for each window. The RMS features are characteristics of a time domain. The CC features are characteristics of the frequency domain [28, 38, 39]. Cepstrum analysis techniques have been used for many years for speech recognition because of their fast response and accurate results. Some recent studies have demonstrated that the techniques are useful also for motion classification based on EMG signals [39–41].

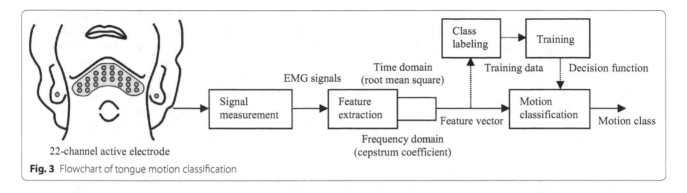

Fig. 3 Flowchart of tongue motion classification

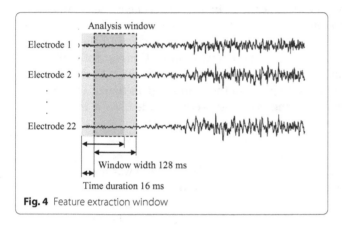

Fig. 4 Feature extraction window

The RMS features provide information related to the amplitude of the EMG signals. Let us denote the EMG signals of the l-th electrode in the n-th sample of the p-th analysis window as $EMG_{l,n}(p)$ ($n = 0, ..., N - 1; l = 1, ..., L$), where N is the number of samples in one analysis window ($N = 256$), and L is the number of electrodes ($L = 22$). The RMS features can be expressed as the following equation:

$$RMS_l(p) = \sqrt{\frac{1}{N} \sum_{n=0}^{N-1} EMG_{l,n}(p)^2} \quad (1)$$

The equation above is useful for calculation of the RMS features of all channels.

To calculate the CC features, the Hanning window procedure was applied to each analysis window of the EMG signals. The Fourier transform $X_l^k(p)$ ($k = 0, ..., N - 1$) of $EMG_{l,n}(p)$ can be expressed as shown below.

$$X_l^k(p) = \sum_{n=0}^{N-1} EMG_{l,n}(p)e^{-j2\pi kn/N} \quad (2)$$

The CC features $CC_l^n(p)$ are calculated from the following equation.

$$CC_l^n(p) = \frac{1}{N} \sum_{k=0}^{N-1} \log\left|X_l^k(p)\right| e^{j2\pi kn/N} \quad (3)$$

Cepstrum analysis enables separation of the power spectrum of the EMG signals into a smooth component (spectral envelope) and a fine fluctuation component (fine structure). Low-order cepstrum coefficients include information about the spectral envelope whereas the high-order coefficients include fine structure information. The low-order coefficients were calculated using formula (3) and by varying n from $n = 0$ to $n = W - 1$. Here, W is a CC feature parameter (order of the cepstrum coefficients).

The feature vector $x(p)$ for classifying tongue motions can be expressed as

$$\begin{aligned} x(p) = (&RMS_1(p), ..., RMS_L(p), \\ &CC_1^0(p), ..., CC_1^{W-1}(p), ..., \\ &CC_L^0(p), ..., CC_L^{W-1}(p))^T \end{aligned} \quad (4)$$

where the dimension of the feature vector $x(p)$ is $L(1 + W)$.

Motion classification

For this study, the support vector machine (SVM) classifier was used to classify tongue motions. The SVM classifier has the following benefits for this classification:

– The SVM classifier offers excellent recognition performance.
– SVM has high generalization capability because it applies a maximum-margin classification function.
– It converges to a global optimal solution and therefore does not fall into a local optimum solution.
– It has extremely short learning time because of the simple procedures used for calculation of the hyperparameters used for training.

SVM is a method for classification of an unknown feature vector $x(p)$ (hereinafter designated as x) into two classes [42]. The decision function is

$$f(x) = sgn\left(\sum_{i=1}^{D} \lambda_i y_i K(x_i, x) + b\right) \quad (5)$$

where D denotes the number of training samples, y_i signifies the class label that corresponds to the i-th training sample x, λ_i is a Lagrangian undetermined multiplier, b is a bias term, and $K(x_i, x)$ denotes a kernel function. For this study, the radial basis function (RBF) was selected as the kernel function to map the input data in a high dimensional feature space. The RBF kernel is expressed as

$$K(x_i, x) = \exp(-\gamma \|x_i - x\|^2) \tag{6}$$

where γ is a kernel parameter. The Lagrangian undetermined multiplier λ_i in the decision function is derived by solving the following equation (quadratic programming).

$$\max_{\lambda_i} \sum_{i=1}^{D} \lambda_i - \frac{1}{2} \sum_{i,j=1}^{D} \lambda_i \lambda_j y_i y_j K(x_i, x) \tag{7}$$

$$\text{subject to} \quad \sum_{i=1}^{D} \lambda_i y_i = 0, \quad 0 \le \lambda_i \le C \tag{8}$$

The SVM classification performance depends on the selection of the kernel parameter γ and the penalty parameter C. The optimal combination of γ and C can be obtained using a grid search.

Usually, the SVM classifier is used for classification of features into two classes. In this study, the SVM algorithm was extended to multi-class classification using the one-against-one method [43]. For the classification of M classes tongue motions, $M(M-1)/2$ decision functions are constructed initially for all combinations of these M classes. The feature vector x is classified against each decision function. The final decision on the class is obtained by majority vote.

Experiments and data acquisition

Subjects

This investigation examined five healthy adult male subjects (22.2 ± 1.3 years old, 169.7 ± 7.4 cm tall, 61.0 ± 11.3 kg weight) who were free of musculoskeletal deficits and neurological impairment and who had normal tongue motor functions. Approval for the tests was obtained in advance by the Ethical Review Board of Iwate University. Before the start of the tests, the study objective, experimental protocol and risks were explained to each subject. Written consent was received from each.

Experimental protocol

First, the skin surface of the underside of the jaw was cleaned with alcohol and electrode paste (Elefix; Nihon Kohden Corp.) was applied to reduce the skin-electrode impedance. The 22-channel active electrode was adhered to the underside of the subject's jaw using film dressing (CATHEREEPLUS; Nichiban Co. Ltd.). A ground

electrode and an active common electrode were attached on the left and right earlobe of the subject using ear clips.

The tongue motion set included five tongue motions (right, left, up, down, and forward) performed with a closed mouth and a saliva swallowing (Fig. 5). During these motions, subjects were asked to position their tongue tips sequentially in the maxillary right second molar tooth, the maxillary left second molar tooth, the hard palate, the floor of the mouth, and near the maxillary central incisor. Saliva swallowing is an unintentional action that is repeated frequently. The saliva swallowing was included in tests to evaluating its effects on tongue motion classification. In the experiment, each tongue motion was executed for 2 s at a subject's comfortable speed. A resting period of 2 s was given to the subject before the start of the next motion. Consequently, all six motions in the set were completed for 22 s. Each subject was asked to perform the motion set 14 times. The EMG signals during each test were recorded. As a result, 14 datasets were produced for each subject.

Data analysis

Matlab (R2013a; The MathWorks Inc.) was used for data analysis. The SVM classification algorithm was designed using an SVM library: LIBSVM [44]. The programs were executed on a PC (Windows 7 64-bit OS, i7-3770 CPU/3.4 GHz, 16 GB RAM).

To justify the selection of the kernel function, it was confirmed that the RBF kernel matrix calculated from the first four datasets is a symmetric, positive semi-definite matrix (i.e., all eigenvalues of the kernel matrix are non-negative). Then the datasets were used as training data of the SVM. The remaining ten datasets were used for tongue motion classification. The feature vector x for tongue motion classification was defined according to Eq. (4). The values of the RMS features and CC features were calculated, respectively, according to Eq. (1) and Eq. (3). The class labels y_i representing the type of motion in Eq. (5) were obtained using threshold triggering of the EMG signals [45]. The relation between the composition of the feature vector and its classification accuracy was evaluated by comparing the classification results when the CC feature parameter W was varied from 0 to 10. For simplicity in these analyses, $W = 0$ expresses the situation when the CC features are not included in the feature vector.

| Right | Left | Up | Down | Forward | Swallowing |

Fig. 5 Definition of the tongue motions included in the tests

As explained in the section describing "Motion classification", the SVM classification performance depends on selection of kernel parameter γ and penalty parameter C. The optimal combination of γ and C was ascertained using a grid search within the training data. The search included 96 combinations of γ and C for $\gamma = \{2^{-10}, 2^{-9}, \ldots, 2^{1}\}$ and $C = \{2^{1}, 2^{2}, \ldots, 2^{8}\}$. The combination with the highest classification rate was defined using fivefold cross validation. Results showed that the optimum values of γ and C differ for each subject. After training of the SVM with the optimized hyperparameters for γ and C, motion classification of the test data was performed. The predicted class was replaced with a "neutral" tongue position when all EMG signals are under the threshold level (i.e., relaxed state). Next, a majority voting technique was applied to reduce the effect of misclassification. Majority voting was applied to a moving window composed of 20 frames that included the present frame and the prior 19 frames. Classification of the tongue motion was determined from the class with the largest number of wins.

The classification accuracy (CA) of the tongue motions was evaluated using the following equation.

$$CA = \frac{\text{number of correct feature vectors}}{\text{total number of feature vectors}} \times 100\,[\%] \quad (9)$$

Results

Effect of feature parameter selection on classification accuracy

The average classification accuracy and the standard deviation of the classification accuracy for all five subjects are presented in Fig. 6. Results reveal the relation between the feature vector and classification accuracy. In cases where the feature vector was composed of RMS features only ($W = 0$), the classification accuracy of the tongue motions was 84.1 ± 1.5 %. The classification accuracy

increased substantially when the CC features were added to the feature vector ($W = 1, \ldots, 10$). The classification accuracy exceeded 95 % and remained almost constant when the CC feature parameter was $W = 5$ or higher. The classification accuracy for $W = 5$ was 95.1 ± 1.9 %. For $W = 10$, the classification accuracy was 95.1 ± 1.3 %. No significant difference was found between the classification accuracies calculated with $W = 5$ and $W = 10$.

The dimension of the feature vector $\boldsymbol{x}$ for tongue motion classification was set to $L(1 + W)$ (see Eq. (4)). Because the computational complexity increases significantly for greater values of W, the smallest possible W that gives comparable classification accuracy should be used. As explained above, no significant difference was found between the classification results with $W = 5$ and $W = 10$, which suggests that satisfying classification results are obtainable with a feature vector based on $W = 5$. For that reason, a more detailed examination of the classification results is given here for the case in which the CC feature parameter was selected as $W = 5$.

Tongue motion classification accuracy

Table 1 presents classification results for all five subjects. The lowest total classification accuracy was 91.9 % (for subject A) and the highest total classification accuracy was 96.7 % (for subject B). The average total classification accuracy for all subjects was 95.1 %. Analysis of the classification results for the separate tongue motions demonstrates that the "left" tongue motion was recognized with the highest classification accuracy (97.6 %). The classification accuracy for the "down" tongue motion was slightly lower (96.7 %), followed by results for "saliva swallowing" (95.3 %), "right" tongue motion (95.0 %), "up" tongue motion (94.5 %), and "forward" tongue motion (91.4 %). Table 2 presents details of the classification errors. The "forward" tongue motions were misclassified as "up", "down", and "saliva swallowing". The "up" tongue motion has the second lowest classification accuracy. Frequently, it has been misclassified as "right" tongue motion.

Short signals at the start and the end of the main motion were often misclassified. By applying majority voting technique, the number of these misclassifications was reduced; 1.0 % of all motions were misclassified as a "neutral" tongue position. However, misclassification as a "neutral" tongue position is less important because the "neutral" tongue position is useful as a stop command when the assistive device is controlled by the developed interface. Misclassification of other motions as a "neutral" tongue position cannot create dangerous situations. It will merely cause the controlled device to stop. Overall, the misclassification errors that might affect the operation of the controlled assistive devices were estimated from the total classification accuracy as about 3.9 %.

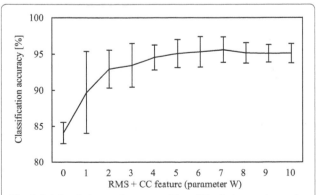

Fig. 6 Relation between the composition of the feature vector and classification accuracy. $W = 0$ means that the CC features are not included in a feature vector

Table 1 Classification accuracy of tongue motions

Subject	Right	Left	Up	Down	Forward	Saliva swallowing	Total
A	96.1	93.8	84.9	98.3	88.3	90.1	91.9
B	94.1	97.8	97.4	95.8	96.4	98.9	96.7
C	99.5	98.8	97.6	99.9	90.0	92.9	96.4
D	89.1	100	96.0	92.4	95.4	96.2	94.9
E	96.2	97.8	96.5	97.2	87.0	98.4	95.5
Mean	95.0	97.6	94.5	96.7	91.4	95.3	95.1

Computer simulation of wheelchair control

A computer simulation model of an electric wheelchair was developed to evaluate the applicability of the developed tongue interface to control assistive devices. The wheelchair model was controlled virtually by operation commands generated from a confusion matrix presented in Table 2. The error of the commands was set to occur according to the possibility in the confusion matrix. In other words, this is a Monte Carlo method. The error timing was determined using a random number with a uniform probability distribution. The virtual trajectory of the wheelchair's center of gravity was used as an indicator to evaluate the effects of misclassification errors on the wheelchair operability.

Simulation model of an electric wheelchair

Figure 7 portrays a simplified model of the electric wheelchair. The angle θ and the center of gravity position P_G (X_G, Y_G) of an electric wheelchair are defined using the following equations.

$$\theta(t) = \frac{1}{T} \int_0^t (R_r\omega_r(t) - R_l\omega_l(t))dt \qquad (10)$$

$$X_G(t) = \frac{1}{2} \int_0^t (R_r\omega_r(t) + R_l\omega_l(t)) \cos\theta(t)dt \qquad (11)$$

$$Y_G(t) = \frac{1}{2} \int_0^t (R_r\omega_r(t) + R_l\omega_l(t)) \sin\theta(t)dt \qquad (12)$$

Therein, R_r and R_l respectively denote the radii of the right and left wheel. $\omega_r(t)$ and $\omega_l(t)$ respectively denote the angular velocity of the right and the left wheel. T is the distance between the right and left wheels. For the wheelchair model, wheels with radius 165 mm were selected. The distance between the wheels was 530 mm.

The maximum velocity of the electric wheelchair model V_{max} was set to 4 km/h. The model is based on a trapezoidal model of acceleration and deceleration. The acceleration time T_a and the deceleration time T_d were set to 1 s. In this simulation, a new operation command is sent to the virtual wheelchair every $T_i = 16$ ms because, in the tongue motion classification experiment, the EMG signals were classified at 16 ms intervals (see Fig. 4). Therefore, velocity commands are sent to the right and the left wheel every 16 ms. These commands are based on the rules presented in Table 3. The wheel velocities $R_r\omega_r(t)$ and $R_l\omega_l(t)$ are defined by the following equations.

$$R_r\omega_r(t) = V_{max}S_r(t)T_i / T_a \qquad (13)$$

$$R_l\omega_l(t) = V_{max}S_l(t)T_i / T_a \qquad (14)$$

Table 2 Confusion matrix for six tongue motions

	Estimated class						
	Right	Left	Up	Down	Forward	Saliva swallowing	Neutral
Actual class							
Right	95.0	0.1	1.6	0.4	2.1	0.0	0.8
Left	1.2	97.6	0.1	0.2	0.0	0.0	0.9
Up	2.6	0.4	94.5	0.4	0.7	0.2	1.2
Down	0.0	0.0	0.7	96.7	1.9	0.3	0.4
Forward	0.4	0.0	2.2	3.4	91.4	1.2	1.4
Saliva swallowing	0.2	0.0	0.9	0.3	2.2	95.3	1.1

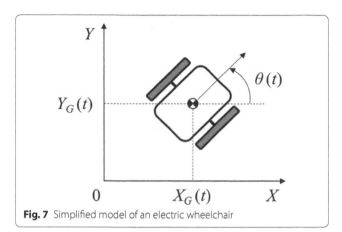

Fig. 7 Simplified model of an electric wheelchair

Table 3 Change amount of velocity commands for right and left wheel

Command	$S_r(t)$	$S_l(t)$
Forward	+1	+1
Back	−1	−1
Right rotation	−1	+1
Left rotation	+1	−1
Brake	$\begin{cases} +1 & \text{if } R_r\omega_r(t) < 0 \\ -1 & \text{if } R_r\omega_r(t) > 0 \end{cases}$	$\begin{cases} +1 & \text{if } R_l\omega_l(t) < 0 \\ -1 & \text{if } R_l\omega_l(t) > 0 \end{cases}$
None	+0	+0

Therein, $S_r(t)$ and $S_l(t)$ respectively represent the commands sent to the right and the left wheel in sequential moments of time. $S_r(t)$ and $S_l(t)$ are defined as follows.

$$-T_a / T_i \leq S_r(t) \leq T_a / T_i \tag{15}$$

$$-T_a / T_i \leq S_l(t) \leq T_a / T_i \tag{16}$$

Linking tongue motions with commands for control of the wheelchair model

Commands for the wheelchair model operation are presented in Table 4. They are based on the confusion matrix of tongue motions, as shown in Table 2. Initially, the "Brake" command is set via the "neutral" tongue position.

Table 4 Definition of operation commands

Operation commands	Tongue motions
Forward	Down
Back	Forward
Right rotation	Right
Left rotation	Left
Brake	Neutral
None	Saliva swallowing, up

It is assumed that the "Brake" command is sent to the wheelchair when all EMG signals are under the threshold level (i.e., relaxed state). The "Brake" command causes the wheelchair to decelerate and stop. The "Forward" command was linked with the "down" tongue motion because the probability for misclassification of the "down" tongue motion as "right" or "left" is nearly zero in the confusion matrix. "Right" and "left" tongue motions were used, respectively, as commands for turning of the wheelchair model to the right and left. Reverse wheelchair movement ("back" command) is initiated by "forward" tongue motion.

The classification accuracy of "forward" tongue motion was lower than that of "up" tongue motion. However, the rate of misclassification of "forward" tongue motion as "right" tongue motion is much lower (0.4 %) than that of "up" tongue motion (2.6 %). Its characteristic means that "forward" tongue motion ensures the straight driving performance. In addition, although "forward" tongue motion is misclassified as "down" tongue motion as about 3.4 %, it does not affect the straight driving performance so much because this misclassification reduces the driving velocity while moving backward.

The remaining "saliva swallowing" and "up" tongue motions were defined as no command.

The driving test consisted of six tasks:

E1	Driving the wheelchair forward 5 m
E2	Driving the wheelchair backward 5 m
E3	Turning the wheelchair 360° to the right
E4	Turning the wheelchair 360° to the left
E5	Swallowing saliva while the wheelchair model is stopped
E6	Swallowing saliva while the wheelchair model is moving straight at maximum velocity

The saliva swallowing times in tests E5 and E6 were set to 1 s.

In this situation, the 100 patterns of velocity commands of right and left wheel S_r, S_l considering that the rate of misclassification as shown in Table 2 was generated using a random function. Then, these resultant trajectories were compared with the ideal trajectory, which was calculated as a classification accuracy of all tongue motions is 100 %.

Simulation results

The simulation results of the angle θ and the center of gravity position P_G (X_G, Y_G) of an electric wheelchair are presented in Fig. 7. In addition, the differences between the ideal trajectory and the trajectory including the effect of misclassification are presented in Fig. 8.

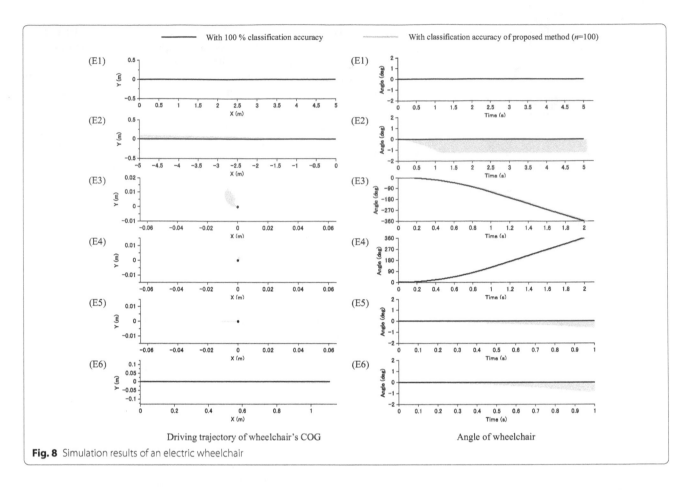

Driving trajectory of wheelchair's COG   Angle of wheelchair

Fig. 8 Simulation results of an electric wheelchair

In test E1, both the angle θ and the driving trajectory in y-direction Y_G were, respectively, 0° and 0 mm. The difference between the maximum time required for 5 m moving of the wheelchair and the time for ideal trajectory was only 40.5 ms. In test E2, the maximum deviations of θ and Y_G for moving the wheelchair backward were, respectively, −1.2° and 99.4 mm. Because the distance of moving backward is about 1 m in daily life, the influence of these errors is believed to present no difficulty. These results suggest that the straight driving performance of an electric wheelchair using the proposed tongue interface is sufficient for practical use.

In test E3, maximum X_G and Y_G while turning the wheelchair 360° to the right were, respectively, −8.2 and 12.7 mm. In test E4, maximum X_G and Y_G for turning to the left were, respectively, 1.6 and 2.3 mm. The deviation of the center of gravity position is slight, which suggests good turning performance.

In test E5, the respective variations of the θ, X_G, and Y_G via swallowing saliva while stopping did not exceed −0.6°, −11.3 and 0.0 mm. In test E6, the respective maximum variations of θ and Y_G via swallowing saliva while moving straight with maximum velocity were −0.9° and −5.7 mm. Moreover, the driving velocity was reduced

from 4.0 km/h of maximum velocity to 3.8 km/h. From these results, it was confirmed that the influence of saliva swallowing on wheelchair operation can be inhibited at most to 11.3 mm.

Discussion

The proposed interface, which has simple appearance, can be attached easily and quickly even by a non-experienced caregiver, as depicted in Fig. 2. The prototype tongue interface was extremely lightweight: just 13.6 g. Because the silicon insulation comprises about 60.7 % of the whole sensor mass, further reduction of the sensor mass can be achieved using thinner silicon insulation sheets. Future studies will explore the optimal electrode unit size and the optimal number and location of electrodes for different categories of individuals. Further improvement might include the development of wireless communication between the sensor and the computer.

Tongue motion classification is based on analysis of the EMG activity of the suprahyoid muscles, which contribute not only voluntary tongue motions but also swallowing motion. Therefore, classification must be done of the large number of voluntary tongue motions that might be used for controlling an electric wheelchair and a PC. Such classification is also necessary for the detection of involuntary motions to inhibit malfunctions of such assistive devices. This study achieved classification accuracy of 95.1 ± 1.9 % using SVM classifier with features of time and frequency domains for five voluntary tongue motions and saliva swallowing. The voluntary tongue motions classified in this study were much more numerous than in our preliminary experiments. They are sufficiently numerous and diverse to control an electric wheelchair and a PC. Our future studies will emphasize further improvement of the tongue motion classification accuracy by optimizing the parameters of classification algorithms such as features and the SVM kernel. In addition, effects of the combination of classifiable tongue motions on the classification accuracy will be clarified.

Computer simulations of driving of an electric wheelchair were conducted to investigate the effectiveness of the proposed classification algorithm. The high performance of straight driving was achieved by finding a voluntary tongue motion that is not misclassified as a "right" or "left" tongue motion from the confusion matrix (Table 2) and by matching this motion with the "forward" command of an electric wheelchair (Table 4). Saliva swallowing during wheelchair driving reduces the velocity slightly. Therefore, this malfunction by saliva swallowing affects driving performance only slightly. However, saliva swallowing while the wheelchair is stopping made the wheelchair back up slightly. To provide a safety margin, some improvement of electric wheelchair control methods must be conducted as future work. As described above, electric wheelchair operation based on the proposed tongue interface has been demonstrated. Future studies will be conducted to evaluate the effects of yawning, talking, drinking, tongue motion speed, muscle fatigue, and head motion on the classification accuracy. The effects of small tongue positioning errors on the classification accuracy of the system will also be assessed. The usability of the tongue interface will be evaluated via new experiments using actual electric wheelchairs, PCs, and other assistive devices, and with testing of people with disabilities.

This study tested the design concept of the new interface through experimentation with five healthy adult male subjects. The results were sufficient to verify the viability of the concept, but a new detailed study will be necessary for evaluation of the developed interface when used by different categories of users. Such a new study will specifically examine the acceptance of the new interface by various users.

Conclusions

This study was conducted to develop and test a novel tongue interface based on the classification of tongue motions from surface EMG signals of the suprahyoid muscles detected at the underside of the jaw. The EMG signals of the suprahyoid muscles were measured via 22 active surface electrodes mounted on a special flexible boomerang-shaped base. The tongue motions were classified from RMS features and CC features of the EMG signals using an SVM classifier. Because the developed interface and this approach require no installation of any sensor into the mouth cavity, the system does not hinder the user's other activities such as eating, chewing, and talking. To verify the effectiveness of the tongue interface, an experiment was conducted with five healthy adult male subjects who had normal motor tongue functions. Results showed that the six tongue motions (i.e., five voluntary tongue motions and saliva swallowing) were classified with high accuracy of 95.1 ± 1.9 %. In addition, the potential of the proposed method was evaluated with a test whereby a computer simulation of an electric wheelchair was controlled using tongue commands. Results from the steering test demonstrated that the computer model was controlled precisely. The developed interface elaborates signals of sufficient number for the control of most assistive devices. This device is therefore useful for people with a high degree of movement paralysis. The tongue control interface can be simplified for use by patients with moderate movement disorders.

Authors' contributions

MS was the main person in charge of conception, design of experiments, acquisition of data, analysis and interpretation of data and drafting of the manuscript. KO took part in the experiments and calculations. DS took part in the conception, interpretation and revision of the manuscript. KK took part in the design of the experimental environment. AN and MY participated in the processing of the experimental data using the SVM. GO took part in the conception and interpretation. All authors read and approved the final manuscript.

Author details

[1] Graduate School of Engineering, Iwate University, Morioka, Iwate, Japan. [2] School of Science and Technology, Middlesex University, London, UK. [3] Pattern Art Laboratory Co., Ltd., Hanamaki, Iwate, Japan. [4] Department of Intelligent Systems Engineering, Ichinoseki National College of Technology, Ichinoseki, Iwate, Japan. [5] Graduate School of Information Science, Nara Institute of Science and Technology, Nara, Japan. [6] Department of Robotics Science and Engineering, Chubu University, Kasugai, Japan.

Acknowledgements

This study was supported in part by the Japan Society of Promotion of Science, Japan (Grants-in-Aid for Scientific Research (C) 24500637 and 15K01450).

Competing interests

The authors declare that they have no competing interests.

References

1. Lau C, O'Leary S (1993) Comparison of computer interface devices for persons with severe physical disabilities. Am J Occup Ther 47:1022–1030

2. Ghovanloo M (2007) Tongue operated assistive technologies. In: Proceedings the IEEE 29th engineering medicine biology conference, pp 4376–4379
3. Jouse3, Compusult limited. http://www.jouse.com
4. Clayton C, Platts RGS, Steinberg M, Hennequin JR (1992) Palatal tongue controller. J Microcomput Applicat 15:9–12
5. Terashima SG, Satoh E, Kotake K, Sasaki E, Uekii K, Sasaki S (2010) Development of a mouthpiece type remote controller for disabled persons. J Biomech Sci Eng 5(1):66–77
6. Kim D, Tyler ME, Beebe DJ (2005) Development of a tongue operated switch array as an alternative input device. Int J Hum Comput Interact 18:19–38
7. Wrench A, McIntosh AD, Watson C, Hardcastle WJ (1998) Optopalatograph: real-time feedback of tongue movement in 3D. In: Proceedings the fifth international conference on spoken language processing, pp 1867–1870
8. Saponas TS, Kelly D, Parviz BA, Tan DS (2009) Optically sensing tongue gestures for computer input. In: Proceedings 22nd annual ACM symposium on user interface software and technology, pp 177–180
9. Sonoda Y (1978) Observation of tongue movements employing a magnetometer sensor. IEEE Trans Magn 10:954–957
10. Huo X, Wang J, Ghovanloo M (2008) Introduction and preliminary evaluation of the tongue drive system: wireless tongue-operated assistive technology for people with little or no upper-limb function. J Rehabil Res Dev 45(6):921–930
11. Yousefi B, Huo X, Kim L, Veledar E, Ghovanloo M (2011) Quantitative and comparative assessment of learning in a tongue-operated computer input device: navigation tasks. IEEE Trans Inf Technol Biomed 15(5):747–757
12. Struijk LNSA (2006) An inductive tongue computer interface for control of computers and assistive devices. IEEE Trans Biomed Eng 53:2594–2597
13. Bentsen B, Gaihede M, Lontis R, Andreasen LNS (2014) Medical tongue piercing—development and evaluation of a surgical protocol and the perception of procedural discomfort of the participants. J Neuroeng Rehabil 11(1):1–11
14. Merletti R, Parker PA (eds) (2004) Electromyography: physiology, engineering, and non-invasive applications. Wiley-IEEE Press, New York
15. Hudgins B, Parker PA, Scott RN (1993) A new strategy for multifunction myoelectric control. IEEE Trans Biomed Eng 40(1):82–94
16. Kermani MZ, Wheeler BC, Badie K, Hashemi RM (1995) EMG feature evaluation for movement control of upper extremity prostheses. IEEE Trans Rehabil Eng 2(4):1267–1271
17. Englehart K, Hudgins B, Parker PA (2001) A wavelet-based continuous classification scheme for multifunction myoelectric control. IEEE Trans Biomed Eng 48(3):302–311
18. Ajiboye AB, Weir RF (2005) A heuristic fuzzy logic approach to EMG pattern recognition for multifunctional prosthesis control. IEEE Trans Neural Syst Rehabil Eng 3(3):280–291
19. Chan ADC, Englehart KB (2005) Continuous myoelectric control for powered prostheses using hidden Markov models. IEEE Trans Biomed Eng 52(1):121–124
20. Chu JU, Moon I, Mun MS (2006) A real-time EMG pattern recognition system based on linear-nonlinear feature projection for a multifunction myoelectric hand. IEEE Trans Biomed Eng 53(11):2232–2239
21. Momen K, Krishnan S, Chau T (2007) Real-time classification of forearm electromyographic signals corresponding to user-selected intentional movements for multifunction prosthesis control. IEEE Trans Neural Syst Rehabil Eng 15(4):535–542
22. Naik GR, Kumar DK, Jayadeva (2010) Twin SVM for gesture classification using the surface electromyogram. IEEE Trans Inf Technol Biomed 14(2):301–308
23. Li G, Schultz AE, Luiken TA (2010) Quantifying pattern recognition-based myoelectric control of multifunctional transradial prostheses. IEEE Trans Neural Syst Rehabil Eng 18(2):185–192

24. Li Z, Wang B, Yang C, Xie Q, Su CY (2013) boosting-based EMG patterns classification scheme for robustness enhancement. IEEE J Biomed Health Inform 17(3):545–552
25. Zhang X, Zhou P (2012) High-density myoelectric pattern recognition toward improved stroke rehabilitation. IEEE Trans Biomed Eng 59(6):1649–1657
26. Stango A, Negro F, Farina D (2015) Spatial correlation of high density EMG signals provides features robust to electrode number and shift in pattern recognition for myocontrol. IEEE Trans Neural Syst Rehabil Eng 23(2):189–198
27. Sasaki M, Arakawa T, Nakayama A, Obinata G, Yamaguchi M (2011) Estimation of tongue movement based on suprahyoid muscle activity. In: Proceeding the 2011 IEEE international symposium on micro-nanomechatronics and human science, pp 433–438
28. Sasaki M, Onishi K, Arakawa T, Nakayama A, Stefanov D, Yamaguchi M (2013) Real-time estimation of tongue movement based on suprahyoid muscle activity. In: Proceeding the IEEE 35th engineering medicine biology conference, pp 4605–4608
29. Ide Y, Koide K (eds) (2004) Fundamental of functional anatomy for chairside evaluation of stomatognathic functions. Ishiyaku Publishers, Tokyo
30. Norton NS (2012) Netter's head and neck anatomy for dentistry, 2nd edn. Elsevier, London
31. Pittman LJ, Bailey EF (2009) Genioglossus and intrinsic electromyographic activities in impeded and unimpeded protrusion tasks. J Neurophysiol 101:276–282
32. Tsukada T, Taniguchi H, Ootaki S, Yamada Y, Inoue M (2009) Effects of food texture and head posture on oropharyngeal swallowing. J Appl Physiol 106(6):1848–1857
33. Coriolano MG, Belo LR, Carneiro D, Asano G, Oliveira AL, Silva DM, Lins G (2012) Swallowing in patients with parkinson's disease: a surface electromyography study. Dysphagia 27:550–555
34. Balata PMM, Silva HJ, Nascimento Moraes KJR, Pernambuco LA, Freitas MCR, Lima LM, Braga RS, Souza SR, Moraes SRA (2012) Incomplete swallowing and retracted tongue maneuvers for electromyographic signal normalization of the extrinsic muscles of the larynx. J Voice 26(6):813.e1–813.e7
35. Yoon WL, Khoo JKP, Liow SJR (2014) Chin tuck against resistance (CTAR): new method for enhancing suprahyoid muscle activity using a shaker-type exercise. Dysphagia 29:243–248
36. Carlo JDL (2002) Surface electromyography: detection and recording. DelSys Incorporated, Boston
37. Englehart K, Hudgins B (2003) A robust, real-time control scheme for multifunction myoelectric control. IEEE Trans Biomed Eng 50(7):848–854
38. Zecca M, Micera S, Carrozza MC, Dario P (2002) Control of multifunctional prosthetic hands by processing the electromyographic signal. Crit Rev in Biomed Eng 30(4–6):459–485
39. Yoshikawa M, Mikawa M, Tanaka K (2007) A myoelectric interface for robotic hand control using support vector machine. In: Proceedings the 2007 IEEE/RSJ international conference on intelligent robots and systems, pp 2723–2728
40. Kang WJ, Shiu JR, Cheng CK, Lai JS, Tsao HW, Kuo TS (1995) The application of cepstral coefficients and maximum likelihood method in EMG pattern recognition. IEEE Trans Biomed Eng 42(8):777–785
41. Lee SP, Kim LS, Park SH (1996) An enhanced feature extraction algorithm for EMG pattern classification. IEEE Trans Rehab Eng 4(4):439–443
42. Cortes C, Vapnik (1995) Support-vector networks. Mach Learn 20:273–297
43. Hsu CW, Lin CJ (2002) A comparison of methods for multi-class support vector machines. IEEE Trans Neural Netw 13(2):415–425
44. Chang CC and Lin CJ (2013) LIBSVM—a library for support vector machines. http://www.csie.ntu.edu.tw/~cjlin/libsvm
45. Micera S, Vannozzi G, Sabatini AM, Dario P (2001) Improving detection of muscle activation intervals. IEEE Eng Med Biol Mag 20(6):38–46

13

"Leg-grope walk": strategy for walking on fragile irregular slopes as a quadruped robot by force distribution

Yuichi Ambe[*] and Fumitoshi Matsuno

Abstract

Problems can often occur when a legged robot attempts to walk on irregular or damaged terrain, such as in search and rescue missions during natural and man-made disasters. In some cases, the ground beneath the robot will collapse because of the pressure of its weight, causing the machine to lose its foothold and topple over. This is a point to which we as designers must pay careful attention when designing a robot. Thus, in such irregular areas, the robot should walk carefully so as not to collapse its footholds. To attempt to solve this problem, we proposed the "leg-grope walk" method which allows a quadruped robot to avoid stumbling or causing a large collapse of the surrounding area on weak horizontal planes. Specifically, when the robot puts its foot on the ground, it applies some excess force on the ground and confirms whether the foothold is likely to collapse, so as to choose a foothold will not collapse. In this study, we extended this method to weak and irregular slopes, where slippage needs to be considered. A new walking method was designed using a force distribution method. To validate the method, we show simulation results from force distribution and robotic experiments in various environments. These results demonstrate that our method allows a robot to walk carefully without slipping or stumbling, even when its foothold is lost.

Keywords: Quadruped robot, Irregular terrain, Fragile environment, Force distribution, Walking strategy

Background

Search and rescue workers face a dangerous and difficult task when they attempt to rescue survivors after a disaster, because they are at risk of getting caught in a secondary disaster. Despite this, they must search quickly because the survival rate drastically drops over time. This is why recently many organizations have begun to use robots in search and rescue missions to decrease the risk to human life. The terrain in rescue scenarios is often very rough, giving legged robots an advantage over wheeled and tracked vehicles. That advantage comes from legged robots' redundancy; therefore, we focused our research on this type of robot.

To walk competently on irregular terrain, stability is a key issue for quadruped robots. The first research on quadruped robots focused on static walking, where the

center of the gravity (COG) is always in the supporting leg polygon [1]. Hirose et al. [2] built a series of quadruped robots (TITAN) that could stably climb up a set of stairs. A stability criterion, the Normalized Energy Stability Margin, was proposed to evaluate the stability of walking [3]. A walking gait with a large stability margin was also proposed [4]. Estremera and Santos proposed a free gait, which allows the quadruped robot (SILO4) to have a statically stable gait by searching for optimal footholds [5, 6]. Many researchers have also suggested the force distribution method to prevent slippage on irregular terrains in simulations [7–10]. Currently, there is a real robot that can avoid slippage by distributing contact forces optimally using joint torque control [11].

However, to walk on irregular terrain continuously, it is also important to generate the path where the robot is to walk, as well as footholds based on geometric information. Path planning on irregular terrain has been much improved through the Learning Locomotion program conducted by the Defense Advanced Research Project

---
*Correspondence: amby.yu@gmail.com
Department of Mechanical Engineering and Science, Kyoto University, Katsura, Nishikyo-ku, Kyoto, Japan

Agency (DARPA). In that project, several researchers showed that a quadruped robot, LittleDog [12], could climb over rough terrain by searching for optimal footholds if the geometrical information about the environment and the robot position were known [13–16]. A team at the Florida Institute for Human and Machine Cognition has proposed many algorithms, such as a fast foothold planning method and a new parametrized gait generator, which can generate static and dynamic walking [14]. Similarly, a team from the University of Southern California proposed the terrain template concept to teach the robot what consists of suitable terrain for footholds [15]. Finally, the Stanford LittleDog has many learning algorithms installed, focusing on recovery and stabilization methods to combat problems such as unexpected slippage [16].

It is important to obtain further information about the environment, including the relevant geometric information, to achieve stable walking. Some researchers have focused on terrain classification based on haptic feedback [17–20]. Hoepfinger et al. [17] estimated surface friction by applying forces on the foothold. This haptic feedback is associated with the foothold shape and can be used to estimate the friction of an untouched foothold using geometrical data. Tokuda et al. [19] proposed a method to estimate fragile footholds using the foot's center of force and pressure changes. Although their quadruped robot could detect when a foothold was collapsing, they did not propose how to make the robot walk on fragile terrain without stumbling.

Thus, in this study, we propose a stable walking method for fragile irregular terrain. We focus on how to detect fragile footholds with haptic information, and how to walk stably using this information. We do not focus on the path planning algorithm, because this is not one of our main aims.

Previously, we proposed a walking method named the "leg-grope walk" method, and discussed the validity of this strategy based on our experiments on a fragile horizontal plane [21]. According to this method, when the robot puts its foot on the ground, it applies some excess force and confirms whether the foothold is stable, and then chooses a foothold that does not collapse. In addition, the robot walks slowly so as not to apply force over probed reaction, avoiding foothold collapse. This algorithm allowed the robot to walk safely while avoiding stumbling on horizontal planes.

In this paper, the environment is extended to an irregular slope, where slippage must be considered. Hence, in the proposed strategy, tip-point forces in the x-y-z directions are distributed using a standard Quadratic Programing method such that the friction and leg-grope constraints (explained later) are satisfied. The simulation results of the force distribution on various terrains are shown to evaluate the validity of the method. We also carried out walking experiments with the robot, not only on a slope but also on irregular terrain, to evaluate the validity of our method. Our results indicate the validity of the leg-grope walk method. This paper is the extension of our published conference paper [22], and extend our previous findings to include: (1) the simulation results of the force distribution; (2) walking experiments with the robot; and (3) a detailed explanation of the method.

Methods
Quadruped robot and model

The developed robot (Fig. 1a) consists of a body and four legs, each of which has three active joints with servomotors. A three-axis force sensor is installed on each toe to sense a resultant force vector. An attitude sensor and an accelerometer are equipped on the center of the robot body. The parameters of the robot are presented in Table 1.

Figure 1b, c shows the leg and the front view of the quadruped model of the robot. The body and the links of the legs are rigid. We name the legs of the robot L_1, L_2, L_3, and L_4, starting clockwise from the left front leg. Each leg i has three links and joints, and we name them Links $i1$, $i2$ and $i3$ and Joints $i1$, $i2$ and $i3$ starting from the root of leg. Joint $i1$ of Leg i is a yaw joint that allows the leg to move from back to front. Joints $i2$ and $i3$ are pitch joints that allow the leg to be lifted up and down. The coordinate frames and variables of the robot are described as follows (see also Fig. 1b, c).

Σ_G:	$O_G - x_G y_G z_G$. A base coordinate frame fixed at the environment. z_G axis: opposite direction of gravity.
Σ_R:	$O_R - x_R y_R z_R$. A robot coordinate frame fixed at the center of the robot body. z_R axis: vertical direction of the robot. x_R axis: forward direction of the robot.
Σ_{iS}:	$O_{iS} - x_{iS} y_{iS} z_{iS}$. A contact coordinate frame fixed at the contact point of L_i. z_{iS} axis: direction of normal reaction. x_{iS} axis: direction of gradient of the contact plane.
M:	Total mass of the robot
g:	Gravitational acceleration
θ_i:	Angle between z_G and z_{iS} axis (i.e. angle of gradient of the slope where L_i contacts)
r_R:	Position vector of the origin of Σ_R
$\phi_{r,p,y}$:	roll ϕ_r, pitch ϕ_p and yaw ϕ_y angles of the robot
q_B:	$= [r_B^T \ \phi_r \ \phi_p \ \phi_y]^T \in R^{6 \times 1}$
θ_{ij}:	Angle of the Joint ij
q_{Li}:	$= [\theta_{i1} \ \theta_{i2} \ \theta_{i3}]^T \in R^{3 \times 1}$

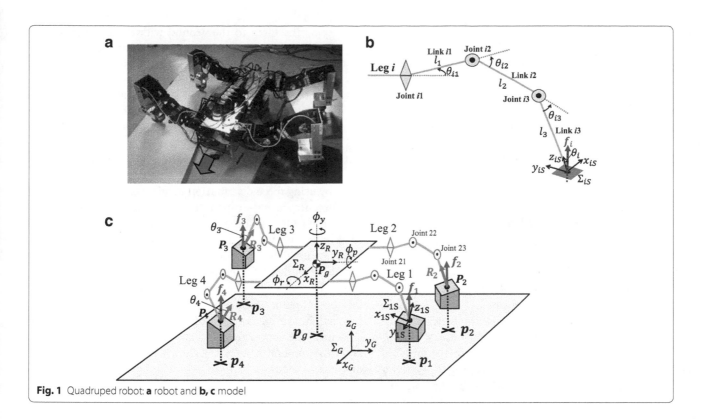

Fig. 1 Quadruped robot: **a** robot and **b, c** model

Table 1 The parameters of the robot

Parameters (m)	Value	Parameters (kg)	Value
Body width	0.15	Body mass	4.54
Body length	0.29	Link1 mass	0.03
Link1 length	0.072	Link2 mass	0.35
Link2 length	0.109	Link3 mass	0.25
Link3 length	0.172	Whole mass	7.06

τ_{ij} :   Torque that is input to the joint j of L_i

τ_i : $= [\tau_{i1}\ \tau_{i2}\ \tau_{i3}]^T \in R^{3\times1}$

f_i :   Resultant force vector of L_i applied by ground

R_i :   Normal reaction vector of the leg L_i

P_g :   Position vector of the COG of the robot

P_i :   Position vector of the contact point of L_i

$p_g \in R^2$:   Vector projected P_g on $O_G - x_G y_G$ plane

$p_i \in R^2$:   Vector projected P_i on $O_G - x_G y_G$ plane Unless otherwise noted, the vectors are defined in the base coordinate frame Σ_G.

Strategy of leg-grope walk

In this section, we describe the strategy of the leg-grope walk as described in [22]. First, we define the type of fragile irregular terrain used in this study; next, we outline the basic strategy of the leg-grope walk; and finally, we

explain the consequent one-cycle walking movement for a leg.

Definition of fragile terrain

For the purpose of this study, we used a fragile and uneven environment for the target area in which our legged robot walks. This environment is defined as to be like an area with scattered debris and collapsed buildings, on which surfaces may collapse when put under external forces such as the pressure from a robot's leg. We define the threshold of normal reaction as $R_{break} \in R^1(> 0)$ to an area of the environment, and assume that this area collapses if the external normal force is over R_{break}. When a robot moves on such areas, it is necessary for it to find strong footholds so as to avoid stumbling and falling. To check for fragile areas, the robot applies some excess force to the environment to confirm whether it collapses or not. A dangerous foothold for robot locomotion is defined as a region that satisfies $R_{break} \leq R_{max}$, where $R_{max} \in R^1(> 0)$ is the maximum value of the normal reaction that is applied to all legs during one walking cycle, except for the leg-grope movement, which will be explained in the next section Walking methods. For simplicity, we assume that the contact area of any leg is a point and that the contact point of any leg is on a smooth surface where a normal reaction can be defined. In addition, we assume that the robot has a geometrical 3D map of the environment.

Walking method

The following two walking strategies are proposed to achieve safe locomotion for a legged robot on fragile terrain.

(1) Examine whether a foothold candidate, which a robot will use for its locomotion, can stand up to a certain value of external force $R_{ref} \in R^1$. In addition, it must be guaranteed that the robot will not fall down even if the foothold collapses.

(2) Satisfy the condition that the maximum normal reaction for all legs R_{max} needed for walking is less than R_{ref} set in the walking strategy 1 while the robot walks on fragile terrain.

In particular, we call walking strategy 1 a leg-grope movement. By using this movement, the robot can distinguish a safe region for its locomotion.

The leg-grope movement is an action by which a robot checks whether a targeted region will collapse, statically; that is, the robot applies force gradually to the targeted region until the normal reaction of a groping leg is over a given value R_{ref} ("grope-reaction") when standing on four legs. If the targeted region collapses in this movement, the robot can remain standing on the other three legs without falling down. When the robot is performing the leg-grope movement, we let the movement of the COG of the robot be negligible for a simple formulation.

The following relation is satisfied if the targeted region does not collapse during the leg-grope movement.

$$R_{ref} < R_{break}. \tag{1}$$

In addition, if walking strategy 2 is satisfied, the robot can walk without causing those footholds that have been already probed to collapse.

Leg-grope walk

On the basis of the above leg-grope movement, the concrete one-leg cycle walking strategy (leg-grope walk) of a quadruped robot is explained in four steps (Fig. 2). Figure 2a represents the status of the robot in following Steps A–D in the case of groping using the right front leg. Figure 2b represents the time response of the normal reaction of the groping leg in each step.

A Move the COG of the robot standing on four legs.
B Reduce the force of the groping leg to 0 gradually without any movement.
C Swing the groping leg to the point of the leg-grope and make the leg touch down.

D Apply the force to the ground with the groping leg gradually, up to R_{ref} (leg-grope movement) with a movement small enough to ignore the movement of the COG. Even if the foothold collapses during this step, the robot can still keep its pose stable by standing on the other three legs. Thus, the robot can repeat this procedure from Step C to find a stable foothold.

It is guaranteed that the robot will not slip or apply normal force over the grope-reaction R_{ref} to the environment by using force distribution in all steps.

To execute the leg-grope walk, the admissible region to which the COG can be moved in Step A and the admissible region on which the groping leg can be placed in Step D should be considered. Furthermore, the way to distribute optimal forces of the legs should be formulated. The geometrical regions of the COG's position and the contact point of the groping leg are shown in the next subsection, the formulation of force distribution is then shown in the following subsection, and the simulation and experimental results are shown in the "Results and discussion" section.

Geometrical relation of leg-grope

In this section, an admissible region of the position of the COG and that of the contact point of groping leg are derived. For easy derivation, we assume that the force vector f_i is parallel to the direction of gravity; in other words, the friction force is determined uniquely. We only consider

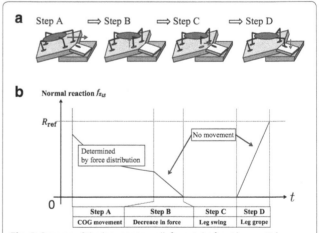

Fig. 2 Process of the leg-grope walk for a right front leg. **a** stick figures of the robot and **b** time response of normal reaction of the right front leg. Step A: the robot moves COG standing on four legs. Step B: the robot reduces the force of the groping leg without any movement. Step C: the robot swings the groping leg to the point of the leg-grope. Step D: the robot applies the force to the ground gradually up to R_{ref}

the static equilibrium because the leg-grope is carried out without any movement as in Fig. 2. First, we show the equilibrium of force and moment of the robot system, and next, we show the admissible geometrical regions of the position of the COG and the contact point for the groping leg.

Equilibrium of force and moment

Under the above assumptions, in the case where three legs L_i, L_j, L_k are on the ground (which we represent as $\Delta(L_i, L_j, L_k)$), the equilibrium of force and moment of the robot is written as

$$1 - h_i - h_j - h_k = 0,$$
$$\boldsymbol{p}_g - h_i \boldsymbol{p}_i - h_j \boldsymbol{p}_j - h_k \boldsymbol{p}_k = 0, \qquad (2)$$

where $h_n = |\boldsymbol{f}_n|/Mg$ ($n = 1, 2, 3, 4$). Note that these equations consist of the projected vectors and h_n. The relationship between $R_n = |\boldsymbol{R}_n|$ (magnitude of normal reaction of leg L_n) and $f_n = |\boldsymbol{f}_n|$ (magnitude of force which the robot applies) is described as follows, because of the assumption about friction:

$$R_n = f_n \cos \theta_n. \qquad (3)$$

Hence, the confirmation of condition (whether the foothold collapsed or not) by applying normal force to the targeted area up to R_{ref}, is the same as the confirmation by applying vertical force to the targeted area up to $f^n_{\text{ref}} = R_{\text{ref}}/\cos \theta_n$ ("grope-force"). We need to select R_{ref} to fulfill the inequality $Mg/3 \leq f^n_{\text{ref}} \leq Mg$. When the robot stands statically on three legs, the largest magnitude of force f_i on those three legs is larger than $Mg / 3$. Hence, the lower bound of R_{ref} should be $Mg / 3$ to satisfy walking strategy 2. The upper bound means that the maximum force that a robot can apply statically in the leg-grope movement should be Mg.

Admissible region of COG and contact point of groping leg

To employ the walking method, we need to determine the position to which the robot moves its COG in Step A of Fig. 2, and also determine the position where the groping leg can apply force to the ground in Step D of Fig. 2.

The admissible region of the position of the COG is determined such that the magnitude of the vertical force of each leg does not exceed that of the grope-force when the robot stands on three legs (Step B of Fig. 2). The admissible region of the contact point for the groping leg is determined such that the magnitude of the vertical force of the groping leg larger than that of the grope-force, and the magnitude of the vertical force for the other three legs less than that of the grope-force. In fact, the vertical force of one of the three legs is assumed to be zero (we call this leg the "float leg"), because the groping leg can apply the maximum force when one of the other legs is floating.

Let us consider the state $\Delta(L_i, L_j, L_k)$ in Step A of Fig. 2, and the residual leg is described as the groping leg L_{grp}. Let leg L_k be the float leg in Step C of Fig. 2 without loss of generality. With this situation, we calculate the admissible region of the position of the COG and that of the contact point for the groping leg on $O_G - x_G y_G$ plane.

Admissible region of COG In the state $\Delta(L_i, L_j, L_k)$, the admissible region of the position of the COG $\pi_g(L_i, L_j, L_k)$ is calculated as follows.

The magnitude of the vertical force of each leg f_n needs to be less than the grope-force f^n_{ref} and this condition is represented as

$$\begin{cases} 0 < h_i \leq h^i_{\text{ref}} \\ 0 < h_j \leq h^j_{\text{ref}} \\ 0 < h_k \leq h^k_{\text{ref}}, \end{cases} \qquad (4)$$

where $h^n_{\text{ref}} \equiv f^n_{\text{ref}}/Mg$. Using these constraints (Eq. 4) and Eq. 2, the projected position of the COG can be represented as follows.

$$\begin{cases} 0 < h_i \leq h^i_{\text{ref}} \\ 0 < h_j \leq h^j_{\text{ref}} \\ 0 < 1 - h_i - h_j \leq h^k_{\text{ref}}, \end{cases} \qquad (5)$$

$$\boldsymbol{p}_g = \{h_i \boldsymbol{p}_i + (1 - h_i)\boldsymbol{p}_k\} + h_j(\boldsymbol{p}_j - \boldsymbol{p}_k). \qquad (6)$$

By changing the parameters h_i and h_j under the constraints (Eq. 5), an admissible region of the COG $\pi_g(L_i, L_j, L_k)$ can be calculated based on Eq. 6. The region $\pi_g(L_i, L_j, L_k)$ is classified into eight geometrical patterns (Fig. 3a) under the relations of variables $h^i_{\text{ref}}, h^j_{\text{ref}}$ and h^k_{ref} (see Table 2). We also represent the region $\pi_g(L_i, L_j, L_k)$ as the gray triangle in Fig. 4a for a specific example ($h^i_{\text{ref}} = h^j_{\text{ref}} = h^k_{\text{ref}} = 1/2$). This example is a special case of **a**-(1) in Table 2, where all conditions satisfy the equality.

Admissible region of groping leg for fixed COG with a particular float leg The region $\pi_{\text{grp},g}(L_i, L_j, L_{\text{grp}})$, which is the admissible region of the contact point of the groping leg for a fixed COG, is calculated as follows.

Because the groping leg L_{grp} can apply the maximum force when one of the legs is floating, we consider leg L_k as the float leg in the leg-grope movement. Then, we consider the state $\Delta(L_i, L_j, L_{\text{grp}})$. From Eq. 2, the equilibrium of force and moment is represented as

$$1 - \hat{h}_i - \hat{h}_j - h_{\text{grp}} = 0,$$
$$\boldsymbol{p}_g - \hat{h}_i \boldsymbol{p}_i - \hat{h}_j \boldsymbol{p}_j - h_{\text{grp}} \boldsymbol{p}_{\text{grp}} = 0, \qquad (7)$$

where the variables of Eq. 7 are distinguished from the ones used before by using a hat "$\hat{}$". The conditions where the magnitude of the vertical force of the groping

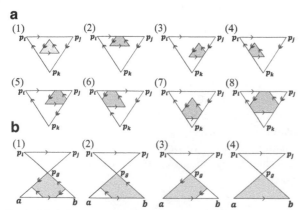

Fig. 3 Admissible region patterns. **a** The region for the COG when three legs (position $\boldsymbol{p}_{i,j,k}$) are on the ground and **b** The region for the groping leg when the COG (position $\boldsymbol{p}_g$) is fixed on $O_G - x_G y_G$ plane. Admissible regions are colored *gray*. These patterns are classified depending on the relation of variables h^i_{ref}, h^j_{ref}, h^k_{ref} and $h^{\mathrm{grp}}_{\mathrm{ref}}$ as shown in Table 2. On figure **b**, $\triangle \boldsymbol{p}_g \boldsymbol{p}_i \boldsymbol{p}_j$ and $\triangle \boldsymbol{p}_g \boldsymbol{b} \boldsymbol{a}$ are similar and the relation is $(1 - h^{\mathrm{grp}}_{\mathrm{ref}})|\boldsymbol{p}_g - \boldsymbol{p}_j| = h^{\mathrm{grp}}_{\mathrm{ref}}|\boldsymbol{p}_g - \boldsymbol{a}|$

Table 2 The relations of variables h^i_{ref}, h^j_{ref}, h^k_{ref} and $h^{\mathrm{grp}}_{\mathrm{ref}}$ in Fig. 3

Number	Conditions
a-(1)	$h^i_{\mathrm{ref}} + h^j_{\mathrm{ref}} \leq 1,\ h^j_{\mathrm{ref}} + h^k_{\mathrm{ref}} \leq 1,\ h^k_{\mathrm{ref}} + h^i_{\mathrm{ref}} \leq 1$
a-(2)	$h^i_{\mathrm{ref}} + h^j_{\mathrm{ref}} > 1,\ h^j_{\mathrm{ref}} + h^k_{\mathrm{ref}} \leq 1,\ h^k_{\mathrm{ref}} + h^i_{\mathrm{ref}} \leq 1$
a-(3)	$h^i_{\mathrm{ref}} + h^j_{\mathrm{ref}} \leq 1,\ h^j_{\mathrm{ref}} + h^k_{\mathrm{ref}} > 1,\ h^k_{\mathrm{ref}} + h^i_{\mathrm{ref}} \leq 1$
a-(4)	$h^i_{\mathrm{ref}} + h^j_{\mathrm{ref}} \leq 1,\ h^j_{\mathrm{ref}} + h^k_{\mathrm{ref}} \leq 1,\ h^k_{\mathrm{ref}} + h^i_{\mathrm{ref}} > 1$
a-(5)	$h^i_{\mathrm{ref}} + h^j_{\mathrm{ref}} > 1,\ h^j_{\mathrm{ref}} + h^k_{\mathrm{ref}} > 1,\ h^k_{\mathrm{ref}} + h^i_{\mathrm{ref}} \leq 1$
a-(6)	$h^i_{\mathrm{ref}} + h^j_{\mathrm{ref}} > 1,\ h^j_{\mathrm{ref}} + h^k_{\mathrm{ref}} \leq 1,\ h^k_{\mathrm{ref}} + h^i_{\mathrm{ref}} > 1$
a-(7)	$h^i_{\mathrm{ref}} + h^j_{\mathrm{ref}} \leq 1,\ h^j_{\mathrm{ref}} + h^k_{\mathrm{ref}} > 1,\ h^k_{\mathrm{ref}} + h^i_{\mathrm{ref}} > 1$
a-(8)	$h^i_{\mathrm{ref}} + h^j_{\mathrm{ref}} > 1,\ h^j_{\mathrm{ref}} + h^k_{\mathrm{ref}} > 1,\ h^k_{\mathrm{ref}} + h^i_{\mathrm{ref}} > 1$
b-(1)	$h^i_{\mathrm{ref}} + h^{\mathrm{grp}}_{\mathrm{ref}} < 1,\ h^j_{\mathrm{ref}} + h^{\mathrm{grp}}_{\mathrm{ref}} < 1$
b-(2)	$h^i_{\mathrm{ref}} + h^{\mathrm{grp}}_{\mathrm{ref}} \geq 1,\ h^j_{\mathrm{ref}} + h^{\mathrm{grp}}_{\mathrm{ref}} < 1$
b-(3)	$h^i_{\mathrm{ref}} + h^{\mathrm{grp}}_{\mathrm{ref}} < 1,\ h^j_{\mathrm{ref}} + h^{\mathrm{grp}}_{\mathrm{ref}} \geq 1$
b-(4)	$h^i_{\mathrm{ref}} + h^{\mathrm{grp}}_{\mathrm{ref}} \geq 1,\ h^j_{\mathrm{ref}} + h^{\mathrm{grp}}_{\mathrm{ref}} \geq 1$

leg f_{grp} is larger than that of the grope-force $f^{\mathrm{grp}}_{\mathrm{ref}}$, and the magnitudes of the vertical forces of the other legs are less than those of the grope-force f^i_{ref} and f^j_{ref}, are described as

$$
\begin{cases}
h^{\mathrm{grp}}_{\mathrm{ref}} \leq h_{\mathrm{grp}} < 1 \\
0 < \hat{h}_i \leq h^i_{\mathrm{ref}} \\
0 < \hat{h}_j \leq h^j_{\mathrm{ref}},
\end{cases}
\tag{8}
$$

where $h^{\mathrm{grp}}_{\mathrm{ref}} = f^{\mathrm{grp}}_{\mathrm{ref}}/Mg$. Using Eqs. 7 and 8 yields

$$
\boldsymbol{p}_{\mathrm{grp}} = \boldsymbol{p}_g + \frac{\hat{h}_i}{1 - \hat{h}_i - \hat{h}_j}(\boldsymbol{p}_g - \boldsymbol{p}_i) + \frac{\hat{h}_j}{1 - \hat{h}_i - \hat{h}_j}(\boldsymbol{p}_g - \boldsymbol{p}_j),
\tag{9}
$$

$$
\begin{cases}
0 < \dfrac{\hat{h}_i}{1 - \hat{h}_i - \hat{h}_j} \leq \dfrac{h^i_{\mathrm{ref}}}{h^{\mathrm{grp}}_{\mathrm{ref}}} \\[2mm]
0 < \dfrac{\hat{h}_j}{1 - \hat{h}_i - \hat{h}_j} \leq \dfrac{h^j_{\mathrm{ref}}}{h^{\mathrm{grp}}_{\mathrm{ref}}}.
\end{cases}
\tag{10}
$$

The region $\pi_{\mathrm{grp},g}(L_i, L_j, L_{\mathrm{grp}})$, which is represented by Eqs. 9 and 10, is classified into four geometrical patterns (Fig. 3b) under the relations of the variables h^i_{ref}, h^j_{ref} and $h^{\mathrm{grp}}_{\mathrm{ref}}$ (see Table 2). Based on the specific example, as shown in Fig. 4a, we can represent the region $\pi_{\mathrm{grp},g}(L_i, L_j, L_{\mathrm{grp}})$ as the *dark gray triangle* in Fig. 4b for a fixed $\boldsymbol{p}_g$ represented in Fig. 4b as an example.

Admissible region of the groping leg for all admissible COG positions with a particular float leg Since leg L_k is the float leg, the admissible region $\pi_{\mathrm{grp}}(L_i, L_j, L_{\mathrm{grp}})$ of the contact point for the groping leg considering an admissible region of COG is calculated as follows. This region

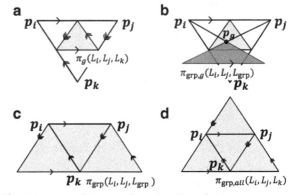

Fig. 4 Process to determine the region where the groping leg can be set. This figure shows the process in the case $(h^i_{\mathrm{ref}} = h^j_{\mathrm{ref}} = h^k_{\mathrm{ref}} = h^{\mathrm{grp}}_{\mathrm{ref}} = 1/2)$ on $O_G - x_G y_G$ plane. In figure **a**, when three legs (position $\boldsymbol{p}_{i,j,k}$) are on the ground, the admissible region of COG can be calculated as in the *gray triangle*, where each top point of the *gray triangle* is the middle point of the side of $\triangle \boldsymbol{p}_i \boldsymbol{p}_j \boldsymbol{p}_k$. Then, in figure **b**, the admissible region of the groping leg for fixed COG (position $\boldsymbol{p}_g$) and a particular float leg (L_k) can be calculated as in the *dark gray triangle*, where the *dark gray triangle* and $\triangle \boldsymbol{p}_i \boldsymbol{p}_j \boldsymbol{p}_g$ are congruent. In figure **c**, the admissible region of the groping leg for all admissible COG positions (*grey triangle* in figure **a** for a particular float leg (L_k) can be calculated as in the *gray trapezoid*. Finally, in figure **d**, by repeating the same procedure for the other float legs ($L_{i,j}$), the admissible region of groping leg can be calculated as in the *gray triangle*

is obtained as the union of the regions $\pi_{\mathrm{grp},g}(L_i, L_j, L_{\mathrm{grp}})$ for all $\boldsymbol{p}_g$ in $\pi_g(L_i, L_j, L_k)$. The region π_{grp} is represented as the *gray trapezoid* in Fig. 4c for the specific example related to Fig. 4a.

Admissible region of groping leg The case where leg L_k is assumed to be a float leg in the leg-grope movement was explained above. Here, the same process is done for legs L_i and L_j. The region $\pi_{\mathrm{grp,all}}(L_i, L_j, L_k)$, which is the whole admissible region of the contact point for the groping leg, is obtained as the union of the regions π_{grp} of the three potential float legs together. The region $\pi_{\mathrm{grp,all}}$ is represented as the *gray triangle* in Fig. 4d for the specific example related to Fig. 4a.

As we explained above, the robot can place the groping leg on the region $\pi_{\mathrm{grp,all}}$, and the position of the COG in π_g should be chosen to realize the desired position for the groping leg. Practically, we can locate the positions of the COG and the groping leg inside of the regions (i.e., apart from the boundaries) to tolerate modeling errors and the COG shift in the leg-grope movement.

Force distribution problem

The geometrical relations were calculated to conduct the leg-grope movement. Here, the force distribution method is proposed based on these relations, and guarantees slippage avoidance.

Robot dynamics

The respective dynamic equations of the robot body and its legs are represented as follows.

$$M_B(\boldsymbol{q})\ddot{\boldsymbol{q}} + \boldsymbol{h}_B(\boldsymbol{q},\dot{\boldsymbol{q}}) + \boldsymbol{g}_B(\boldsymbol{q}) + J_B \boldsymbol{f} = \boldsymbol{0} \in R^{6 \times 1}, \quad (11)$$

$$M_L(\boldsymbol{q})\ddot{\boldsymbol{q}} + \boldsymbol{h}_L(\boldsymbol{q},\dot{\boldsymbol{q}}) + \boldsymbol{g}_L(\boldsymbol{q}) + J_L \boldsymbol{f} = \boldsymbol{\tau} \in R^{12 \times 1}. \quad (12)$$

$\boldsymbol{q}$: $= [\boldsymbol{q}_B^T \ \boldsymbol{q}_{L1}^T \ \boldsymbol{q}_{L2}^T \ \boldsymbol{q}_{L3}^T \ \boldsymbol{q}_{L4}^T]^T \in R^{18 \times 1}$;

$\boldsymbol{f}$: $= [\boldsymbol{f}_1^T \ \boldsymbol{f}_2^T \ \boldsymbol{f}_3^T \ \boldsymbol{f}_4^T]^T \in R^{12 \times 1}$;

$\boldsymbol{\tau}$: $= [\boldsymbol{\tau}_1^T \ \boldsymbol{\tau}_2^T \ \boldsymbol{\tau}_3^T \ \boldsymbol{\tau}_4^T]^T \in R^{12 \times 1}$;

$M_B(\boldsymbol{q})$: Inertia matrix of the body [6 × 18];

$M_L(\boldsymbol{q})$: Inertia matrix of the legs [12 × 18];

$\boldsymbol{h}_B(\boldsymbol{q},\dot{\boldsymbol{q}})$: vector defining centrifugal and Coriolis effects of the body [6 × 1];

$\boldsymbol{h}_L(\boldsymbol{q},\dot{\boldsymbol{q}})$: vector defining centrifugal and Coriolis effects of the legs [12 × 1];

$\boldsymbol{g}_B(\boldsymbol{q})$: vector of the gravity terms of the body [6 × 1];

$\boldsymbol{g}_L(\boldsymbol{q})$: vector of the gravity terms of the legs [12 × 1];

J_B: Jacobian matrix of the body [6 × 12].;

J_L: Jacobian matrix of the legs [12 × 12]

Unless the leg is in the singular configuration ($\theta_{i3} = n\pi$ (where n is an integer)), J_L is a non-singular matrix. Let the kinematic motion be designed to avoid the singular condition and to fulfill the geometrical relation

of leg-grope (that is, the $(\boldsymbol{q}, \dot{\boldsymbol{q}}, \ddot{\boldsymbol{q}})$ are given at each time step); the above equations can be represented as follows.

$$\boldsymbol{b} = A\boldsymbol{\tau}, \quad (13)$$

$$\boldsymbol{f} = J_L^{-1}(\boldsymbol{\tau} - \boldsymbol{\tau}_o), \quad (14)$$

where $\boldsymbol{b} \in R^{6 \times 1}$, $A \in R^{6 \times 12}$ and $\boldsymbol{\tau}_o \in R^{12 \times 1}$ are calculated from (Eqs.11 and 12) with designed $(\boldsymbol{q}, \dot{\boldsymbol{q}}, \ddot{\boldsymbol{q}})$ (see Additional file 1: Appendix S1 for detail). The vector $\boldsymbol{\tau}$ consists of 12 components and fulfills six linear equality constraints (Eq. 13) (which consist of the equilibrium of force and moment). $\boldsymbol{\tau}$ has a one-to-one relation with the force vector $\boldsymbol{f}$ as Eq. 14. Hence, at each time step, we need to determine the optimal vector $\boldsymbol{\tau}$ that fulfils the six equality constraints (Eq. 13), avoids slippage, and also fulfills the constraints for the leg-grope.

To date, various methods for force distribution problems have been proposed. For example, methods based on a pseudo inverse matrix method [9], a linear programming method (LP method) [7], and a quadratic programming method (QP method) [23, 24] were proposed. In this study, the standard QP method is applied to consider inequality constraints and a quadratic evaluation function as follows.

Constraints

Slippage avoidance For a leg L_i that stands on the ground, a normal force must satisfy the following inequality to assure definite foot contact:

$$^{iS}f_{iz} \geq 0, \quad (15)$$

and horizontal force elements also need to satisfy the following inequality constraints for preventing slippage:

$$\sqrt{(^{iS}f_{ix})^2 + (^{iS}f_{iy})^2} \leq \mu \, |^{iS}f_{iz}|, \quad (16)$$

where μ is the coefficient of static friction, and $(^{iS}f_{ix}, {}^{iS}f_{iy}, {}^{iS}f_{iz})$ are the components of $\boldsymbol{f}_i$ on the contact coordinate frame Σ_{iS}. To apply the QP method, Eq. 16 is changed to linear inequality constraints that are tighter than the original one as follows.

$$-\frac{\mu}{\sqrt{2}} {}^{iS}f_{iz} - \frac{1}{\sqrt{2}} {}^{iS}f_{ix} - \frac{1}{\sqrt{2}} {}^{iS}f_{iy} \leq -s,$$
$$-\frac{\mu}{\sqrt{2}} {}^{iS}f_{iz} + \frac{1}{\sqrt{2}} {}^{iS}f_{ix} + \frac{1}{\sqrt{2}} {}^{iS}f_{iy} \leq -s, \quad (17)$$

$$-\frac{\mu}{\sqrt{2}} {}^{iS}f_{iz} + \frac{1}{\sqrt{2}} {}^{iS}f_{ix} - \frac{1}{\sqrt{2}} {}^{iS}f_{iy} \leq -s,$$
$$-\frac{\mu}{\sqrt{2}} {}^{iS}f_{iz} - \frac{1}{\sqrt{2}} {}^{iS}f_{ix} + \frac{1}{\sqrt{2}} {}^{iS}f_{iy} \leq -s, \quad (18)$$

where $s \geq 0$ is defined as the safety margin on the friction constraints, and represents the minimum safety margin within the friction constraints pyramid. Hence, by maximizing s, slippage avoidance may be further enhanced.

Constraints for leg-grope To ensure that the magnitude of the normal force is less than R_{ref}, the following inequality must be satisfied for each stance leg L_i.

$$^{iS}f_{iz} \leq R_{ref}. \tag{19}$$

In addition, in Step B of the leg-grope walk, $^{kS}f_{kz}$ for the groping leg L_k is constrained to decrease linearly to zero as shown in Fig. 2b. In Step D of the leg-grope walk, $^{kS}f_{kz}$ for the groping leg L_k is constrained to increase linearly to R_{ref} as shown in Fig. 2b. We derive the geometrical relations by assuming that the force vector $\boldsymbol{f}_i$ is parallel to the direction of gravity. Hence, the normal reaction of the groping leg can be distributed to be R_{ref} by making the force vector $\boldsymbol{f}_i$ parallel to the direction of gravity.

Minimization problem

Adding the safety margin s to the primary variable $\boldsymbol{\tau}$, the QP formulation is represented as follows for each Step i ($i =$A, B, C and D) of the leg-grope walk.

$$\hat{\boldsymbol{\tau}} = \begin{bmatrix} \boldsymbol{\tau} \\ s \end{bmatrix}_{13 \times 1}, \quad \hat{\boldsymbol{b}}_i = \hat{A}_i \hat{\boldsymbol{\tau}}, \quad \hat{G}_i \hat{\boldsymbol{\tau}} \leq \hat{\boldsymbol{d}}_i, \tag{20}$$

where $\hat{A}_i$ and $\hat{\boldsymbol{b}}_i$ represent the equality constraints of Eq. 13 and the leg-grope, $\hat{G}_i$ and $\hat{\boldsymbol{d}}_i$ represent the inequality constraints (see Additioanl file 1: Appendix S1 for detail). These matrixes and vectors are determined by designed kinematic motion ($\boldsymbol{q}$, $\dot{\boldsymbol{q}}$, $\ddot{\boldsymbol{q}}$) at each time step.

The minimized evaluation function is composed of three terms.

$$\Phi(\hat{\boldsymbol{\tau}}) = C\hat{\boldsymbol{\tau}} + \frac{1}{2}\hat{\boldsymbol{\tau}}^T W_\tau \hat{\boldsymbol{\tau}} + \frac{1}{2}(\hat{\boldsymbol{\tau}} - \hat{\boldsymbol{\tau}}_b)^T W_c(\hat{\boldsymbol{\tau}} - \hat{\boldsymbol{\tau}}_b), \tag{21}$$

$$C = [\boldsymbol{0}_{1 \times 12} \mid h_s]_{1 \times 13},$$

$$W_\tau = \begin{bmatrix} \mathrm{diag}[h_{\tau 1}, h_{\tau 2}, \ldots, h_{\tau 12}] & \boldsymbol{0}_{12 \times 1} \\ \boldsymbol{0}_{1 \times 12} & 0 \end{bmatrix}_{13 \times 13},$$

$$W_c = \begin{bmatrix} \mathrm{diag}[h_{c1}, h_{c2}, \ldots, h_{c12}] & \boldsymbol{0}_{12 \times 1} \\ \boldsymbol{0}_{1 \times 12} & 0 \end{bmatrix}_{13 \times 13}, \tag{22}$$

where $\boldsymbol{\tau}_b$ is the vector of the input torque of the previous time step. C is a weight vector for maximizing the safety margin s, W_τ is a weight matrix for minimizing the norm of the torque, and W_C is a weight matrix for evaluating the continuity of the torque. Note that $h_s < 0$, $h_{\tau 1 \ldots 12} > 0$, $h_{c1 \ldots 12} > 0$. Then, W_τ and W_C are positive definite.

By solving the QP formulation for each time step, the input torque of each joint can be calculated, and by using this torque, the optimal force distribution can be achieved. The simulation results for the force distribution are shown in the next section.

Results and discussion

Simulation

In this section, the simulation results for the force distribution are shown.

Setting

In the simulation, the robot walks on various slopes in various directions using the proposed one-cycle leg-grope walk. The inclination angle of the slope and the angle of walking direction are represented as θ [rad] and ψ [rad], respectively, as shown in Fig. 5. We solve the force distribution problem for various $\theta = (-\pi/2, \pi/2)$ and $\psi = [-\pi/2, \pi/2]$ with the following conditions.

Conditions for the geometrical relations of leg-grope The grope-reaction is set as $R_{ref} = \frac{1}{2}Mg\cos\theta$ depending on θ. The robot swings its four legs L_2, L_1, L_3 and L_4 in sequence using the explained leg-grope walk method. The contact point of each groping leg (L_2, L_1, L_3 and L_4) and the COG are represented on $O_G - x_G y_G$ in Fig. 6.

Conditions for the force distribution We designed the robot body and leg movement to fulfill the above geometrical relations, while the maximum acceleration and velocity of the robot body are set as $a_{max} = 0.15$ [m/s^2] and $v_{max} = 0.1$ [m/s], respectively. In addition, the designed movement keeps the robot body parallel to the surface. The detail of this kinematic motion is explained in Additional file 1: Appendix S2. Based on this kinematic motion, we solve the force distribution problem formulated in the "Methods" section.

The parameters for the evaluation function are set as $h_s = -2$, $h_{\tau 1 \ldots 12} = 1$ and $h_{c1 \ldots 12} = 80$. As $-h_s$ and $h_{\tau 1 \ldots 12}$ become larger, slippage avoidance and energy saving are further enhanced, respectively. However, the torque output changes abruptly when a leg touches down or lifts off. Additionally, as $h_{c1 \ldots 12}$ becomes larger, smooth torque output is further enhanced. In this simulation, we use a larger value for $h_{c1 \ldots 12}$ to ensure a smooth torque output.

The coefficient of static friction and time step of force distribution are set as $\mu = 0.45$ and $dt = 15$ [ms], respectively. We used the MATLAB function "quadprog" with a

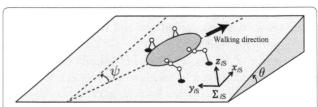

Fig. 5 Definition of the environment where the robot walks in the simulation. The robot walks on a simple slope whose inclination angle is θ. The angle between the walking direction and the gradient vector of the slope is defined as ψ

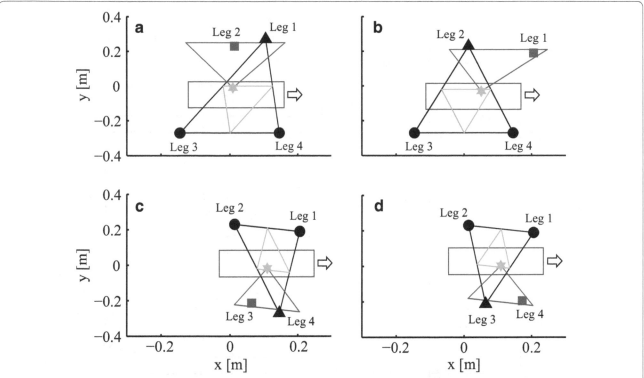

Fig. 6 Geometrical relations of leg-grope for the simulation and experiments. Each figure shows the relation in the case of the groping leg **a** L_2, **b** L_1, **c** L_3 and **d** L_4 on $O_G - x_G y_G$. For one walking cycle, the robot moves its COG and swings four groping legs L_2, L_1, L_3 and L_4 in sequence by following these geometrical relations. In each graph (**a–d**), an *arrow* and a *red rectangle* represent the moving direction of the robot and the shape of the robot body, respectively. The *blue square* point is the targeted point of the groping leg, and the other three points (*two black circles and black triangle*) represent the contact points of the other three legs. The *biggest black triangle* region represents the supporting leg polygon, except for the groping leg. The *green triangle* region represents the admissible region of the COG π_g, and the *green asterisk* point represents the targeted position of the COG. The *blue triangle* region represents the admissible region of the position of the groping leg for the COG $\pi_{grp,g}$, where the float leg is shown by the *black triangle* point

computer (CPU: Core i7 4 GHz; Memory: 16 GB) for the calculation.

Results

In simulation, the robot performs the leg-grope walk successfully when the magnitude of the inclination angle θ is less than around 0.40 [rad]. If the inclination angle is less than that critical value, the robot performs well irrespective of the walking direction ψ. When the magnitude of the inclination angle θ is over the critical value, the robot cannot avoid slippage and fails in the leg-grope walk.

If a rigid body is static on the slope, the maximum absolute inclination angle to avoid slippage is calculated as $\theta = \arctan(\mu) = 0.42$ [rad]. This value is close to the critical inclination angle for the leg-grope walk, which means that the force distribution method works well. The critical inclination angle of the leg-grope walk is slightly smaller than that of the rigid body because the robot applies additional forces to accelerate its body. We also confirmed that the computational time to solve this force distribution problem is less than the period of one walking cycle in all cases.

As an example of one leg-grope walk cycle, Fig. 7 shows the time response of the elements of the force vector f_i on the contact coordinate Σ_{iS} at $(\theta, \psi) = (\pi/12, 0)$. *Dotted horizontal lines* represent the value of R_{ref}. As time goes by, the robot moves its COG by standing on four legs and decreasing the normal reaction of the groping leg (Steps A and B), swings the groping leg to the point to be probed (Step C), and probes the foothold by applying the reference force R_{ref} (Step D). The robot repeats this procedure for four groping legs L_2, L_1, L_3 and L_4 in order. The areas that the robot successfully applies the normal force R_{ref} in the leg-grope step (Step D) are marked with *black circles*. However, the magnitude of the normal forces except for Step D are less than R_{ref}. Figure 8 represents the time response of the safety margin of the friction s. This result shows that the safety margin s is assured and is never negative, which means that slippage does not occur. Figure 9 shows the time response of the torque inputs. The torque inputs depend smoothly on time, as our design of the minimized evaluation function of the QP formulation intended.

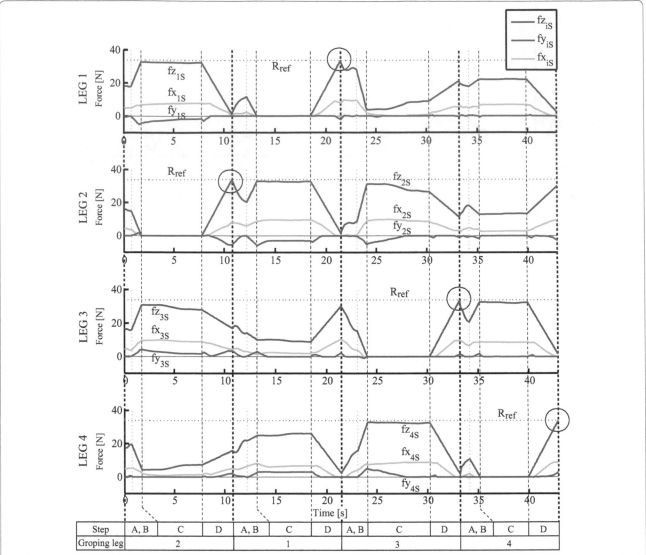

Fig. 7 Time response of the force distributions on Σ_{iS} of the simulation. Each leg applies normal reaction R_{ref} in the leg-grope movement as marked with the *black circles*. Aside from that, each normal reaction is less than R_{ref}. The *red line*, the *blue line* and the *green line* represent the z_{iS}, y_{iS} and x_{iS} elements of the force with respect to Σ_{iS}, respectively. Each *dotted horizontal line* represents R_{ref}

As a conclusion, the proposed force distribution method achieves suitable torque inputs, taking account of slippage for the leg-grope walk in various environments.

Experiments

Setting

To demonstrate the effectiveness of the proposed method, we also carried out some experiments with the real robot. However, the results of the force distribution could not be used, because the joints of the robot were controlled not by torque inputs but by position inputs. Hence, we only consider the geometrical relations of the leg grope walk by following the assumption about

friction. The leg grope movement (Step D) is replaced by two steps: Steps D′-1 and D′-2 as in the following description. The modified leg-grope walk sequence consists of the following five steps.

A′ Move the COG of the robot inside of the admissible region of COG while standing on four legs. The COG is placed more with in the leg supporting polygon than the COG position for the leg-grope.

B′ Move the groping leg up gradually until the normal reaction becomes 0 without any other movement.

C Swing the groping leg to the point of the leg-grope and make the leg touch down.

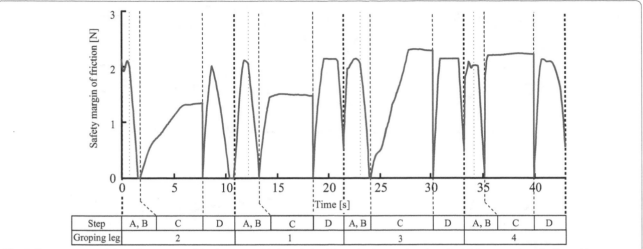

Fig. 8 Time response of the safety margin of the friction *s* of the simulation. The value is never less than 0, which means that distributed forces prevent slippage successfully

D′-1   Move the COG to the position for the leg-grope standing on four legs. As a result, the normal reaction of the groping leg increases gradually.

D′-2   Put down the groping leg gradually until the normal reaction is over R_{ref} without any other movement.

Figure 10 shows an example of this leg-grope walk. Note that Steps A′ and D′-1 allow the robot to get a large stability margin in a swing movement (the COG is placed more within the leg supporting polygon than the COG position for the leg-grope) (Fig. 10). Note that Steps B′, D′-1 and D′-2 allow the robot to apply the force using position control.

We demonstrated one cycle of walking with the proposed method on a simple slope and an irregular slope to validate the robot being able to apply the force over R_{ref} to the foothold to probe the environment. We also demonstrated that even if the robot's foothold collapsed during the leg-grope movement, the robot did not stumble.

We conducted three trials for each experiment, and show one of them as a typical result. In these experiments, the angle of the slope and the grope-reaction are set as $\pi/12$ [rad] and $R_{\text{ref}} = \frac{1}{2}Mg\cos(\pi/12)$, respectively. The robot swings its four legs L_2, L_1, L_3 and L_4 in sequence, and the contact point of each groping leg (L_2, L_1, L_3 and L_4) and the COG position for groping are represented on $O_G - x_G y_G$ in Fig. 6, as in the simulation.

Result of walking on a simple slope

The robot climbs a simple slope ($\theta = \pi/12$, $\psi = 0$ [rad]) using the proposed leg-grope walk, as in the simulation result. Figure 11 shows the time response of the

elements of the force vector f_i on the contact coordinate Σ_{iS} of one walking cycle of experiments. In Fig. 11, *dotted horizontal lines* represent the value of R_{ref}. As time goes by, the robot moves its COG while standing on four legs, decreases the normal reaction of the groping leg (Steps A′ and B′), swings the groping leg to the point to be probed (Step C), and probes the foothold by applying the grope reaction R_{ref} (Step D′). We repeat this procedure for four groping legs L_2, L_1, L_3 and L_4 in order. We find that the robot applies the normal force to the ground over R_{ref} in the leg-grope step (Step D′), as marked with *black circles*. However, the magnitude of each normal force is less than R_{ref}, except for the leg-grope step D′. The other four trials also have the same properties. Hence, we can say that the leg-grope walk is achieved successfully, as we expected.

Result of walking on an irregular slope

We also carried out the experiment on an irregular slope. The environment consists of slopes whose inclination angle is $\theta = \pi/12$ [rad], but the directions of the gradient vectors are not the same, as shown in Fig. 1a. Figure 12 shows the time response of the elements of the force vector f_i on the contact coordinate Σ_{iS}. The representation of the figure is the same as in Fig. 11. We find that the magnitude of the normal force is less than R_{ref}, except for the leg-grope step (Step D′). However, the grope-reaction R_{ref} can be applied in the leg-grope step, as shown with *black circles*. The other four trials also have the same properties. Hence, we conclude that the robot also performs well on the irregular slope. A video of this experiment is contained in Additional file 2.

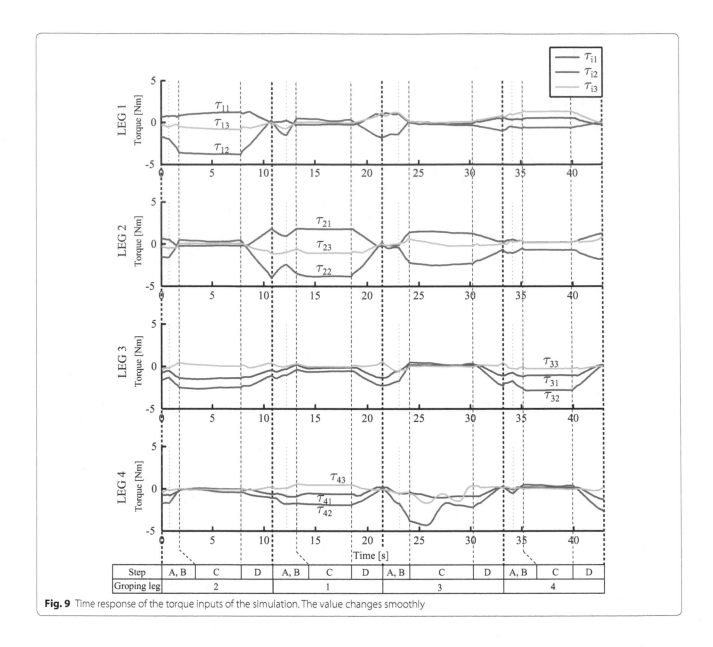

Fig. 9 Time response of the torque inputs of the simulation. The value changes smoothly

Result in the case of foothold collapse

In this experiment, the robot climbs a simple slope that is the same as the previous one. The grope-reaction R_{ref} and contact points of the legs are set to the same way as the previous ones. We set the foothold of leg L_1 as fragile enough to collapse while walking. The robot stops walking after the detection of the foothold collapse. Figure 13 shows the time response of the attitude and the z_R-axis acceleration of the robot body. At the marked time (around 35 [s]), the foothold of the leg L_1 collapsed. We found that the robot attitude changed by approximately 2 degrees only, and it never fell when and after the environment collapse. Figure 14 shows the time response of the elements of the force vector f_i on the contact coordinate

Σ_{iS}. The magnitude of the normal force of each leg ($f_{z_{iS}}$ on Fig. 14) is almost less than R_{ref}, although that of leg L_2 is larger than R_{ref} at very short moments near the collapse (a *blue circle* on Fig. 14). The sudden loss of one foothold induces a sudden change in body attitude (Fig. 13) because the leg is not rigid (back-lash of joints, flexibility of joints induced by PD controller, etc.). This sudden attitude change causes non-negligible acceleration and forces over R_{ref} (Figs. 13 and 14). Although this is the limitation caused by design failure, the method is practical enough to allow the robot to walk without stumbling. The other four trials also have the same properties. A video of this experiment is also included and can be found in Additional file 3.

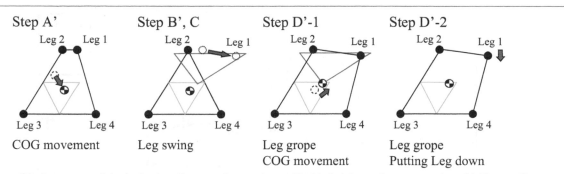

Fig. 10 Process of the leg-grope walk for the leg 1 on $O_G - x_G y_G$ for experiments. The *black circles* are the contact points of the leg toe. The *green triangle* and *blue triangle* represent the admissible region of the COG and the groping leg, respectively. Step A': the robot moves the COG inside of the admissible region of the COG while standing on four legs. Step B',C: the robot moves the groping leg up and swings it to the point of the leg-grope, and the leg touches down. Step D'-1: the robot moves the COG to the position for the leg-grope. Step D'-2: the robot pushes the groping leg down gradually until the normal reaction is over R_{ref}

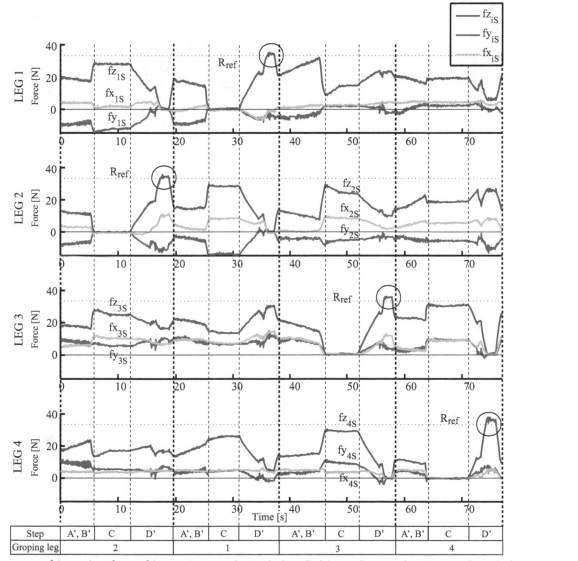

Fig. 11 Time response of the resultant forces of the experiment on the simple slope. Each leg applies normal reaction over R_{ref} in the leg-grope movement as marked with the *black circles*. Aside from that, each normal reaction is less than R_{ref}. The *red line*, the *blue line* and the *green line* represent the z_{iS}, y_{iS} and x_{iS} elements of the resultant force with respect to Σ_{iS}, respectively. Each *dotted horizontal line* represents R_{ref}

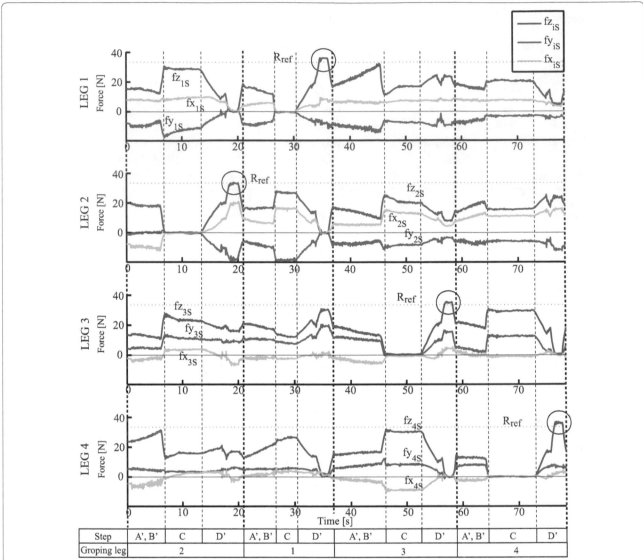

Fig. 12 Time response of the resultant force of the experiment on the irregular slope. Each leg applies normal reaction over R_{ref} in the leg-grope movement as marked with the *black circles*. Aside from that, each normal reaction is less than R_{ref}. The method of representation is the same as in Fig. 11

Conclusion

We propose the leg-grope walk on fragile irregular terrain considering slippage by force distribution. In simulation, the proposed method successfully derives the torque inputs to distribute the forces appropriately considering slippage avoidance. We also conducted various robotic experiments, and show the effectiveness of the method. The robot can walk stably by probing footholds step by step. Even if the foothold collapses, the robot can keep its posture and never stumbles. Thus, we conclude that the proposed method is useful for robots to walk safely on fragile irregular terrain.

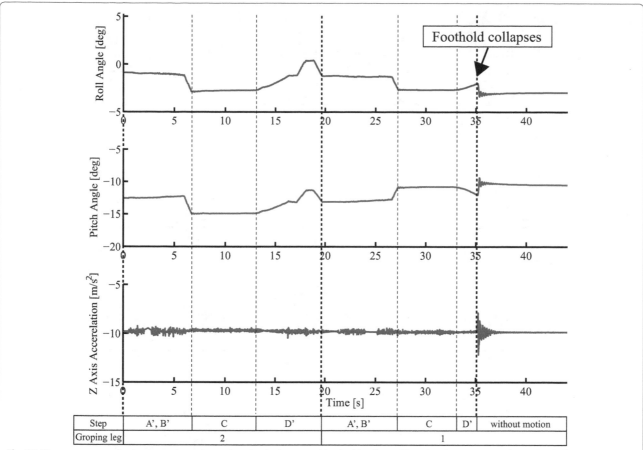

Fig. 13 Time responses of attitudes and z_R-axis acceleration in the case of foothold collapse. The robot keeps its attitude angles (roll and pitch angle) and never stumbles when and after the foothold of leg L_1 collapses

As limitations, foothold collapse based on slippage is not considered in this study, although we ensure that the robot fulfils friction cone constraints. The force distribution method is not demonstrated with the robot, because the joints of the robot are designed to be controlled by position inputs. Thus, we carried out the robotic experiments based on the geometric relation by following the assumption about the friction. However, the experiments show that the method is still practical. In the future, we need to design a robot whose joints are controlled by torque inputs to demonstrate the effectiveness of the force distribution method.

Practically, the robot cannot walk fast, because the method is designed based on static equilibrium and the

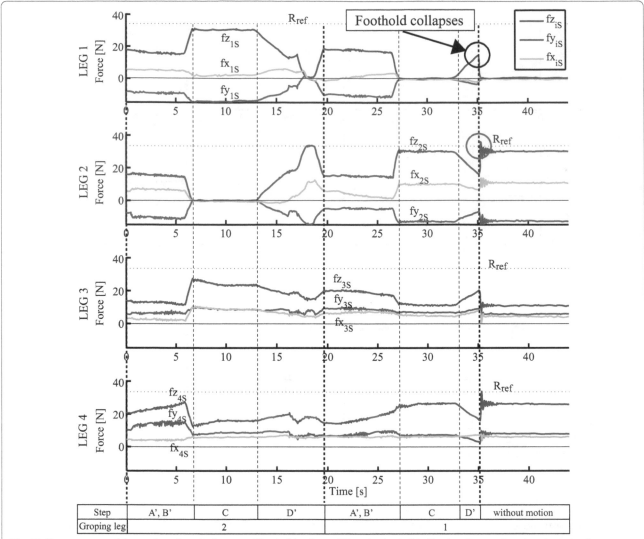

Fig. 14 Time response of the resultant force in the case of foothold collapse. When the foothold of leg L_1 collapses, the ground reaction becomes zero, as shown with the *black circle*. Conversely, the normal reactions of the other legs are still less than R_{ref} after the collapse, except for the impulse as shown with the *blue circle*. The method of representation is the same as in Fig. 11

region for the leg-grope is not so large. Recent dynamical walking strategies for legged robots [25–29] may outshine the proposed strategy in terms of walking speed. However, our walking strategy must be useful in a situation where scattered debris or a fragile environment should not be further compromised. This is the only method that allows robots to walk safely by making the magnitude of the normal reaction as small as possible. We think that a robot should change its walking strategy depending on the terrain, as LittleDog does [14–16]. If the terrain is flat, the robot can use a fast gait. However, if the terrain is fragile, we believe that our method will be useful.

It would be interesting in future work to combine this method and the terrain classification methods [17]. For example, terrain that is found to be fragile using the leg-grope walk can be used as learning data for terrain classification to estimate fragile footholds in advance, which compensates for the slow walking speed of the leg-grope walking method.

Additional files

Additional file 1: Appendixes. Appendix S1 derives the formulation of force destribution. Appendix S2 explains how to design the kinematic motion in the simulation.

Additional file 2. Leg-grope walking on the irregular terrain. This is the movie of the leg-grope walking on the irregular terrain of Fig. 12. The robot swings each leg and probes the foothold.

Additional file 3. Leg-grope walking in the case of foothold collapse. This is the movie of the leg-grope walking in the case of foothold collapse, as in Figs. 13 and 14. When the robot probes the foothold of the left front leg, the foothold collapses. The movie shows that the robot can keep its attitude after the foothold collapses.

Authors' contributions
All authors conceived and designed the algorithms and experiments. YA carried them out and wrote the paper. Both authors read and approved the final manuscript.

Acknowledgements
The authors thank Yuji Sato for his useful comments and help on the development of the robot, and Shinya Aoi for his useful comments on the manuscript.

Competing interests
The authors declare that they have no competing interests.

References
1. McGhee RB, Frank AA (1968) On the stability properties of quadruped creeping gaits. Math Biosci 3:331–351. doi:10.1016/0025-5564(68)90090-4
2. Hirose S, Yoneda K, Arai K, Ibe T (1991) Design of prismatic quadruped walking vehicle TITAN VI. In: The 5th International Conference on Advanced Robotics, vol. 1, pp. 723–728. doi:10.1109/ICAR.1991.240685
3. Hirose S, Tsukagoshi H, Yoneda K (2001) Normalized energy stability margin and its contour of walking vehicles on rough terrain. In: IEEE International Conference on Robotics and Automation (ICRA), pp 181–186. doi:10.1109/ROBOT.2001.932550
4. Konno A, Ogasawara K, Hwang Y, Inohira E, Uchiyama M (2003) An adaptive gait for quadruped robots to walk on a slope. In: IEEE/RSJ International Conference on Intelligent Robots and Systems (IROS), pp 589–594. doi:10.1109/IROS.2003.1250693
5. Estremera J, de Santos PG (2002) Free gaits for quadruped robots over irregular terrain. Int J Robot Res 21(2):115–130
6. Estremera J, de Santos PG (2005) Generating continuous free crab gaits for quadruped robots on irregular terrain. IEEE Trans Robot 21(6):1067–1076. doi:10.1109/TRO.2005.852256
7. Klein CA, Kittivatcharapong S (1990) Optimal force distribution for the legs of a walking machine with friction cone constraints. IEEE Trans Robot Autom 6(1):73–85
8. Zhou D, Low KH, Zielinska T (2000) An efficient foot-force distribution algorithm for quadruped walking robots. Robotica 18:403–413
9. Martins-Filho LS, Prajoux R (2000) Locomotion control of a four-legged robot embedding real-time reasoning in the force distribution. Robot Auton Syst 32(4):219–235. doi:10.1016/S0921-8890(99)00128-1
10. Li Z, Ge SS, Liu S (2014) Contact-force distribution optimization and control for quadruped robots using both gradient and adaptive neural networks. IEEE Trans Neural Netw Learn Syst 25(8):1460–1473. doi:10.1109/TNNLS.2013.2293500
11. Marco H, Christian G, Michael B, Mark H, Roland S (2013) Walking and running with StarlETH. In: The 6th International Symposium on Adaptive Motion of Animals and Machines (AMAM)
12. Murphy MP, Saunders A, Moreira C, Rizzi AA, Raibert M (2011) The LittleDog robot. Int J Robot Res 30(2):145–149
13. Byl K, Shkolnik A, Prentice S, Roy N, Tedrake R (2008) Reliable dynamic motions for a stiff quadruped. In: The 11th International Symposium on Experimental Robotics (ISER)
14. Neuhaus PD, Pratt JE, Johnson MJ (2011) Comprehensive summary of the institute for human and machine cognition's experience with LittleDog. Int J Robot Res 30(2):216–235
15. Kalakrishnan M, Buchli J, Pastor P, Mistry M, Schaal S (2011) Learning, planning, and control for quadruped locomotion over challenging terrain. Int J Robot Res 30(2):236–258
16. Kolter JZ, Ng AY (2011) The Stanford LittleDog: A learning and rapid replanning approach to quadruped locomotion. Int J Robot Res 30:150–174
17. Hoepflinger MA, Hutter M, Gehring C, Bloesch M, Siegwart R (2013) Unsupervised identification and prediction of foothold robustness. In: IEEE International Conference on Robotics and Automation (ICRA), pp 3293–3298. doi:10.1109/ICRA.2013.6631036
18. Hoepflinger MA, Remy CD, Hutter M, Spinello L, Siegwart R (2010) Haptic terrain classification for legged robots. In: IEEE International Conference on Robotics and Automation (ICRA), pp 2828–2833. doi:10.1109/ROBOT.2010.5509309
19. Tokuda K, Toda T, Koji Y, Konyo M, Tadokoro S, Alain P (2003) Estimation of fragile ground by foot pressure sensor of legged robot. In: IEEE/ASME International Conference on Advanced Intelligent Mechatronics, vol 1. pp 447–453. doi:10.1109/AIM.2003.1225137
20. Degrave J, van Cauwenbergh R, Wyffels F, Waegeman T, Schrauwen B (2013) Terrain classification for a quadruped robot. In: 12th International Conference on Machine Learning and Applications (ICMLA), 2013, vol 1. pp 185–190. doi:10.1109/ICMLA.2013.39
21. Kamegawa T, Suzuki T, Otani K, Matsuno F (2010) Detection of footholds with Leg-grope and safety walking for quadruped robots on weak terrain. J Robot Soc Japan 28(2):215–222 in Japanese
22. Ambe, Y, Matsuno, F (2012) Leg-grope-walk–walking strategy on weak and irregular slopes for a quadruped robot by force distribution. In: IEEE/RSJ International Conference on Intelligent Robots and Systems (IROS), pp 1840–1845. doi:10.1109/IROS.2012.6385798
23. Chen JS, Cheng FT, Yang KT, Kung FC, Sun YY (1999) Optimal force distribution in multilegged vehicles. Robotica 17:159–172
24. Erden MS, Leblebicioglu K (2007) Torque distribution in a six-legged robot. IEEE Trans Robot 23(1):179–186. doi:10.1109/TRO.2006.886276
25. Estremera J, Waldron KJ (2008) Thrust control, stabilization and energetics of a quadruped running robot. Int J Robot Res 27(10):1135–1151
26. Hyun DJ, Seok S, Lee J, Kim S (2014) High speed trot-running: Implementation of a hierarchical controller using proprioceptive impedance control on the MIT Cheetah. Int J Robot Res 33(11):1417–1445
27. Semini C, Barasuol V, Boaventura T, Frigerio M, Focchi M, Caldwell DG, Buchli J (2015) Towards versatile legged robots through active impedance control. Int J Robot Res 34(7):1003–1020
28. Hutter M, Gehring C, Hopflinger MA, Blosch M, Siegwart R (2014) Toward combining speed, efficiency, versatility, and robustness in an autonomous quadruped. IEEE Trans Robot 30(6):1427–1440. doi:10.1109/TRO.2014.2360493
29. Raibert M, Blankespoor K, Nelson G, Playter R, the BigDog Team (2008) BigDog, the rough-terrain quadruped robot. In: The 17th IFAC World Congress, pp 10822–10825

14

BBot, a hopping two-wheeled robot with active airborne control

Huei Ee Yap[*] and Shuji Hashimoto

Abstract

Most two-wheeled robots have algorithms that control balance by assuming constant contact with the ground. However, such algorithms cannot confer stability in robots deployed on non-continuous ground terrain. Here, we introduce BBot, a robot that can hop as well as move over stepped terrains. BBot has a two-wheeled lower body platform and a spring-loaded movable upper body mass. Hopping results from the impact force produced by release of pretensed springs. An inertia measurement unit detects the angle of body tilt, and an ultrasonic distance sensor records the height above ground. An accelerometer in the inertia measurement unit measures the impact force to determine the beginning and end of the phases of hopping and landing. Torque generated from rotation of the drive wheels controls the airborne robot's body angle. Sensors detect the impact of landing, and controls immediately switch to ground balance mode to stay upright. Experiment results show that BBot is capable of traversing down a 17 cm step, enduring manual toss landing and hopping 4 cm above ground.

Keywords: Two-wheeled robot, Step traverse, Attitude control, Inverted pendulum, Hopping

Background

Mobile robots often need to navigate non-continuous terrains with obstacles. Crawler-type robots can move over spacious uneven terrains but are usually heavy and slow. Biped robots that can navigate through natural environments are currently expensive to make and deploy. Lunar rovers such as the NASA Soujourner and the Ecole Polytechnique Federale de Lausanne (EPFL) Shrimp robot combine adaptive legs with the efficiency of wheels to traverse stepped terrains. This approach relies on redundancy of static points of support on the robot's base to stabilize climbing and descending steps. These robots, albeit efficient, have complex structural designs and driving systems. Currently, there are many robots capable of jumping. Examples include 7g [1], Grillo [2], MSU [3] and mowgli [4]. These robots use linkage leg systems and springs to leap over large obstacles. However, the horizontal movements of such electro-mechanical machines are inefficient and inaccurate. Mini-Whegs [5], Jollbot [6], Scout [7] integrate mobility and jumping to cover large areas and negotiate stair-like obstacles. But, jumping is passive, and there is no control of airborne attitude.

In 2008, Kikuchi et al. [8] introduced a wheeled-based robot that climbs up and down stairs dynamically. A spring-loaded movable upper body mass allows their robot to land softly and double-hop in midair. Kikuchi robot consists of a statically stable wheel base. One of the drawbacks of this is that the robot has to land with minimum body tilt angle to ensure a successful landing. Safe landing is not guaranteed if there exists any external disturbances during airborne. Another jumping robot named iHop[9] is a transformable two wheeled robot. In hoping mode, it uses both wheels as weights and has a lockable hopping mechanism. iHop pushes its wheels upward while balancing on the central chassis. iHop exhibits hoping capability but it is not shown that the robot is capable of climbing up or down step terrains.

The tradeoff between mobility and system complexity is the main challenge faced by many mobile robots. For example, Boston Dynamics' BigDog [10] moves easily over rough terrains, but complexity and construction costs limit the wide adoption of such machines. By contrast, simpler designs such as the 7g have mobilities that are too restricted for practical use. Unlike existing robots,

---

*Correspondence: huei@shalab.phys.waseda.ac.jp
Department of Applied Physics, Graduate School of Advanced Science and Engineering, Waseda University, Tokyo, Japan

humans and animals negotiate stepped terrains dynamically and efficiently. For example, when jumping up or down a step, we tend to bend our knees to absorb the impact during landing and use the momentum to push ourselves forward and continue our next step.

In this research, we introduce a prototype robot named BBot that utilizes the movement momentum to achieve hopping motion. BBot is an improved version of our previous prototype [11] with additional hopping capability. Our goal is to develop a mobile robot that is able to traverse common terrains such as steps and stairs in an efficient manner. BBot consists of a lower body with wheels and a movable upper body connected to the main chassis by springs. Previous prototype suffers from impact recoil upon landing due to the rigid body structure. The movable upper body acts as an absorber to mitigate the impact force during landing. In addition, BBot uses the spring loaded movable upper body to generate lifting force to hop. We have chosen the two-wheeled structure due to the dynamic nature and simple design. During airborne, BBot uses the drive wheels to generate torque to control the angle of body tilt. As a result, it is resistant to external disturbances when traversing step terrains. Here we describe the hopping mechanism, motion characteristics, sensors, and control algorithms. We discuss the experimental results compare to the theoretical findings, as well as the limitations of the current model.

Mathematical model

In this section, we develop the mathematical models and analyze the behavior of the robot during a hopping motion. The hopping cycle has three distinctive phases: (1) ground balance; (2) pre-airborne impact; and (3) airborne balance. The robot goes from ground to air and back in cycles. Figure 1 shows the three stages during a hop. To simplify the problem, we derived and analyzed separate models for each phase.

Ground balance phase

Figure 2 depicts the dynamic behavior of the robot on the ground and in the air in simplified two-dimensional models. Our 2D model consists of an upper body with mass m_b and moment of inertia I_b as well as a lower body of mass (wheel) m_w and moment of inertial I_w. The upper body rotates about the center axis of the wheel. A motor connected to the wheel generates torque τ_g. The equations of motion on the ground are as follows:

$$(I_w + (m_b + m_w)R^2)\ddot{\theta}_w + m_b l_w R cos\theta_b \ddot{\theta}_b$$
$$- m_b l_w R \dot{\theta}_b^2 sin\theta_b = \tau_g \tag{1}$$

$$(I_b + m_b l_w^2)\ddot{\theta}_b + m_b l_w R cos\theta_b \ddot{\theta}_w - m_b g l_w sin\theta_b = -\tau_g \tag{2}$$

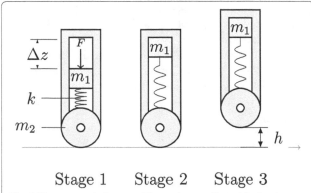

Fig. 1 Three steps during pre-airborne impact. *Stage 1* Spring is compressed to accumulate potential energy. *Stage 2* Spring is released and impact on outer body frame. *Stage 3* Conservation of momentum causes both upper and lower body to lift off from ground

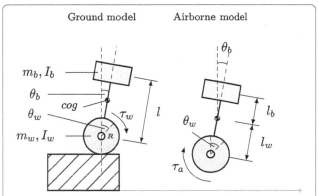

Fig. 2 Two dimensional models of the robot on ground and in air

Pre-airborne impact phase

Three steps during pre-airborne impact. Stage (1) Compression of the spring increases potential energy. (2) Release of the spring impacts the frame of the outer body. (3) Conservation of momentum causes the upper and lower body to lift off from ground. Hopping results from conversion of the potential energy stored in the pre-compressed spring into vertical kinetic energy. The transition from ground to air takes place in three stages.

1. External force F pushes down on the upper body mass to increase potential energy in the spring. Stored potential energy is directly proportional to the spring constant k and the square of the compressed distance of the spring Δz. F is proportional to k and Δz.
2. Release of the upper body mass allows upward acceleration by converting spring potential energy into kinetic energy. Motion of the upper body mass is brought to a stop upon impact with the body frame (Δz), unlike the free motion of the upper body mass

proposed in [8]. This design limits the magnitude of change in the center of mass that might affect the balance algorithm.

3. A perfectly inelastic collision occurs at the moment of impact. The Law of Conservation of Momentum mandates changes in the velocities of the upper body mass and the lower body according to the following equation:

$$m_1 v_1 = (m_1 + m_2) v_2 \qquad (3)$$

If the initial velocity of the upper body generates enough energy, the lower body lifts off the ground. When the wheels rise above ground, the pre-airborne phase ends and the airborne phase begins.

Jumping height is directly proportional to the initial potential energy stored in the spring. Assuming potential energy is zero for the uncompressed spring ($\Delta z = 0$), the potential energy of the system is

$$E_0 = \frac{1}{2} k \Delta z^2 - m_1 g \Delta z \qquad (4)$$

Neglecting energy lost from friction, the total energy just before impact is

$$E_1 = \frac{1}{2} m_1 v_1^2 \qquad (5)$$

The Law of Conservation of Energy dictates that the velocity of m_1 prior to impact is

$$v_1 = \sqrt{\frac{2}{m_1} \left(\frac{1}{2} k \Delta z^2 - m_1 g \Delta z \right)} \qquad (6)$$

Assuming a perfectly inelastic collision occurs after impact, the Law of Conservation of Momentum (e.g. (3)) requires that the velocity of the robot obeys the following equation:

$$v_2 = \frac{m_1}{m_1 + m_2} v_1 \qquad (7)$$

The energy right after impact is

$$E_2 = \frac{m_1}{m_1 + m_2} E_0 \qquad (8)$$

$$= \frac{m_1}{m_1 + m_2} \left(\frac{1}{2} k \Delta z^2 - m_1 g \Delta z \right) \qquad (9)$$

The maximum height of the jump is

$$h = \frac{E_2}{(m_1 + m_2) g} \qquad (10)$$

$$= \frac{m_1}{(m_1 + m_2)^2 g} \left(\frac{1}{2} k \Delta z^2 - m_1 g \Delta z \right) \qquad (11)$$

Jump height increases in direct proportional to k and Δz.

Airborne phase

We modelled the airborne robot as a downward pointing reaction wheel pendulum with a pivot at its center of mass (Fig. 1).

$$(I_b + m_b l_b^2 + m_w l_w^2) \ddot{\theta}_b = -\tau_a \qquad (12)$$

$$I_w \ddot{\theta}_w = \tau_a \qquad (13)$$

These two equations show that wheel acceleration is directly proportional to the torque τ_a generated by the motor. Rearranging and solving the differential equations reveals that the wheel angle and the angle of body tilt are related as follows:

$$\theta_b = - \frac{I_w}{I_b + m_b l_b^2 + m_w l_w^2} \theta_w \qquad (14)$$

Accordingly, the body tilts in the direction opposite to the rotation of the wheels. Increasing the wheel's moment of inertia permits effective control over body tilt.

Design and implementation

The jumping mechanism of the stair climbing robot developed by Kikuchi et al. [8] uses an upper body mass that is free to oscillate along the vertical axis to create momentum for lift. Maximum extension of the attached spring limits the movement of the upper mass. This design requires a tall body frame to accommodate the full movement of the mass. Large movement of the upper body mass would result in a major shift in the center of mass. Such a shift makes it harder to keep the robot stable and requires a strict control scheme to maintain balance. By contrast, our design limits the vertical movement of the upper body mass to reduce the magnitude of the shift in the center of mass. Figure 3 shows the 2D CAD model of BBot. Figure 4 shows the actual prototype of BBot. The upper body incorporates two 11.1 V lipo batteries for powering the motors and electronic components respectively. The upper body can slide vertically along a bar, and it is connected to the lower body frame with tension springs that pull upwards.

Tension springs instead of compression springs embedded in vertical slider eliminate possible force from friction. Spring tensioning mechanism is not implemented in this version of the prototype. In this prototype, the upper body is manually compressed and locked in place to prepare the robot in a jump-ready state. A servo activated latch controls the locking and releasing of the upper body. Weights ranging from 0.5 to 1 kg are attached to the upper body to test their effect on the jumping ability.

The lower body consists of two differential wheels powered separately by dc motors, a main electronic

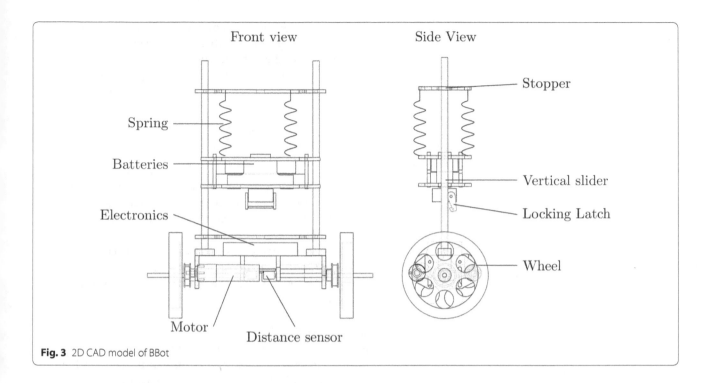

Fig. 3 2D CAD model of BBot

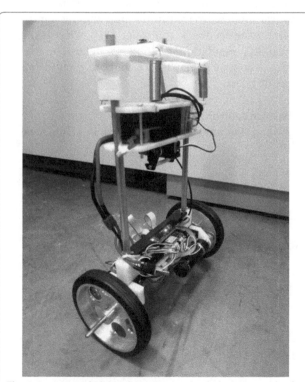

Fig. 4 Prototype of BBot, a dynamic two wheeled robot capable of hopping motion

control board, and an ultrasonic distance sensor. An inertia measurement unit (IMU), consisting of a 3-axis accelerometer and a 3-axis gyroscope, measures the angle of tilt. An ultrasonic distance sensor attached to the bottom measures the height the robot attains above ground. Lower body mass is minimal to make the hops as large as possible. We calculated conversion efficiency η as the ratio of the kinetic energy at takeoff to the energy stored in the compressed spring before takeoff [12]. The equation for conversion efficiency is

$$\eta = \frac{E_2}{E_0} = \frac{1}{1+r} \tag{15}$$

where $r = m_2/m_1$.

Figure 5 plots the conversion efficiency for various upper and lower body masses. Increasing upper body mass m_1 increases the efficiency of energy conversion and height of jumping. For a fixed m_1, lower body mass m_2 is inversely related to the efficiency of energy conversion. Consequently, increasing m_1 while minimizing m_2 leads to a higher height of jumping that is directly proportional to spring constant k.

Figure 6 shows a simplified 2D model constructed in Working Model 2D simulation software to simulate the effect of upper body weight on jump height. In this model, upper body mass m_1 is constrained to move vertically inside m_2. m_2 can moves freely horizontally and vertically above ground. m_2 and spring constant k are held constant. Gravitational force is set to 9.8 m/s^2 and air resistance is neglected. For each simulation spring extension Δz is initialized to 0.15 m. Upon released the jump height of the system is recorded. Figure 7 plots the time variation of the

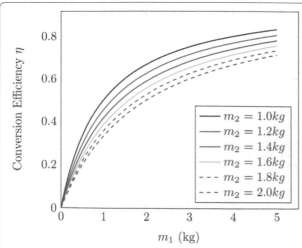

Fig. 5 Conversion efficiency plot for various upper body masses m_1 and lower body masses m_2. Efficiency increases with increasing m_1. For a fixed m_1, decreasing m_2 increases efficiency

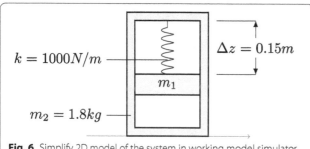

Fig. 6 Simplify 2D model of the system in working model simulator

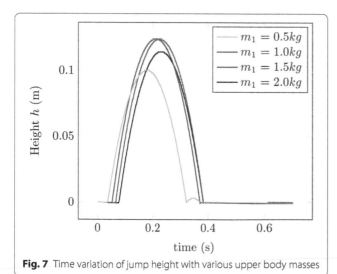

Fig. 7 Time variation of jump height with various upper body masses

jump height with upper body masses varying from 0.5 to 2.0 kg. The results reveal that jump height is not linearly proportional to upper body mass. An upper body mass

larger or smaller than the optimal decreases the height of the jump. Larger upper body masses delay the transition from ground to the airborne state.

Figure 8 plots jump height h versus upper body mass m_1. Optimum m_1 occurs at the maximum point of the curve:

$$m_{1optimal} = \frac{m_2 k \Delta z}{4 m_2 g + k \Delta z} \qquad (16)$$

The 12 V dc motor including gearing produces a maximum torque of 0.2 Nm. Body height l is 250 mm. Assuming the robot can recover from an angle of 30° when balancing on the ground, the maximum allowable weight of m_1 is 1.6 kg. Our design uses four separate springs arranged in parallel with a spring constant of 250 N/m and m_2 of 1.8 kg. m_1 is set at 1.2 kg, close to the optimum of 1.22 kg. The H × D × W dimensions are 300 × 160 × 420 mm.

Balance and attitude control

Static instability requires active control of body position during the airborne phase to ensure an upright landing. Two independent control schemes balance the robot on the ground and in the air (Fig. 9). We linearized the mathematical models derived in the previous section to design a full state-feedback controller [11]. We use the linear quadratic regulation method to determine the state-feedback control gain matrix K for both ground and airborne controllers.

Ground balance control

On the ground, the linearized systems are in state-space form:

$$\dot{\mathbf{x}}_g = A_g \mathbf{x}_g + B_g u_g \qquad (17)$$

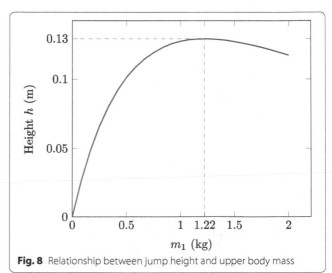

Fig. 8 Relationship between jump height and upper body mass

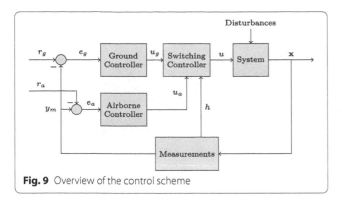

Fig. 9 Overview of the control scheme

where $\mathbf{x}_g = \begin{bmatrix} \theta_w & \dot{\theta}_w & \theta_b & \dot{\theta}_b \end{bmatrix}^T$ represents the states of the system, A_g is the state transition matrix, B_g is the input matrix and u_g is the input control torque from the motor:

$$u_g = \tau_g = K_g(\mathbf{r}_g - \mathbf{x}_g) \qquad (18)$$

where $\mathbf{r}_g = \begin{bmatrix} \theta_{wref} & \dot{\theta}_{wref} & \theta_{bref} & 0 \end{bmatrix}^T$ represents the target references to the feedback controller and K_g is a vector of controller gains. The quadratic cost function to be minimized is defined as

$$J(u) = \int_0^\infty (\mathbf{x}^T Q \mathbf{x} + u_g' R u_g) dt \qquad (19)$$

The values of weight matrix Q and R are manually tuned to yield a satisfactory response of a mathematical simulation constructed in Matlab.

Figure 10 shows the feedback control block diagram detailing the input and output of the system. BBot's differential drive requires separate controllers for left and right wheels. The input to the system consist of four state references, left wheel velocity $\dot{\theta}_{wleft}$, right wheel velocity $\dot{\theta}_{wright}$, body tilt angle θ_b and body tilt angular velocity $\dot{\theta}_b$. Reference θ_b and $\dot{\theta}_b$ is set zero to keep the robot upright. Varying $\dot{\theta}_{wleft}$ and $\dot{\theta}_{wright}$ reference controls the forward, backward and turning movement on ground.

The yaw PD feedback controller controls yaw movement:

$$u_{yaw} = K_{yaw}(\mathbf{r}_{yaw} - \mathbf{x}_{yaw}) \qquad (20)$$

$$\mathbf{r}_{yaw} = \begin{bmatrix} \theta_{yawref} & \dot{\theta}_{yawref} \end{bmatrix}^T \qquad (21)$$

$$\mathbf{x}_{yaw} = \begin{bmatrix} \theta_{yaw} & \dot{\theta}_{yaw} \end{bmatrix}^T \qquad (22)$$

where the yaw angle θ_{yaw} and yaw velocity $\dot{\theta}_{yaw}$ are the differences between left and right wheel angles and wheel velocities. The calculated yaw output is added to left and

right wheel input torque to yield the final input torque to the system.

Airborne attitude control

When in the air, the robot behaves like a reaction wheel pendulum with a pivot at its center of mass. The drive wheels act as reaction wheels to create reaction torque that changes the angle of tilt. Equation 14 shows the angle of tilt θ_b is directly proportional to the wheel angle θ_w. Manipulating torque generated by the motor directly controls θ_b. The controller is designed in a similar manner as the ground controller. The following control law controls the attitude in the air:

$$u_a = \tau_a = K_a \begin{bmatrix} \theta_{bref} - \theta_b & -\dot{\theta}_b & -\dot{\theta}_w \end{bmatrix}^T \qquad (23)$$

where K_a is the vector of feedback gains and θ_{bref} is the reference body tilt angle in the air. Angular positions of the wheels θ_w are not controlled, so they are not included in the feedback loop.

Phase transition control

Data from the sonar distance sensor and the accelerometer determine the phase transition. Details of the approach are in the next section. A switching controller activates the corresponding feedback controller based on the current phase of the robot.

$$u = \begin{cases} u_g & \text{(if current phase = ground phase)} \\ u_a & \text{(if current phase = airborne phase)} \end{cases} \qquad (24)$$

Sensors and sensing approach

To have full-state feedback control for balance, the robot must know its state at every point in time in the control loop. An IMU measures body angle, and a combination of an ultrasonic sensor and an accelerometer measures height and state transition.

Body angle detection

The IMU sensor comprises a 3-axis accelerometer and a 3-axis gyroscope to measure the angle of body tilt. In principle, for a known initial state, direct integration of the gyroscope measurement gives the angle of tilt. In reality, however, the MEMS gyroscope is subject to white noise and fluctuating bias over time. So, errors accumulate if direct integration is performed on gyro measurements. This is known as gyro "drift". One solution is to implement a sensor fusion scheme using an accelerometer as an additional reference measurement. Schemes such as a complementary filter [13], Kalman filter [14], or Direct Cosine Matrix (DCM) filter [15] can help estimate

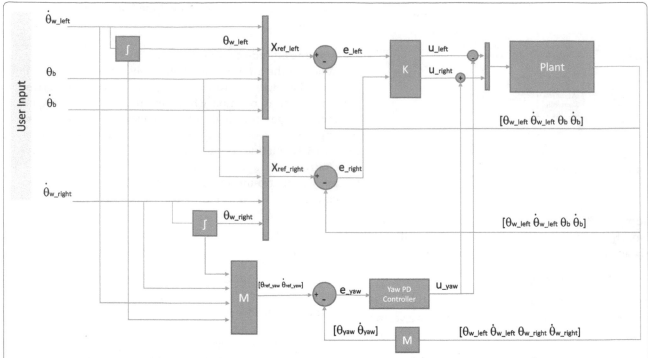

Fig. 10 Detailed feedback controller scheme for BBot. 4 user reference input left wheel velocity $\dot{\theta}_{wleft}$, right wheel velocity $\dot{\theta}_{wright}$, body tilt angle θ_b and body tilt angular velocity $\dot{\theta}_b$ are used to control the motion of BBot

rotational information of an IMU. Each type of filter has its pros and cons regarding speed, accuracy and complexity of implementation. We found that a Kalman filter sensor fusion scheme was more resilient and performed better during hopping and at the impact of landing, where accelerometer measurements tend to peak and overshoot. The Kalman filter also performed better under the influence of linear acceleration, especially when the robot platform was in motion.

Figure 11 compares the estimated angles for the Kalman, DCM, and complementary filters when the IMU sensor is accelerated back and forth in a fixed direction at a horizontal angle of zero degrees. The lower graph shows the corresponding raw accelerometer data. We found that a complementary filter that estimates the angle based on weighted angles from accelerometer and gyro sensor performed poorly during acceleration. The DCM filter fluctuated less that the complementary filter, but the influence from linear acceleration was still obvious. By contrast, the Kalman filter showed little fluctuation in angle estimation.

Height and phase transition detection
We have mounted an ultrasonic distance sensor on the bottom of the body, facing downwards towards the ground to measure the height of the robot from the ground. Detection of the transition from ground

to airborne phase and vice versa is crucial to maintain balance throughout the jumping motion. A simple approach is to set a threshold to determine the phase transition. This approach, however, has its drawback. Specifically, the sensor reading increases when the robot leans forward during acceleration. Increased sensor reading triggers the threshold settings and causes a false detection in phase transition. For larger tilt angle (>60°), the ultrasonic waves reflects away from the sensor and causes incorrect readings.

In order to build a more reliable phases detection scheme, we use the vertical axis accelerometer data to detect impact indicating a phase transitions. Figure 14 shows the accelerometer data at the instant of an impact. We can determine an impact event by detecting a sudden spike in accelerometer data. Figure 12 shows a flow diagram of a more reliable approach in detecting phase transition using both ultrasonic sensor data and accelerometer data.

Experiments and discussion
BBot is a self contained, with electronics and power supply encased in the chassis. A host pc controls and communicates with BBot via bluetooth connection. Realtime sensor data is streamed to the host pc and logged at a rate of 100 Hz.

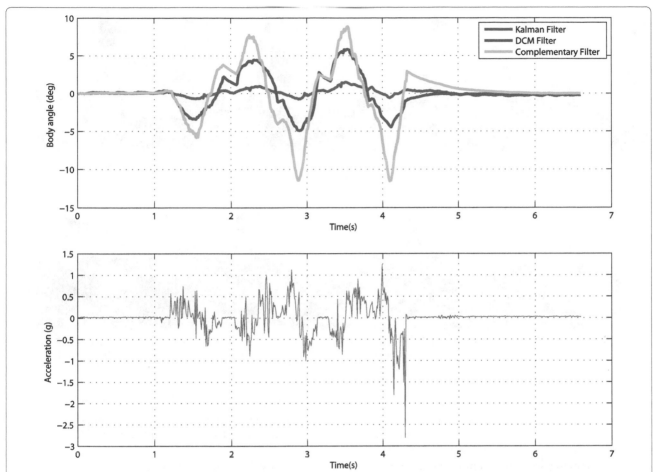

Fig. 11 Angle estimation comparison between Kalman filter, DCM filter and complementary filter. The sensor is accelerated back and forth in a fix direction at tilt angle of 0°. Kalman filter shows little fluctuation in angle estimation under the influence of linear acceleration noise. The *lower graph* shows the raw linear acceleration observed by the sensor

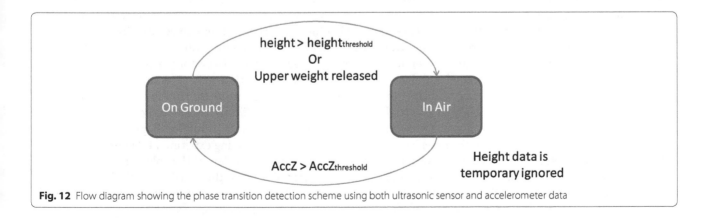

Fig. 12 Flow diagram showing the phase transition detection scheme using both ultrasonic sensor and accelerometer data

Step traversing experiment

In this experiment, the robot travels down a 17 cm step. The step height is equivalent to the average height of a stair step. Additional file: 1 (Bbot-StepTraverseExperiment) shows the video of the step traversing experiment. Figure 13 shows the snap shots of the motion. Figure 14

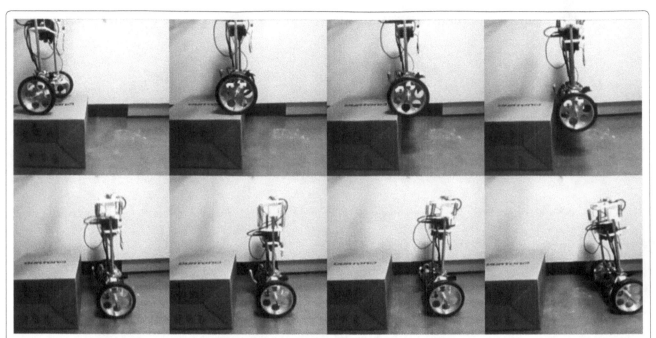

Fig. 13 The robot is driven down a step terrain of height 17 cm. While in air, the robot uses the drive wheels to maintain its body angle close to vertical. Using this approach, the robot prevents its motor from saturation and hence possesses sufficient torque to balance itself upon landing

shows the corresponding raw sensor data and the body tilt angle plot against time. Height data plots the height calculated from the ground to the sonar sensor. The height data has an offset of 5 cm above ground when the robot is balancing still. AccZ graph plots the acceleration in the robot frame z axis (pointing up) with gravity subtracted. At time $t = 1.86$ s, sudden increased in height measurement indicates that the robot is currently airborne. The controller switches into airborne control mode and applies balancing torque generated through the rotating wheels. We have set the reference body tilt angle for airborne controller to a small positive value (+6 degrees), to tilt the robot backwards during airborne. The reason for this is to compensate the forward momentum during landing and reduce the torque needed to balance upon landing. This is analogous to landing with feet in front and uses momentum to bring the body to a neutral position. At time $t = 2.02$ s, the wheel hits the ground and causes large fluctuation in the accelerometer readings. Phase changed is detected and the robot switches into ground mode to keep its balance. From the figure, the fluctuation in accelerometer reading does not affect the body angle estimation due to robustness of the Kalman filter. At the moment of impact upon landing, we observe the upper body mass moves downward. This motion acts as an absorber to absorb the impact force from the ground and mitigate any recoil effect which is observed in the previous rigid prototype. From the height vs time plot, there is a false positive indication of increase

height right after landing. This is the effect explained in the previous section where the ultrasonic waves bounds away from the ground when the body tilt angle is large.

Toss landing experiment

We have devised another experiment to confirm the effectiveness of attitude control during airborne. In this experiment, the robot is manually tossed with an initial angular rotation to test the capability to recover from such situation and land safely. Without attitude control, initial rotation causes a tilt away from the vertical. Upon landing the robot is not be able to recover from large tilt angle. With attitude control turned on, the controller constantly generates recovery torque to keep the body angle close to the vertical axis. The small tilt angle on landing prevents the motor from saturating. The motor is able to generate sufficient torque to keep the robot in balance. Additional file: 2 (BBot-TossExperiment) shows the video of the toss landing experiment. Figure 15 shows the snapshots of the motion of the robot in this experiment. Figure 16 shows the height, vertical acceleration, body angle and motor pwm duty plot against time. Red solid line in the graph shows the instant when the robot is released. At time $t = 4.5$ s the robot reaches peak height and starts to free fall. The airborne controllers maintains a positive body tilt angle in air. Landing impact happens at time $t = 4.7$ s. The toss experiment introduces large rotational torque to the robot. From time $t = 4.7$ s ~ 5.5 s, the robot rocks back and forth to

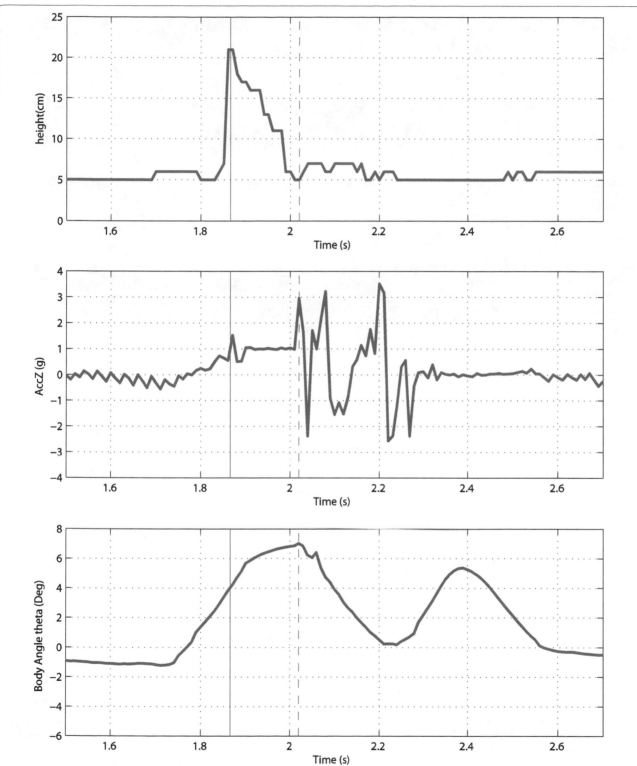

Fig. 14 Height, raw accelerometer plot against time during step traverse event. The graph body angle (filtered) against time shows the robustness of Kalman filter fusion scheme towards fluctuating accelerometer measurements. *Solid red line* and *dotted red line* indicate the beginning and the end of an airborne phase

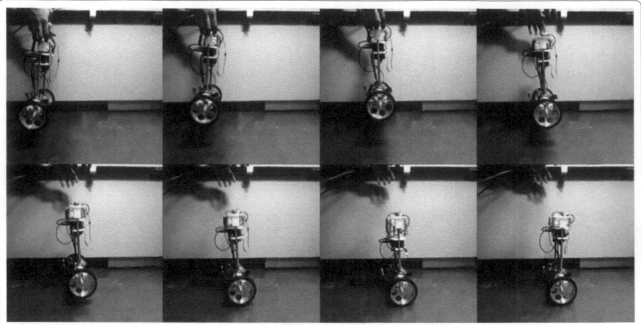

Fig. 15 Snapshot of the robot manually tossed with an initial angular rotation

keep balance before settling down on a stable upright position at $t > 6$ s. The motor duty graph shows that the pwm duty for the left motor. Pwm duty is proportional to the torque apply to the motor. The pwm duty is kept below 50 % at any instant of time indicates that the motor is below saturation limit.

Hopping experiment

In the last experiment, we present the hopping action using our proposed method. The springs are pre-tensioned manually by pushing down the upper body weight. A catch and release mechanism locks the upper body weight in place. Remote command from the host pc unlocks the locking mechanism to release the upper body weight. Additional file: 3 (BBot-HoppingExperiment) shows the video of the hopping experiment. Figure 17 shows the continuous snapshots of a hopping action. Figure 18 shows the sensor data plots against time. When the lock is released, pre-tensed springs accelerate the upper body weight. Impact on the stopper, occurs at time $t = 2.38$ s, converts into lifting force and causes the robot to jump. The robot detects the impact event and switches into airborne balancing mode. The rest of the action is similar to step traversing. At time $t = 2.55$ s, landing is detected and the controller switches back to ground balance mode. The robot successfully performs

a hopping motion. From Fig. 18, the maximum hopping height of this model is roughly 4 cm. The actual jump height is less than the value calculated in the simulation due to the following reasons: (1) Friction from the manually constructed sliding joint connecting the upper and lower body contributes to energy lost. (2) The springs used in the robot do not have the exact spring constant as assumed in the simulation. (3) Contrast to the assumption, the impact collision is not fully inelastic. The impact force does not convert fully into lifting force. (4) The simulation does not take into account friction encountered in physical world. Nevertheless, the prototype robot demonstrated the possibility to achieve hopping motion using proposed approach.

From these experiments, we observed a few short comings on the current prototype. In order to improve the hopping height of the robot, the current sliding joint has to be replaced by a lubricated ball bearing sliding joint to further reduced the friction. Jump height of the current prototype is too small to have practical application in real world environment. In order to be able to jump up a step, the robot needs to be able to have a jump height of at least 17 cm. To achieve higher jump height, we plan to scale up the prototype to accommodate larger springs. We observed recoil effect during impact of the upper body mass and the body frame. The partially inelastic collision differs from the

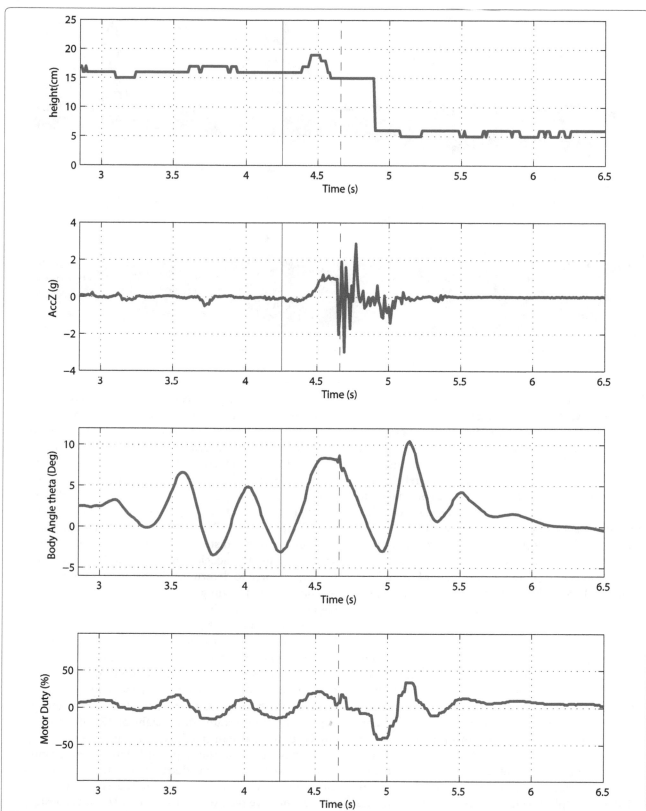

Fig. 16 Data plot of robot tossing experiment. *Red solid* line above shows the instant when the robot is released. The robot generates correcting torque to prevent it body angle to tilt away from vertical axis so that the body angle upon landing is within recoverable range. *Dashed line* shows the instant when the robot hits the ground

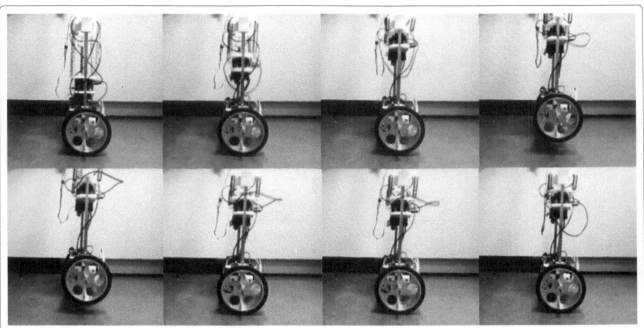

Fig. 17 Hopping motion demonstrated by the prototype robot

simulation assumption where the impact is perfectly inelastic. For the next version of the prototype, we plan to use a latching mechanism to enable locking of the upper body mass upon impact with the body frame. The reference body tilt angle is manually tuned and hard coded to enable a more efficient landing. As for future work, the relationship between velocity, jump height and body tilt angle needs to be investigated so that the robot can dynamically calculate the optimal tilt angle for different jump height. Current sonar sensor used to detect jump height suffers from false positive readings when the body is tilted away from the vertical. We are investigating the possibility of using laser scanner or small radar chips which are capable of measuring distance at a higher accuracy and reliability.

Conclusions

Tradeoff between robustness and increased system complexity is the main challenges faced by many mobile robots. Highly dynamic and robust robots tend to have increased complexity as well as cost of development and manufacturing. Simpler robots, on the other hand, have limited capability to be put into practical use. In order to tackle this problem, we have created BBot, a dynamically stable two wheeled robot capable of hopping and negotiating step terrains. The dynamic nature and relatively simple mechanical construction enables

the robot to have a good compromise between robustness and complexity. We have presented the theoretical analysis, design and mechanism of the robot, choice of sensors and sensing approach as well as the experiment results of a working prototype. The robot consists of a two wheeled lower body platform and a spring loaded movable upper body mass. Hopping motion is achieved through the impact force produced by releasing pretensed springs. An IMU detects the body tilt angle, along with an ultrasonic distance sensor to detect the height above ground. The accelerometer in the IMU measures the impact force to determine the beginning and the end of a hopping and landing phases. Due to statically unstable property of the robot, the attitude has to be actively controlled during airborne to ensure stability upon landing. While in air the body angle is controlled using the torque generated through the rotation of the drive wheels. Upon landing impact detection through onboard sensors, the controller switches to ground balance mode to maintain balance. Current prototype is capable of negotiating step terrains of a height of 17 cm, which is similar to normal stair height, and is capable of hopping 4 cm above ground. The hopping height is significantly less than simulated results due to mechanical constraints and perfect world assumption in the simulation. For future work, we plan to improve the mechanical design,

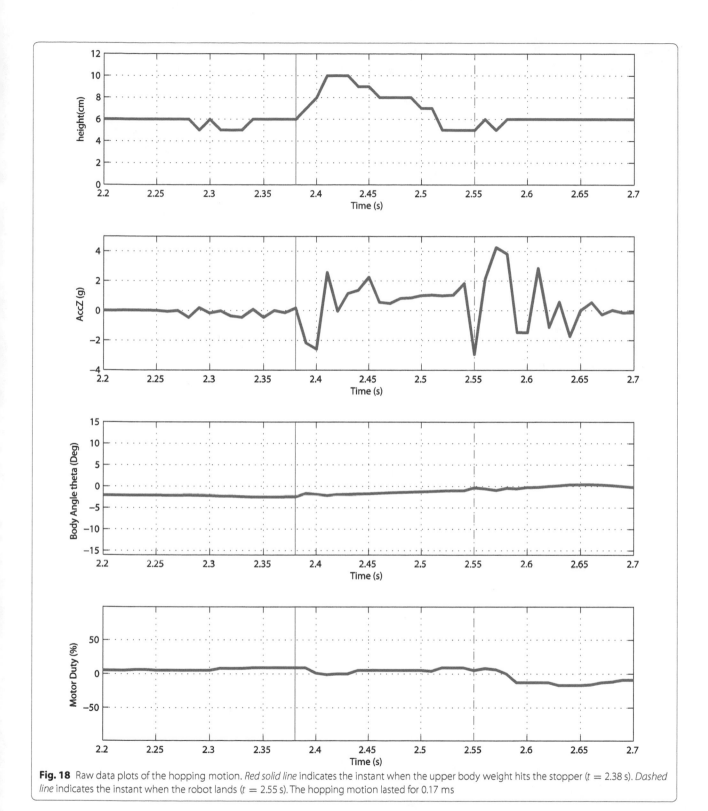

Fig. 18 Raw data plots of the hopping motion. *Red solid line* indicates the instant when the upper body weight hits the stopper ($t = 2.38$ s). *Dashed line* indicates the instant when the robot lands ($t = 2.55$ s). The hopping motion lasted for 0.17 ms

increase the size of the prototype to accommodate larger springs to increase the lifting force and enable the robot to jump at a higher height. We plan to take advantage of the dynamic nature of the robot to travel up and down step terrains, leap over gap obstacles in a swift and reliable manner. We are also investigating laser scanners or radar sensors in replace of sonar sensor for height detection.

Abbreviations

IMU: inertia measurement unit; DCM: direct cosine matrix.

List of symbols

m_b	mass of body
m_w	mass of wheels
I_b	inertia of body around body center of mass
I_w	inertia of wheels around wheel center of mass
R	wheel radius
l_b	length of upper body's center of mass to robot's center of mass
l_w	length of wheel axis to robot's center of mass
l	length of upper body's center of mass to wheel axis
g	gravity
θ_b	tilt angle of robot body
θ_w	rotational angle of wheels
τ_a	motor torque in air
τ_g	motor torque on ground

Additional files

Additional file 1. BBot-StepTraverseExperiment Description: Movie showing step traverse experiment of BBot.

Additional file 2. BBot-TossExperiment Description: Movie showing tossing experiment of BBot.

Additional file 3. BBot-HoppingExperiment Description: Movie showing the hopping experiment of BBot.

Authors' contributions

HEY developed the methodology, performed the analysis, constructed the prototype and wrote the manuscript. Both authors read and approved the final manuscript.

Competing interests

The authors declare that they have no competing interests.

References

1. Kovac M, Fuchs M, Guignard A, Zufferey J-C, Floreano D (2008) A miniature 7g jumping robot. In: 2008 IEEE international conference on robotics and automation. ICRA 2008. pp 373–378. doi:10.1109/ROBOT.2008.4543236
2. Scarfogliero U, Stefanini C, Dario P (2006) A bioinspired concept for high efficiency locomotion in micro robots: the jumping robot grillo. In: Proceedings 2006 IEEE international conference on robotics and automation. ICRA 2006. pp. 4037–4042. doi:10.1109/ROBOT.2006.1642322
3. Zhao J, Xi N, Gao B, Mutka MW, Xiao L (2011) Development of a controllable and continuous jumping robot. In: 2011 IEEE international conference on robotics and automation (ICRA), pp 4614–4619. doi:10.1109/ICRA.2011.5980166
4. Niiyama R, Nagakubo A, Kuniyoshi Y (2007) Mowgli: a bipedal jumping and landing robot with an artificial musculoskeletal system. In: 2007 IEEE international conference on robotics and automation, pp 2546–2551. doi:10.1109/ROBOT.2007.363848
5. Lambrecht BGA, Horchler AD, Quinn RD (2005) A small, insect-inspired robot that runs and jumps. In: Proceedings of the 2005 IEEE international conference on robotics and automation. ICRA 2005, pp 1240–1245. doi:10.1109/ROBOT.2005.1570285
6. Armour R, Paskins K, Bowyer A, Vincent J, Megill W (2008) Jumping robots: a biomimetic solution to locomotion across rough terrain. Bioinspir Biomim 3(3):039801
7. Stoeter SA, Papanikolopoulos N (2006) Kinematic motion model for jumping scout robots. IEEE Trans Robot 22(2):397–402. doi:10.1109/TRO.2006.862483
8. Kikuchi K, Sakaguchi K, Sudo T, Bushida N, Chiba Y, Asai Y (2008) A study on a wheel-based stair-climbing robot with a hopping mechanism. Mech Syst Signal Process 22(6):1316–1326. doi:10.1016/j.ymssp.2008.03.002 (Special Issue: Mechatronics)
9. Schmidt-Wetekam C, Bewley T (2011) An arm suspension mechanism for an underactuated single legged hopping robot. In: 2011 IEEE international conference on robotics and automation (ICRA). pp 5529–5534. doi:10.1109/ICRA.2011.5980339
10. Railbert M, Blankespoor K, Nelson G, Playter R (2008) Bigdog, the rough-terrain quaduped robot. In: The international federation of automatic control
11. Yap HE, Hashimoto S (2012) Development of a stair traversing two wheeled robot. In: 2012 IEEE/RSJ international conference on intelligent robots and systems (IROS). pp 3125–3131. doi:10.1109/IROS.2012.6385767
12. Burdick J, Fiorini P (2003) Minimalist jumping robots for celestial exploration. Int J Robot Res 22(7–8):653–674. doi:10.1177/02783649030227013. http://ijr.sagepub.com/content/22/7-8/653.full.pdf+html
13. Batista P, Silvestre C, Oliveira P, Cardeira B (2010) Low-cost attitude and heading reference system: filter design and experimental evaluation. In: 2010 IEEE international conference on robotics and automation (ICRA), pp 2624–2629. doi:10.1109/ROBOT.2010.5509537
14. Lin Z, Zecca M, Sessa S, Bartolomeo L, Ishii H, Takanishi A (2011) Development of the wireless ultra-miniaturized inertial measurement unit wb-4: preliminary performance evaluation. In: 2011 Annual international conference of the IEEE Engineering in Medicine and Biology Society, EMBC, pp 6927–6930. doi:10.1109/IEMBS.2011.6091751
15. Madgwick SOH, Harrison AJL, Vaidyanathan R Estimation of imu and marg orientation using a gradient descent algorithm. In: 2011 IEEE international conference on rehabilitation robotics (ICORR), pp 1–7 (2011). doi:10.1109/ICORR.2011.5975346

15

TITAN-XIII: sprawling-type quadruped robot with ability of fast and energy-efficient walking

Satoshi Kitano[1][*][†] , Shigeo Hirose[2][†], Atsushi Horigome[1][†] and Gen Endo[1][†]

Abstract

In this paper, we discuss development of a sprawling-type quadruped robot named TITAN-XIII which is capable of high speed and energy efficient walking. We consider a sprawling-type quadruped robot is practical, because of its high stability which comes from the large supporting leg polygon and the low center of gravity. However in previous researches, the speed and the energy efficiency of a sprawling-type quadruped robot is lower than a mammal-type quadruped robot. Since cost of transport (COT) can be reduced by increase of walking velocity, we decided to design a fast walking sprawling-type quadruped robot. As a demonstrator, we developed the sprawling-type quadruped robot named TITAN-XIII. For a lightweight and compact leg, the right-angle type wire driven mechanism is adopted to the robot. To confirm its performance, several experiments were carried out and the robot walked at 1.38 m/s and COT of 1.76 was achieved. Finally, we compared the performance of TITAN-XIII with other quadruped robots, and confirm that its performance is almost same level as mammal-type quadruped robots.

Keywords: Quadruped robot, Wire driven, Dynamic walking

Background

There has been increasing interest in walking robots which can walk over uneven terrain where it is difficult for wheeled or crawler type robots to travel. Among the various types of walking robots, the four-legged walking robot has the minimum number of legs establishing static walking, and thus is regarded as highly practical configuration.

Quadruped robots can be classified into two groups, a mammal-type and a sprawling-type, according to its leg configuration (Fig. 1). Here, a mammal-type means the robot which locates its foot vertically downward from the base of the leg as a standard posture (Fig. 1a). A sprawling-type means the robot whose first leg segment (thigh) is in horizontal direction and second leg segment (shank) is in vertical direction as a standard posture (Fig. 1b).

There are several advantages in a mammal-type quadruped robot. First, a mammal-type quadruped robot can walk faster than a sprawling-type quadruped robot by utilizing two actuators (e.g., hip and knee) in each leg. Second, a mammal-type quadruped robot required small torque on each joint by straighting its leg, especially when the robot stands. Third, because of the its small footprint, the robot can walk through narrow space or side of a cliff like a mountain goat doing.

Recently, as a represented by BigDog [1] developed by BostonDynamics, mammal-type quadruped robots is studied actively. HyQ [2] which is a hydraulic quadruped robot, is capable to walk over rough terrain using the active compliance of the joints. MIT Cheetah [3] which is an electric quadruped robot, achieves energy efficient walking and dynamic jumping by developing a high torque density electromagnetic motor. StarlETH [4] which equips the series elastic actuator (SEA) on each joint, can walk on rough terrain interacting with ground dynamically. CheetahCub [5] and Tekken [6] also equip springs in its leg structure and utilize the springs for walking on rough terrain. Additionally,

*Correspondence: kitano.s.ac@m.titech.ac.jp
[†]All authors contributed equally.
[1] Department of Mechanical and Aerospace Engineering, Tokyo Institute of Technology, Meguro-ku, Okayama 2-12-1, 152-8550 Tokyo, Japan
Full list of author information is available at the end of the article

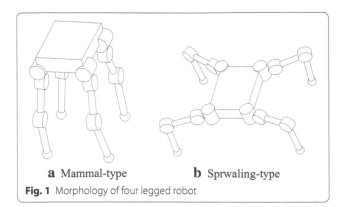

a Mammal-type   **b** Sprwaling-type

Fig. 1 Morphology of four legged robot

in Learning Locomotion Program held by Defense Advanced Research Projects Agency (DARPA), LittleDog is developed by BostonDynamics and several US universities studied about learning and planning of locomotion with the developed robot [7–9].

On the other hand a sprawling-type quadruped robot has another features. First, the robot has high stability, because the robot can locate its center of gravity at low position and have a wider supporting leg polygon. Second, the robot has wide range of motion because of its proximal yaw axis, therefore it can choose foot placement widely. Third, since its center of gravity is low, even if the robot falls down, the damage to the robot is considered to be relatively small. Additionally, it is easy to use the body of the robot as a fifth foot depending on the terrain.

Considering about a practical legged robot, the robot have to be capable of carrying objects or doing some operation rather than just moving. Otherwise, just for moving over rough terrain, another type of robot such as a flying robot also can perform. To carry objects, lowering center of gravity reduces the risk of a falling and damages. For doing some operation, wide range of motion helps to establish a stable platform like an outrigger. Thus, we consider that a sprawling-type quadruped robot is a practical quadruped robot.

So far, various types of sprawling-type quadruped robot has been developed and their high rough terrain adaptability is reported. TITAN III [10] walked over stair case autonomously using "the whisker sensor" equipped on each foot. SILO4 [11] has a compact leg using the spiroid gear, and walk over a step and a slope using a force sensor on each foot. MRWALLSPECT IV [12] succeeded to walk over various type of rough terrain autonomously by introducing the body workspace concept. Compact rescure robot (CRR) [13] achieves big payload more than its weight by using a pneumatic actuator on each joint.

Although a sprawling-type robot is highly practical, it is generally said that a sprawling-type quadruped robot has following disadvantages. First, because of the sprawling

first segment, the proximal pitch axis always has to generate the torque to support its own weight, therefore its energy efficiency seems to be low. Second, mainly the walking velocity is generated by only proximal yaw actuator, its walking speed would be limited compared to mammal-type. Although the walking velocity and the energy efficiency are very important parameter for a moving machine, so far there is almost no research about walking velocity and energy efficiency of a sprawling-type quadruped robot.

Thus, the purpose of this research is to investigate and improve walking velocity and energy efficiency of a sprawling-type quadruped robot. Contribution of this paper is proving that a sprawling-type quadruped robot can walk fast and efficiently against common belief of the sprawling-type quadruped robot, and proposing a mechanical design of a sprawling-type quadruped robot to achieve fast walking and energy efficiency.

This paper first discuss how to improve walking velocity and energy efficiency. Then, we explain in detail about the sprawling-type quadruped robot named TITAN-XIII which developed as a demonstrator. Finally, we carried out experiments to verify its walking velocity and energy efficiency, and compare its performance with other quadruped robots.

Design concept

In this section, we discuss how to improve walking speed and energy efficiency of the sprawling-type quadruped robot.

Walking velocity and energy efficiency

First of all, walking velocity and energy efficiency are not independent parameter, and there is a strong relationship between two parameters. As a criterion of walking efficiency of a legged robot, cost of transport (COT), also known as specific resistance [14] is used broadly. COT is defined as follows:

$$COT = \frac{E}{mgd} = \frac{P}{mgv},\qquad(1)$$

where E: consumed energy [J], m: mass of the robot [kg], g: gravitational acceleration [m/s^2], d: moving distance [m], P: consumed power [W], v: walking velocity [m/s]. Smaller COT means higher energy efficiency. The equation shows that increase of velocity causes decreasing COT. However, normally increase of velocity also causes increase of the mechanical energy which is required to move forward.

In case of robots, generally consumed energy means all of the energy consumption of the robot, including electric circuits such as microcontrollers and sensor devices which does not directly contribute to generate mechanical work. Additionally in case of legged robots, the robot have to support gravitational force which wheeled or tracked type robot does not need to support. These energy consumption is not small in reality. For example, in case of the sprawling-type quadruped robot TITAN-VIII [15], the electric energy consumption is almost 40 % of its total energy consumption when the robot walks at 0.1 m/s [16]. However, unlike the mechanical energy, these consumed energies are proportional to a running time. Therefore by reducing a running time, e.g. increase of walking velocity, total COT can be reduced because of decrease of the electric energy and the gravitational support energy.

Increase of walking velocity

To increase the walking velocity, the foot velocity in the moving direction should be as fast as possible. In case of a legged robot, the leg is always reciprocated. Hence, the inertia of the leg should be small as much as possible to achieve fast leg movement and high energy efficiency.

In addition to the foot velocity, walking gait is also important for increasing walking velocity. In the previous researches of sprawling-type quadruped robots, the crawl gait is mainly focused [12, 15], because of its static stability. However the walking speed of the crawl gait is relatively slower than other dynamic walking gaits. On the other hand, in researches of mammal-type quadruped robots, the trot gait is mainly used.

To compare each gait, we estimate the walking velocity of legged robot which can be defined as follows:

$$V = \frac{(1 - \beta)U}{\beta}, \quad (2)$$

where U: a foot velocity, β: walking duty factor ($\beta > 0.5$). According to the equation, the maximum walking speed of the crawl gait is $\frac{1}{3}U$ at minimum duty factor of 0.75. Similarly, the maximum walking speed of the trot gait is U at minimum duty factor of 0.5. The trot gait is theoretically three times higher than the crawl gait. Thus, it is very effective to choose the low duty factor gait such as the trot gait to increase walking velocity.

Additionally, we consider the trot gait is also having high stability on rough terrain. Assuming the robot walking with the trot gait and it starts to fall because of unexpected disturbances or errors, the robot rotates around diagonal supporting line. However in both rotating directions, the swinging legs exist, and one of the swinging legs hits the ground and forms a supporting leg polygon with the supporting legs. Because of the newly formed supporting polygon, the robot will maintain standing posture if the tumbling velocity is low.

Thus, to achieve fast walking on a rough terrain, we believe the trot gait is most suitable.

Reducing gravitational support energy

Although the electric and the gravitational support component of COT can be decreased by increase of velocity, still the electric energy consumption and gravitational support energy consumption remain in the system. Especially, in case of the sprawling-type quadruped robot, gravitational support energy tend to be bigger than the mammal-type quadruped robot, because of its horizontally extended leg. To decrease gravitational support energy, it is considered to use non-backdrive mechanism such as a worm gear and a screw mechanism. However this kind of mechanisms has disadvantages which is heaviness and low transmission efficiency. Because of these disadvantages of non-backdrive mechanisms, it does not suit for a fast walking quadruped robot. As a practical solution, the reduction ratio of an actuator supporting gravitational power should be as high as possible, within achieving required velocity for walking motion.

Summarize above discussion, the important points to achieve high speed and energy efficiency are below.

- COT can be reduced by increasing walking velocity.
- By using a low duty factor gait such as the trot gait, the walking velocity can be increased.
- To reduce gravitational support energy, the actuator which supports gravitational force should have large reduction ratio.

Hardware design

As a demonstrator model, we developed the new sprawling-type quadruped robot named TITAN-XIII in accordance with the previous design concept. Overall view of the robot is shown in Fig. 2 and Table 1 shows the specification of the robot. We also define the standard posture as shown in Fig. 3 and the size of Table 1 is based on this standard posture.

The weight, reduced by decreasing the component count as far as possible and using carbon fiber reinforced plastic for main structure of each leg segment, is only 5.65 kg including the battery. As an actuator, customized DC brushless motors (FX1206-11 made by Nippo Denki Co., Ltd., max. power 68 W) is used for all of the joint. As a result, the power-weight ratio which is summation of maximum output of actuators divided by weight of the robot is 144.4 W/kg, which is about 4 times higher than the previously developed sprawling-type quadruped robot TITAN-VIII [15] with similar dimension whose power-weight ratio is 40 W/kg.

Additionally a modularized design was adopted for maintainability, and the robot is composed of four identical leg unit. This makes it possible to easily maintain each unit, and if a malfunction occurs, function can be restored simply by replacing the malfunctioning unit.

Fig. 2 Sprawling-type quadruped robot TITAN-XIII

Table 1 Specification of the TITAN-XIII

Characteristic	Value
Size (L × W × H)	213.4 × 558.4 × 340.0 mm
Weight (w/o battery)	5.29 kg
Weight (with battery)	5.65 kg
Payload	5.0 kg
Battery	LiFe 26.4V 1100mAh
Battery run time	approx. 20 min.

With regard to the power source, power is supplied to the actuators and microcontrollers by an LiFePO4 battery (26.4 V, 1100 mAh) made by A123 and installed in the center of the body.

Mechanism of leg unit

Figure 4 shows the overall view of the leg unit and Fig. 5 shows a schematic and detailed diagram of the leg unit. Table 2 shows the specification of the leg unit.

The leg unit has three degrees of freedom and consists of a planar mechanism with two degrees of freedom employing two pitch axes (Axis 2 and Axis 3), and a yaw axis (Axis 1) which rotates the planar mechanism. To reduce the inertia of the leg around the yaw axis, the actuators for pitch axes are placed close to the yaw axis and transmit power to each joint by using wires.

Two wires each are used to drive one joints. Since the diameter of idler pulley on Axis 2 and the diameter of the pulley fixed on Axis 3 are the same, this planar mechanism comprises a parallel mechanism.

For the transmission of the yaw axis (Axis 1), a timing belt and a pulley are used. Although a pulley is made with metal material usually, but this time the pulley made with poly-acetal plastic is used as an output pulley for lightweight. Since it is difficult to have small teeth on a plastic

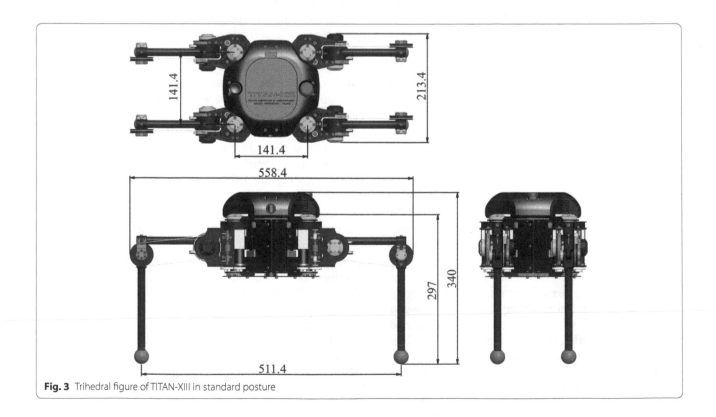

Fig. 3 Trihedral figure of TITAN-XIII in standard posture

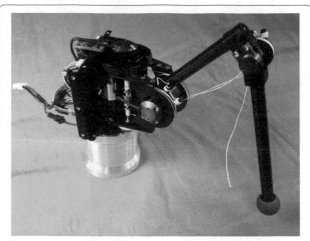

Fig. 4 Leg unit of TITAN-XIII

pulley, the pulley has a dent to fasten a timing belt with a screw via tensioner plate as show in Fig. 6.

The screw fixing the timing belt to the pulley also work as tensioner which is usually achieved by additional idler pulley. This pulley with dent can not be used in case of requiring rotation of 360° or more, but can be regarded as effective at case of where only 180° of range of motion

is required, such as the Axis 1. As shown in Fig. 7, the motor is fixed to the back side of the leg base section, and the timing belt is connected to the output axis via an opening provided in the leg base.

The gear reduction ratio of each axis is shown in Table 2. As we discussed above, the reduction ratio of Axis 1 is set to low to achieve high walking speed. The reduction ratio of Axis 2 is set to high, as the leg can support whole weight of the robot. In case of Axis 3, we consider it does not require the so much torque as Axis 2.

Therefore Axis 3 uses a smaller pulley and lower gear reduction ratio than Axis 2.

The range of motion of the planer mechanism is shown in Fig. 8. All of the joint have range of motion of ± 90°, and makes it possible to completely fold up a leg to the inside.

Wire driven mechanism

As we mentioned above, the wire driven mechanism is used for the Axis 2 and the Axis 3. Figure 9 shows the wire driven mechanism for the Axis 2. For the Axis 3, almost symmetric design is used with extra relay pulley to transmit the Axis 3.

There are several reasons, aside from transmitting power to a remote location, why a wire driven mechanism

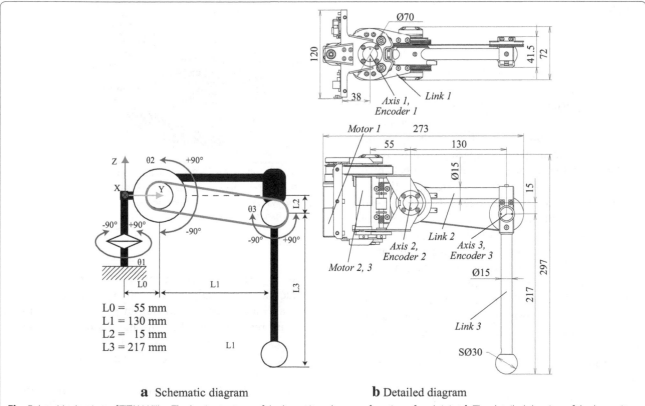

a Schematic diagram　　　　　**b** Detailed diagram

Fig. 5 Leg Mechanism of TITAN-XIII. **a** The basic structure of the leg unit and range of motion of each joint. **b** The detailed drawing of the leg unit and indicates the position of the motors and the encoders

Table 2 Specification of the leg unit

Weight	1.2 kg
Actator	Nippo Denki: FX1206-011 brushless DC motor (max. output:68W)
Motor driver	Hibot: 1BLDC power module
Reduction mechanism for the axis1	Planetary gear, timing belt
Reduction mechanism for the axis2, 3	Planetary gear, spur gear, wire and pulley
Reduction ratio (each reduction ratio of mechanism)	Motor1: 115 (23.04, 5)
	Motor2: 273 (23.04, 1.4, 8.5)
	Motor3: 163 (23.04, 1, 7.1)
Sensor	Sastinable robotics: 16 bit absolute magnet encoder
Power weight ratio	170 W/kg

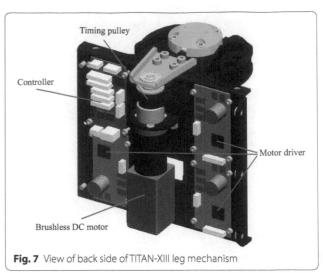

Fig. 7 View of back side of TITAN-XIII leg mechanism

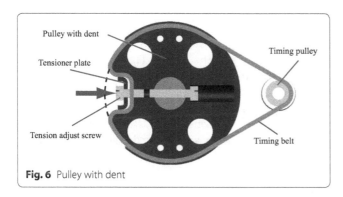

Fig. 6 Pulley with dent

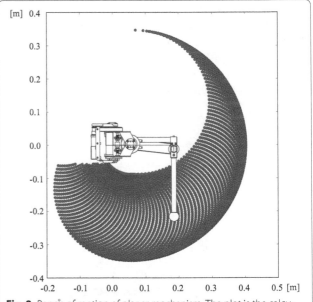

Fig. 8 Range of motion of planer mechanism. The plot is the calculated by forward kinematic inputting angle of Axis 2 and Axis 3. Input angle are changed from $-\pi/2$ rad to $\pi/2$ rad by 0.05 rad

is used instead of usual gears for the final output stage of the joint mechanism.

The most important feature is that the wire driven mechanism can achieve higher allowable torque density (allowable torque of reduction mechanism [Nm] / weight of reduction mechanism [kg]) comparing to other reduction mechanisms. In case of a gear reduction mechanism, the power can be reliably transmitted. However the output torque is supported by one or two of the gear teeth, and in order to output high torque, the module of the gear must be large or material of the gear need to have high yield strength which generally makes a mechanism heavy such as steel. The same is true in the case of a timing belt, which can transmit power to a separate axis just like wire. In the case of the wire driven mechanism, strength of wire that transmits torque between pulleys, limits allowable torque. On the side of pulley in which wire wound, since wire tension is attenuated along with winding onto the pulley surface, even plastic material can endure wire tension.

Additionally the wire driven mechanism does not have backlash which is major problem in gear reduction mechanism.

In case of normal gear reduction mechanism, increasing reduction ratio only affects pitch diameter ratio between an input gear and an output gear. However in case of wire reduction mechanism, increasing reduction ratio also causes increasing number of windings of wire and axial direction length of the pulley if range of motion of output pulley is kept.

As shown in Fig. 10a, in most of cases axes of the output pulley and the input pulley are arranged in parallel. In this configuration the whole mechanism needs to expand

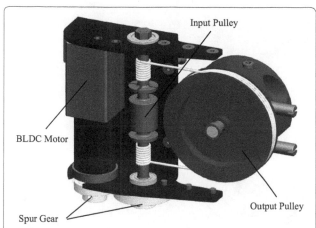

Fig. 9 Wire driven mechanism which is comprised of a brushless DC geared motor, spur gears, two wires, and input and output pulleys

in both direction, radial direction of the output pulley and axial direction of the input pulley, as the reduction ratio is increased. Additionally, when the input pulley is rotated, the point where the wire separates from the input pulley moves in the axial direction, and thus the output pulley must have equivalent length in the axial direction.

Therefore we choose another pulley arrangement, where the input pulley and output pulley are arranged at right angles as shown in Fig. 10b. In this configuration, the axial direction length of the input pulley, which increases as the reduction ratio increases, is matched with the radial direction of the output pulley, and this conserves space.

A space-saving new wire reduction mechanism achieved a high reduction ratio of 8.46 on Axis 2, taking into account an input pulley diameter of 5.5 mm, output pulley diameter of 54.5 mm, and wire diameter of 1.05 mm. In the case of Axis 3 output pulley diameter is 45 mm and the reduction ratio of 7.1 is achieved.

One problem of this wire driven mechanism with right-angle type is that, as the output pulley rotates the point of separation of the wire on the input pulley moves up and down because of the winding of the wire as shown in Fig. 11. This motion changes the path length of the wire and adds extra tension on the wire. Theoretically speaking in order to solve this problem, an input pulley or an output pulley must have sliding degree of freedom along its axis to absorb the up and down motion, but this makes the mechanism large and complicated. Therefore, we examined the elongation of the wire in right-angle type wired driven mechanism. In Fig. 11, L is the length of the wire between the input pulley and the output pulley in the standard posture where theoretically the wire

is perpendicular to the input pulley. Now assuming the output pulley rotates θ, and the input pulley rotates $\phi = reduction\ ratio \times \theta$. Then the point of separation of the wire is moved distance $dl = d \times \phi / 2\pi$, where d is diameter of the wire. Along with the moving of the point of separation of the wire, the length of wire between the input pulley and the output pulley is changed to L'. This new wire length L' can be expressed as $\sqrt{dl^2 + L^2}$. As a result, wire elongation $L' - L$ can be calculated. Since Axis 2 has larger reduction ratio of 8.46 than Axis 3 we will check elongation of Axis 2. Considering L is 40 mm and output pulley rotates $\pm$ 90°, estimated elongation $L' - L$ is only 0.06 mm. Thus we consider this elongation is sufficiently tolerable and does not need a complicated mechanism.

Synthetic fiber rope

For the wire driven mechanism, the property of the wire affects its performance. Conventionally a stainless steel wire rope is commonly used for the wire driven mechanisms including TITAN-VIII [15]. However, it is difficult to wind stainless steel wire onto a pulley which has an extremely small diameter such as the input pulley used for the TITAN-XIII. Thus, we decided to use a synthetic fiber rope due to its expected superior flexibility to stainless steel wire. There are many types of synthetic fiber rope has been developed. Among them, we choose Ultra-high-molecular-weight polyethylene (UHMWPE) a.k.a Dyneema, because of its high tensile strength of 2.6 GPa [17] which is almost five times higher than SUS304 (around 0.5MPa). As a Dyneema rope, we used DB-28HSL (developed by Hayami Indutstry Co., Ltd, shown in Fig. 12) which is braided of eight Dyneema strands. Table 3 shows the basic properties of the Dyneema rope and stainless steel wire rope. Because of braiding, breaking force of the DB-28HSL is low considering tensile strength of the material, but it is still higher than stainless steel wire rope. Thus, we consider Dyneema is suitable for robotic application.

The problem of using the wire driven mechanism is an elongation of a wire, which decreases mechanical bandwidth. Therefore to be able to adjust wire tension easily a co-axial tensioner shaft shown in Fig. 13 was devised as the input pulley. The co-axial tensioner shaft has a dual structure made up of a core shaft which passes through the center of the input axis pulley, and an outer shaft which covers the lower half. Each shaft is connected with a compact roller type one-way clutch (NTN: HF0612, torque capacity 1.76 Nm). Due to the one-way clutch, the two shafts turn relative to each other only in the direction which increases wire tension, and thus it is possible

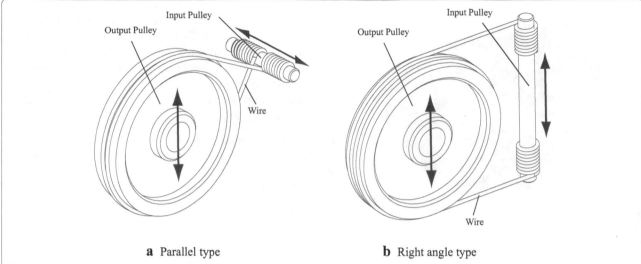

a Parallel type   **b** Right angle type

Fig. 10 Comparison of the pulley arrangement of wire reduction mechanism. The *arrows* show the expanding direction of each pulley if reduction ratio is increased. In case of **a** expanding direction is different, but in case of **b** expanding direction is same

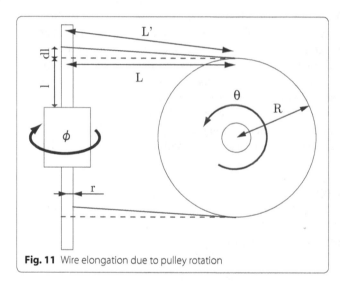

Fig. 11 Wire elongation due to pulley rotation

to adjust tension of the wire by rotating one of the shafts. The wire which is wound to the core shaft transmit power directly to the motor, but the wire wound to the outer shaft transmit power via one-way clutch.

Since the one-way clutch has maximum transmittable torque, the transmittable torque of the outer shaft is limited by torque capacity of the one-way clutch. However walking robots normally require high torque in one direction that gravity works, we deploy wire driven mechanism as the core shaft supports the direction of the gravity force.

Control system

Figure 14 shows an schematic of the control system of TITAN-XIII.

The main control program running on an external PC calculates each joint angle using inverse kinematics from an designed leg trajectory. This control program also provides interface to control a robot motion via a GUI and a joystick. Calculated joint angles are sent each 20 ms via WiFi communication using IEEE802.11ac (5 Ghz) avoiding radio frequency interference most commonly used 2.4 GHz of bandwidth. The robot equips a small WiFi router and the TITech M4 controller (made by Hibot corp.) for communication with the router and the each leg unit.

Each leg unit equips the TITech SH2 Tiny Controller (made by Hibot corp.) and three motor drivers on the back of the leg unit as shown in Fig. 7. The microcontrollers receive joint angle data from the external PC and performs position control with a 1 kHz-cycle PID controller and interpolation of joint angle. Every joint angle is measured by the 16-bit magnet encoders made by Sustainable Robotics (using the AEAT-6600-T16 of AVAGO Technologies) mounted on each joint as an absolute angle sensor.

Experiment

Position tracking of the foot

Firstly, position tracking performance of the leg unit was tested. In the experiment, the leg unit executed walking motions in the air.

As a walking motion, the longitudinal acceleration trot gait which is described in detail in our previous work [18] was used. The commanded swing height and walking stride was kept at 0.05 and 0.18 m. In the experiment, the robot was in the posture shown in Fig. 15 as a normal

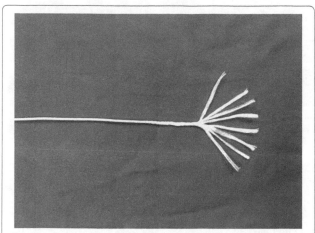

Fig. 12 Dyneema Rope DB-28HSL made by Hayami Industry Co., Ltd

Table 3 Properties of Dyneema rope and stainless steel wire rope

Rope material	DB-28HSL dyneema	Stainless steel wire rope(1x7) SUS304
Diameter	1.05 mm	1.0mm
Breaking force	1.38 kN	1.03 kN
Weight per length	0.66 g/m	5.24 g/m
Specific strength	2060 kNm/kg	197 kNm/kg

Breaking force of DB-28HSL were measured by HAYAMI and weight per Length was measured by authors. Properties of stainless steel wire rope is refereed from JIS G3540(Wire ropes for mechanical control)

posture. The position of the foot was measured by using a motion capture system (Flex 13 by OptiTrack). Since we cannot put a reflection marker on the foot directly, we put the reflection marker on the Link 3 and calculated the position of the foot.

Figures 16 and 17 show the commanded and measured foot trajectory of walking motion at 0.33 m/s of the target velocity (0.94 Hz) and 1.2 m/s of the target velocity (3.1 Hz) respectively. In case of 0.33 m/s, the foot mostly follows the commanded trajectory. However, in case of 1.2 m/s there is big errors especially while the transition between support phase and swing phase where right and left edge of the trajectory. The reason of these errors are considered as error of the magnetic encoder which is used for joint control. Since magnetic encoder has internal averaging function to output stable value, when the joint rotates fast, it outputs different value to actual joint angle position. Therefore, because of this magnetic encoder averaging issue, we were getting inaccurate feedback for the PID controller at high rotational speed. Then, when setting a commanded angle position to the joint, the output position exceeded the commanded position, resulting in a difference between the commanded foot position and the foot position measured by the motion capture system as shown in Fig. 17. However, since the measured stride exceeds commanded stride and its error to commanded stride is 11 %, we consider the robot can walk at target velocity or more.

Walking experiment

Walking experiments were conducted to confirm the walking velocity and the energy consumption of TITAN-XIII.

As a walking algorithm, the same longitudinal acceleration trot gait was used. In the experiment, the robot walked straight 3.0 m on a flat wooden plate. Through the experiments the duty factor was kept at 0.55, but the target velocity and the stride which are control parameters are changed gradually. While the walking experiment, the normal posture of the robot was the posture shown in Fig. 15. Comparing to the standard posture (Fig. 15), the angle of the diagonal supporting line is increased to avoid collision between the forward feet and rear feet when the

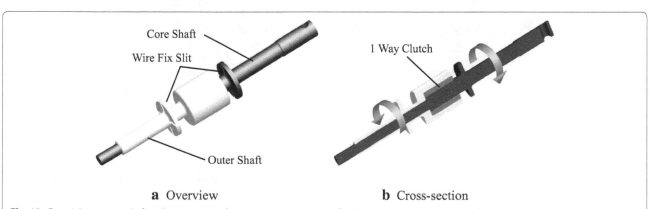

a Overview   **b** Cross-section

Fig. 13 Co-axial tensioner shaft. **a** The overview of the co-axial tensioner shaft. The white part is the outer shaft and the gray part is the core shaft which penetrate the outer shaft. **b** The cross section of the co-axial tensioner shaft. The outer shaft and the core shaft are connected by one-way clutch and only rotate one direction which increases tension of the wire

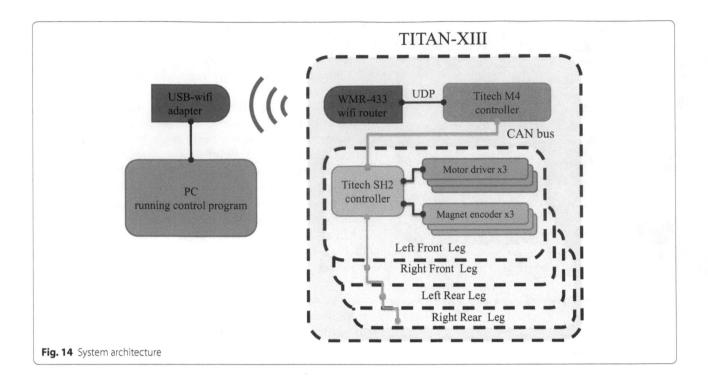

Fig. 14 System architecture

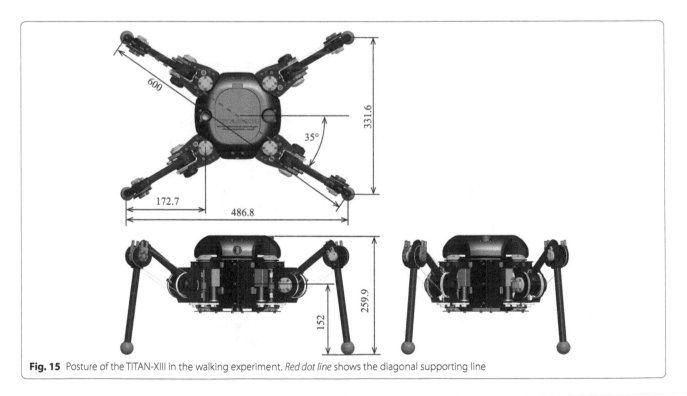

Fig. 15 Posture of the TITAN-XIII in the walking experiment. *Red dot line* shows the diagonal supporting line

stride is increased. Furthermore the height of the robot is decreased to avoid accidental complete tumbling. Relationship between a height of the center of gravity and stability of the legged robot is known as Normalized Energy Stability Margin (NESM) [19] which indicates a required

energy to complete tumble normalized by the weight of a robot. According to the definition of NESM, lower center of gravity means higher stability.

Walking velocity is measured by using the motion capture system which also used in the position tracking of

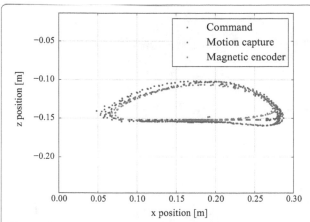

Fig. 16 Foot trajectory at target walking velocity 0.33 m/s with 0.18 m of stride

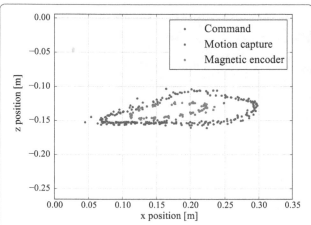

Fig. 17 Foot trajectory at target walking velocity 1.2 m/s with 0.18 m of stride

the foot experiment. A power supply was connected to the robot to measure the consumed current although the battery is still installed in the robot. The consumed power was estimated by multiplying the 26.4 V of supply voltage and the measured current. This power includes not only an energy consumption in the actuators but also one in the electronics e.g. the microcontrollers, the WiFi router and a DC/DC converter. Each experiment was conducted three times with same parameter, and the result is shown as an average of three experiments.

Figure 18 shows side view of the sequence of walking experiment with 0.2 m/s of the target velocity and 0.04 m of the stride and at this time the measured velocity was 0.22 m/s. As shown in the figure, the robot walked stably at constant speed and the body posture was kept in almost horizontal.

In the experiment, 1.38 m/s of walking speed was achieved at target velocity 1.2 m/s and at this speed

the robot consumed 135 W. The reason of higher walking velocity than target velocity is considered as longer stride which is also confirmed in Fig. 17. Figure 19 shows the commanded foot trajectory and measured foot trajectory while walking at 1.38 m/s. As shown in Fig. 19, actually the measured stride is larger than commanded stride.

We also conducted walking experiment with an additional mass. In static payload experiment, the robot can support 20 kg of a weight continuously by four legs. Assuming walking with trot gait, the robot have to support weight by only two leg in most of the sequence and can walk with additional mass of 10 kg. However we set safety factor and conducted experiment with 5 kg of additional mass which is almost same as the weight of the robot.

The weight is fixed on the top of the robot putting the sponge rubber between the robot and the weight. Although the normal position of the foot is same as the Fig. 15, the height of the robot was decreased 0.02m for stable walking. As same as previous experiment, the target velocity and the stride was gradually increased.

As a result, the robot successfully walked with additional mass and the fastest speed in the experiment was 0.9 m/s with 0.15 m of stride at target velocity 1.0 m/s. At this speed the robot consumed 152 W, whereas the robot consumed 103 W without payload at 0.96 m/s.

Increased power is considered to be consumed to support, accelerate and decelerate the additional weight.

Energy efficiency

By using the measured power and the velocity, COT of each walking parameter were calculated. The Fig. 20 shows the COT of the TITAN-XIII at each velocity and also the ratio of power consumption in COT. The shown velocity and COT is average of three trial.

As shown in the Fig. 20, the COT is decreased as the walking velocity increased. As a result, at 1.38 m/s of the walking speed, the COT of 1.76 which is the lowest COT through the experiment was achieved.

To examine each component of power consumption, we define "the electric power consumption" which is the power consumption in the microcontrollers, the WiFi router and the DC/DC converter, "the gravitational support power consumption" as the power consumption of the robot standing still subtracted by the electric power consumption and "the mechanical power consumption" as the power consumption measured while walking subtracted by the electric power consumption and the gravitational support power consumption. In the experiment, electric power consumption is alway around 10 W, and gravitational support power is around 13 W when the robot stands with four legs.

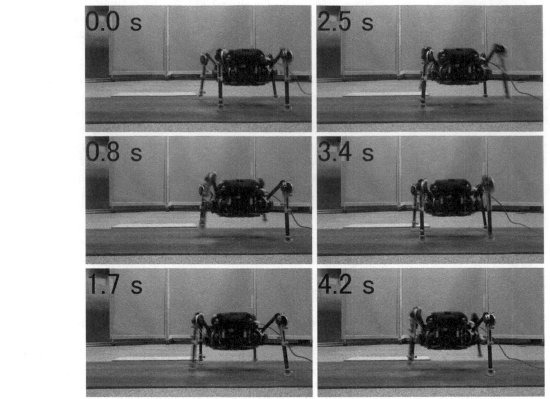

Fig. 18 Walking experiment at target velocity 0.2 m/s with 0.04 m of stride. The measured velocity was 0.22 m/s

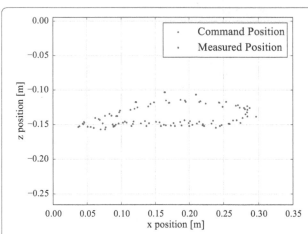

Fig. 19 Foot trajectory while walking on the ground at 1.38 m/s. Each foot trajectory is calculated with respect to the body coordinate system

The ratio of the electric power and the gravitational support power component of COT is decreased as walking velocity increased. At 1.38 m/s of the walking speed, a occupancy of the electric power consumption and the gravitational support power consumption in the total power consumption is 17 % whereas at 0.22 m/s the occupancy is 39 % As a result, the total COT is asymptotic to the mechanical component of COT. Therefore, we can confirm that the total COT is decreased as increase of walking velocity because of reducing the electric power consumption and the gravitational support power.

The mechanical component of COT is almost around 1.75 and it gradually decreased as increase of velocity. The reason of gentle decrease of mechanical COT is considered as the changing of the stride. By increasing the stride, the required number of steps to walk a desired distance is reduced. Hence, the number of acceleration and deceleration to rotate the leg unit is also decreased and causes decrease of mechanical energy consumption.

Discussion
Comparison with other robots
We compared minimum COT and maximum walking velocity of TITAN-XIII with other quadruped robots. Figure 21 shows the graph of minimum COT vs mass. Black markers indicate robots and white markers indicate animals which is adopted from Tucker's graph [20]. As we previously mentioned, in the case of a sprawling-type quadruped robot, there is no research about energy efficiency except our previous robot TITAN-VIII [16].

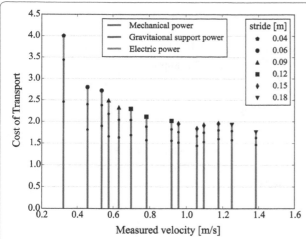

Fig. 20 COT vs velocity of TITAN-XIII: Each *bar* indicates total COT which include all energy consumption of the robot. *Blue bar* indicates the mechanical component of COT, *red bar* indicates gravitational support component of COT and *green bar* indicates electric component of COT. Net COT which exclude electric component is indicated as summation of *blue bar* and *red bar*

Therefore, most of the robots in the graph are a mammal-type quadruped robot.

Looking at other surrounding TITAN-XIII, the Scout II shows lower COT than TITAN-XIII, however Scout II can not select foot placement three dimensionally because of 4 DoF with hopping optimized leg design. Although StarlETH also shows slightly lower COT than TITAN-XIII, the weight of StarlETH is almost four times higher than TITAN-XIII. According to Tucker's graph, in the case of running animals minimum COT is proportional to the 0.3 power of the body weight, which is shown as a line in the Fig. 21. Therefore, a plot which is close or overcome this animal line means high performance and among the plotted robots, only MIT Cheetah overcomes this line.

Estimated minimum COT of the animal which has same weight with TITAN-XIII is around 1.0. This value is not so far from current lowest COT of 1.76. Actually except MIT Cheetah, TITAN-XIII is located closest to the estimated COT line of the animals.

We also plotted the lowest COT of 1.2 of TITAN-XIII with additional 5 kg of weight at 0.83 m/s of the walking velocity in Fig. 21. Comparing COT of the robot without payload, the COT of the robot with payload is close to animal line. Therefore we can say TITAN-XIII is not just optimized for fast walking but possible to achieve payload and energy efficiency at a same time.

Next, we compare walking velocity of quadruped robot by using Froude number. Froude number represents a normalization of the walking velocity for the size of the robot, and higher Froude number means faster walking

performance regardless. Froude number is defined as *Froude number* $= v^2/Lg$, where the $v\,[m/s]$ is walking velocity, the $L\,[m]$ is representative length and g is the gravitational acceleration [21]. According to literature [22] by Alexander, representative length: L is defined as "the height of the hip joint from the ground in normal standing". However, using the height of the hip joint to calculate Froude number is not suit for a sprawling-type quadruped robot, because in case of a sprawling-type robot, the height of the hip joint can be decreased until its belly touches to the ground without decreasing possible stride. We have considered important length to calculate Froude number is the length which actually generate speed. Thus we have extended the Alexander's definition of representative length to "the length between the hip joint axis and the foot projected on a plane orthogonal to the moving direction in normal standing posture"?. By using this extended definition of representative length, Froude number can be calculated for both the mammal-type and the sprawling-type. Especially in case of mammal-type, if the foot is on the same sagittal plane of the hip joint, the extended definition will get same result as the Alexander's definition. According to the extended definition, representative length of TITAN-XIII in walking experiment is 0.172 as shown in Fig. 18.

Figure 22 shows the graph of Froude numbers vs mass of quadruped robots and animals. As shown in the graph, velocity of TITAN-XIII is almost same level as other mammal-type quadruped robots such as Cheetah-cub and Raibert quadruped and much faster than TITAN-VIII.

From the comparison of COT and Froude number, we can say that the walking speed and energy efficiency of the sprawling-type quadruped robot can be the same level as the mammal-type quadruped robot. Especially, the energy efficiency of the TITAN-XIII is closer to animal line than most of mammal-type quadruped robot.

Validity of the sprawling-type quadruped robot

As we mentioned in the previous section, TITAN-XIII can achieve high walking velocity and energy efficiency than previous sprawling-type quadruped robot and its performance is almost same level as a mammal-type quadruped robot. Additionally the payload of TITAN-XIII which is almost same as its own weight, is relatively better than other mammal-type quadruped robots. Considering the nature of the sprawling-type robot which is high static stability and wide range of motion of the foot, we believe the sprawling-type quadruped robot is highly practical.

However we also see trade-off problem about energy efficiency and rough terrain adaptability. Although

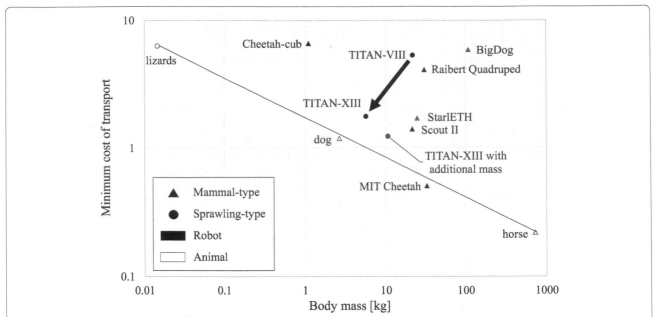

Fig. 21 Minimum Cost of Transport vs body mass of quadruped robots. This plot is based on Tucker's graph [20] and added relatively new quadruped robots [3, 5, 16, 24–27]. The *black line* is drawn by the author

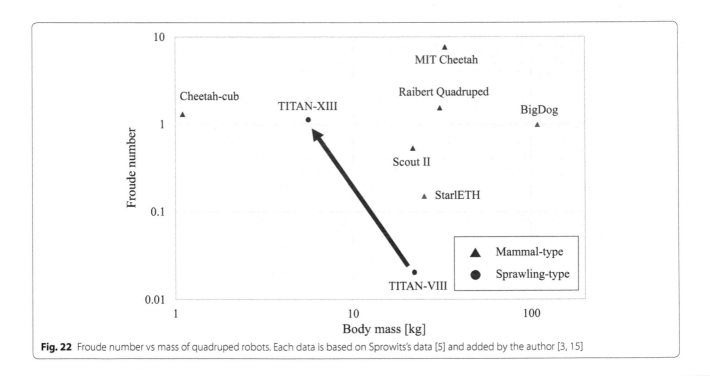

Fig. 22 Froude number vs mass of quadruped robots. Each data is based on Sprowits's data [5] and added by the author [3, 15]

energy efficiency and payload can be increased by increasing gear ratio of the actuator which supports gravitational force, at a same time vertical velocity of the foot is decreased and resulted low tracking performance in vertical direction as shown in Fig. 17. This low vertical velocity of the foot is problematic when the robot overcomes a obstacles, and eventually limit the maximum walking speed on rough terrain.

To solve this problem in a fundamental way, the changing reduction ratio depending on the walking phase is required as TITAN VI equipped [23]. However usually such kind of mechanism makes a robot heavy. Another solution is using the actuator of Axis 3 to generate the vertical velocity of the foot. Currently the Axis 3 joint does not move so much while walking because of parallel link mechanism. Therefore if we design another swing leg trajectory which is not in sagittal plane, the Axis 3 can be utilized and increase vertical velocity of the foot.

Conclusion

In this paper, to improve speed and energy efficiency of a sprawling-type quadruped robot, we focused on three design concepts (increase of walking velocity to reduce cost of transport, using the trot gait as low duty factor and safety gait, and increase of gear ratio of the actuator supporting gravitational force). Based on the concepts, we developed a sprawling-type quadruped robot named TITAN-XIII. To achieve a compact and low inertia leg, right-angle type wire driven mechanism was proposed and adopted to the robot. To confirm the validity of the developed sprawling-type quadruped robot, several experiments was conducted. In the experiment, the robot walked at 1.38 m/s and minimum cost of transport of 1.76 was achieved. We compared the walking speed and minimum COT with other quadruped robots. In comparison, TITAN-XIII shows almost the same level of walking velocity and energy efficiency. Considering advantages of the sprawling-type quadruped robot, which are static stability and wide range of motion, we can say the sprawling-type quadruped robot is highly practical on rough terrain.

In the future work, we will conduct walking experiments with the developed robot on more difficult terrain and evaluate effectiveness of a sprawling-type quadruped robot.

Authors' contributions
All authors equally contributed. All authors read and approved the final manuscript.

Author details
[1] Department of Mechanical and Aerospace Engineering, Tokyo Institute of Technology, Meguro-ku, Ookayama 2-12-1, 152-8550 Tokyo, Japan. [2] Hibot Corp., 5-9-15 Kitashinagawa, Shinagawa-ku, 141-0001 Tokyo, Japan.

Acknowledgements
This work was supported by Grant-in-Aid for Scientific Research (C) 25420214.

Competing interests
The authors declare that they have no competing interests.

References
1. Raibert M, Blankespoor K, Gabriel N, Playter R, the Big-Dog Team (2008) BigDog, the Rough-Terrain Quadruped Robot. In: Proceedings of the 17th World Congress. The International Federation of Automatic Control Seoul, Korea, July 6–11, 2008
2. Semini C, Tsagarakis NG, Guglielmino E, Focchi M, Cannella F, Caldwell DG (2011) Design of hyq, a hydraulically and electrically actuated quadruped robot. Proceedings of the Institution of Mechanical Engineers, Part I: Journal of Systems and Control Engineering 225(6):831–849. doi:10.1177/0959651811402275. URL: http://pii.sagepub.com/content/225/6/831.full.pdf+html
3. Seok S, Wang A, Chuah MYM, Hyun DJ, Lee J, Otten DM, Lang JH, Kim S (2015) Design principles for energy-efficient legged locomotion and implementation on the mit cheetah robot. Mechatronics, IEEE/ASME Transactions on 20(3):1117–1129. doi:10.1109/TMECH.2014.2339013
4. Hutter M, Gehring C, Hopflinger MA, Blosch M, Siegwart R (2014) Toward combining speed, efficiency, versatility, and robustness in an autonomous quadruped. Robotics, IEEE Transactions on 30(6):1427–1440. doi:10.1109/TRO.2014.2360493
5. Sprowitz A, Tuleu A, Vespignani M, Ajallooeian M, Badri E, Ijspeert AJ (2013) Towards dynamic trot gait locomotion: design, control, and experiments with Cheetah-cub, a compliant quadruped robot. Int J Robotics Res 32(8):932–950. doi:10.1177/0278364913489205
6. Fukuoka Y, Kimura H, Cohen AH (2003) Adaptive dynamic walking of a quadruped robot on irregular terrain based on biological concepts. Int J Robotics Res 22(3–4):187–202. doi:10.1177/0278364903022003004. URL: http://ijr.sagepub.com/content/22/3-4/187.full.pdf+html
7. Kalakrishnan M, Buchli J, Pastor P, Mistry M, Schaal S (2011) Learning, planning, and control for quadruped locomotion over challenging terrain. Int J Robotics Res 30(2):236–258. doi:10.1177/0278364910388677
8. Zico Kolter J, Ng AY (2011) The stanford littledog: a learning and rapid replanning approach to quadruped locomotion. Int J Robotics Res 30(2):150–174. doi: 10.1177/0278364910390537. URL: http://ijr.sagepub.com/content/30/2/150.full.pdf+html
9. Neuhaus PD, Pratt JE, Johnson MJ (2011) Comprehensive summary of the institute for human and machine cognition's experience with littledog. Int J Robotics Res 30(2): 216–235. doi: 10.1177/0278364910390538. URL: http://ijr.sagepub.com/content/30/2/216.full.pdf+html
10. Hirose S, Fukuda Y, Kikuchi H (1986) The gait control system of a quadruped walking vehicle. Adv Robotics 1(4):289–323. doi:10.1163/156855386X00193
11. Garcia E, de Santos PG (2006) On the improvement of walking performance in natural environments by a compliant adaptive gait. Robotics, IEEE Transactions on 22(6):1240–1253. doi:10.1109/TRO.2006.884343
12. Loc V-G, Koo I, Tran D, Park S, Moon H, Choi H (2012) Body workspace of quadruped walking robot and its applicability in legged locomotion. J Intell Robotic Sys 67(3–4):271–284. doi:10.1007/s10846-012-9670-0
13. Wait KW (2010) The use of pneumatic actuation to address shortcoming concerning normalized output power in state of the art mobile tobotics. PhD thesis, Graduate School of Vanderbilt University
14. Gabrielli G, von Karman T (1950) What price speed? Mechanical Eng 72(10):775–781
15. Arikawa K, Hirose S (Nov) Development of quadruped walking robot titan-viii. In: Intelligent Robots and Systems '96, IROS 96, Proceedings of the 1996 IEEE/RSJ International Conference On, vol 1, pp. 208–2141. doi:10.1109/IROS.1996.570670
16. Endo G, Hirose S (2012) Study on roller-walker—improvement of locomotive efficiency of quadruped robots by passive wheels. Adv Robotics 26:969–988. doi:10.1163/156855312X633066
17. TOYOBO dyneema basic properties. http://www.toyobo-global.com/seihin/dn/dyneema/seihin/. Accessed Mar 2015
18. Kitano S, Hirose S, Endo G, Fukushima EF (2013) Development of lightweight sprawling-type quadruped robot titan-xiii and its dynamic walking. In: Intelligent robots and systems (IROS), 2013 IEEE/RSJ International Conference On, pp 6025–6030. doi:10.1109/IROS.2013.6697231
19. Hirose S, Tsukagoshi H, Yoneda K (2001) Normalized energy stability margin and its contour of walking vehicles on rough terrain. In: Robotics and Automation, 2001. Proceedings 2001 ICRA. IEEE International Conference On, vol 1, pp 181–1861. doi:10.1109/ROBOT.2001.932550
20. Tucker VA (1975) The energetic cost of moving about: walking and running are extremely inefficient forms of locomotion. much greater efficiency is achieved by birds, fish-and bicyclists. Am Sci 63(4):413–419
21. Alexander RM, Jayes AS (1983) A dynamic similarity hypothesis for the gaits of quadrupedal mammals. J Zool 201(1):135–152. doi:10.1111/j.1469-7998.1983.tb04266.x

22. Alexander RM (2003) Principles of animal locomotion. Princeton University Press, Princeton, pp 58–60

23. Hirose S, Yoneda K, Arai K, Ibe T (1991) Design of prismatic quadruped walking vehicle titan vi. In: Advanced Robotics, 1991. 'Robots in Unstructured Environments', 91 ICAR., Fifth International Conference On, pp 723–7281. doi:10.1109/ICAR.1991.240685

24. Hutter M, Gehring C (2013) Walking and running with StarlETH. In: The 6th international symposium on adaptive motion of animals and machines (AMAM), pp 5–9. URL: http://publications.asl.ethz.ch/files/hutter13dynamic.pdf

25. Poulakakis I, Smith JA, Buehler M (2005) Modeling and experiments of untethered quadrupedal running with a bounding gait: The scout ii robot. Int J of Robotics Res 24(4):239–256. doi:10.1177/0278364904050917. URL: http://ijr.sagepub.com/content/24/4/239.full.pdf+html

26. BigDog overview[Online]. URL: http://www.bostondynamics.com/img/BigDogOverview.pdf

27. Raibert MH, Brown H, Benjamin J, Chepponis M, Koechling J, Hodgins JK, Dustman D, Brennan WK, Barrett DS, Thompson CM, Hebert JD, Lee W, Borvansky L (1989) Dynamically stable legged locomotion. MIT Technical Report, 134

16

Design of a three-segment continuum robot for minimally invasive surgery

Bo Ouyang[1*] ⓘ, Yunhui Liu[2] and Dong Sun[1]

Abstract

Continuum robot, as known as snake-like robot, usually does not include rigid links and has the ability to reach into a confined space by shaping itself into smooth curves. This paper presents the design of a three-segment continuum robot for minimally invasive surgery. The continuum robot employs a single super-elastic nitinol rod as the backbone and concentric disks assembled on the backbone for tendons attachment. Each segment is driven by four tendons and controlled by two linear actuators. The length of each segment is optimized based on the surgical workspace. A visual servo system is designed to assist the surgeon in operating the robot. Simulation experiment is conducted to demonstrate the proposed design.

Keywords: Continuum robot, Dimensional synthesis, Visual servo, Medical robot

Background

A continuum robot is a flexible robot inspired by caterpillars, elephant trunks, octopus arms, and mammalian tongues. The robot can vary its nature shape because of the materials flexibility and is capable of reaching into a complex environment. Therefore, the continuum robot has the potential in single-port access surgery and natural orifice transluminal.

The basic elements of a continuum robot are backbone, actuators, and disks, as shown in Fig. 1. An elephant trunk-like multi-segment continuum robot has been designed by using tendons as the actuators [1]. The multi-backbones continuum robots have been developed for the surgeries in throat and abdomen [2, 3]. This robot has a primary backbone, and other backbones were regarded as the actuators. Active catheter is another type of continuum robot, which employs the tube as the backbone [4].

It is generally assumed that the segment of continuum robot bends with constant curvature [5]. The kinematics of multi-segment continuum robot can be formulated by a Denavit–Hartenberg-type approach [1]. Although there

are various ways for kinematic modeling, the piecewise constant curvature is assumed finally [6]. Variable curvature continuum robot has been also developed [7]. However, the kinematic modeling is extremely hard.

Control of continuum robot possesses a great challenge because of the compliance of continuum robot. The dynamic model of a planar continuum robot has been introduced [8]. The dynamics of a spatial continuum robot has also been reported based on the principle of virtual power [9]. The statics and dynamics of variable curvature continuum robot have been presented by the classical Cosserat rod model [10]. However, the design of a controller is still a difficult issue because of the material flexibility. A neural network controller has been tried, where a hypothesis dynamic model is estimated online [11]. A model-less feedback control has been proposed without using the constant curvature kinematic frameworks [12].

A three-segment continuum robot for minimally invasive surgery has been developed in this study. The robot employs a single super-elastic nitinol rod as the backbone. Twelve tendons passing through the concentric disks are used to operate the robot. These tendons are divided into three groups. Each group has two pairs of tendons and is controlled by two linear actuators. The segment length is determined by the cable nodes in the group, which can be adjusted by varying the position

*Correspondence: boouyang2-c@my.cityu.edu.hk
[1] Department of Mechanical and Biomedical Engineering, City University of Hong Kong, Kowloon, Hong Kong, China
Full list of author information is available at the end of the article

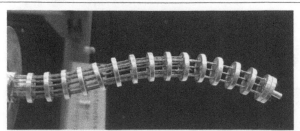

Fig. 1 A three-segment continuum robot prototype. It is shaped by twelve tendons connected to six linear actuators

of the nodes. Moreover, the approximate boundary of reachable workspace is formulated. A unique method is proposed to minimize the length of continuum robot. Finally, a visual servo system is designed.

Mechanical structure

The mechanical structure of the proposed continuum robot is shown in Fig. 2. The backbone material is a super-elastic rod. The robot is shaped by tendons passing through the disks. One segment of the continuum robot has two degrees of freedom (DOFs), i.e., rotation around z-axis (ϕ) and y-axis (θ), based on the constant curvature kinematics frameworks. The number of tendons is at least three for driving one segment, because tendons must be work in tension. This property brings a disadvantage for the kinematic control of continuum robot with three tendons. Because the shape of one segment is determined by two tendons, the tension of the third tendon is extremely large as the translation error of the third tendon is positive. The third tendon would be snapped. The designed continuum robot with each segment driven by four tendons and controlled by two linear actuators is developed to compensate this disadvantage.

One of the differences between the continuum robot developed in this paper and the continuum robots driven by three tendons is arrangement of actuators.

Two common arrangements of actuators are shown in Fig. 3. The second one is selected to arrange the tendons, because the two pairs (H_x and H_y) of tendons are uncoupled with each other. Here, linear motor is selected because it can provide large range of motion. Furthermore, each pair of tendons is connected to a single linear motor. One segment of the continuum robot is driven by six motors. The manufacturing cost is thus lower. The question is whether it is possible to steer four tendons by two linear actuators only. This question will be answered by addressing the kinematic model of continuum robot.

The configuration of each segment robot is defined by three parameters: the curvature ($k(\rho)$), the angle of the plane containing the arc ($\phi(\rho)$), and arc length (l), as shown in Fig. 2, where ρ is the tendon length and $k(\rho) = \theta/l$. By comparing the varied length of tendons in each pair, one can find

$$\Delta l = 2\left(l - 2n\frac{l}{\theta}\sin\left(\frac{\theta}{2n}\right)\right) \geq 0, \tag{1}$$

where n is the number of disks. Equation (1) indicates that one tendon works in tension and the other one is slack in each pair, which is exactly the requirement of the continuum robot control. Therefore, one just needs to change the moving direction of linear motor based on ϕ for the operation of the continuum robot. On the other side, multiple segments are employed to provide sufficient DOF for accomplishing complex surgical tasks. Twelve tendons and six linear motors are used to operate the robot finally. The tendons are divided into three groups (H_1, H_2, and H_3), and each group has two pairs (H_{ix}, H_{iy}, and $i = 1, 2, 3$). The length of each segment (l_i) is defined by the cable nodes. Thus, the length of each segment can be adjusted based on the surgical requirement.

Dimensional synthesis

The three-segment continuum robot developed in this paper has six DOFs. The workspace of each segment is determined by three parameters: ϕ_i, θ_i, and l_i, $i = 1, 2, 3$.

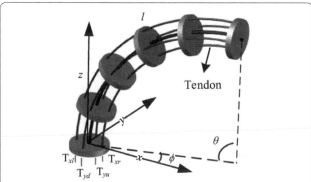

Fig. 2 One segment of continuum robot. The mechanical structure of a continuum robot usually contains tendons, disks, and backbone

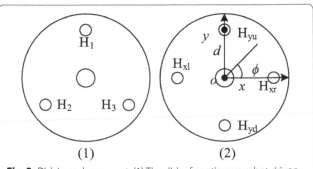

Fig. 3 Disk in each segment. (1) The disk of continuum robot driven by three tendons; (2) the disk of continuum robot driven by four tendons

The range of rotation angle ϕ_i is 0°–360°. Note that the rotation angle θ_i is limited because of the constant curvature kinematics. In general, the range of rotation angle θ_i is ranged 0°–90° or 0°–120°. The entire attitude space is independent of the arc length l_i. However, the position of end effector depends on l_i. The arc length of each segment should be optimized based on surgical workspace.

In this paper, the range of θ_i is set as 0°–90°. The reachable workspace of the continuum robot is a circular symmetry in geometry, so it can be defined by the cross section in plane. The approximate boundary of the cross section can be formulated. For the three-segment continuum robot, its left approximate boundary of the cross section can be divided into four sections:

$$s_1 = \mathbf{T}_1(\phi_1,\theta_1)[00l_2 + l_3 1]^{\mathrm{T}}, \tag{2}$$

$$s_2 = \mathbf{T}_1(\phi_1,\theta_1)\mathbf{T}_2\left(\phi_2,\frac{-\pi}{2}\right)\mathbf{T}_3\left(\phi_3,\frac{-\pi}{2}\right)[0001]^{\mathrm{T}}, \tag{3}$$

$$s_3 = \mathbf{T}_1\left(\phi_1,\frac{-\pi}{2}\right)\mathbf{T}_2(\phi_2,\theta_2)[00l_3 1]^{\mathrm{T}}, \tag{4}$$

$$s_4 = \mathbf{T}_1\left(\phi_1,\frac{-\pi}{2}\right)\mathbf{T}_2\left(\phi_2,\frac{-\pi}{2}\right)\mathbf{T}_3(\phi_3,\theta_3)[0001]^{\mathrm{T}}, \tag{5}$$

where $\phi_1,\phi_2,\phi_3 = 0$, $\theta_1,\theta_2,\theta_3 \in (0,\pi/2]$, $\mathbf{T}_i$ is the transform matrix of the ith segment, and $x \leq 0$. If ϕ_1 changes from 0 to 2π, the cure of each section turns into a surface, which forms the approximate boundary of workspace.

Now the dimensional synthesis of continuum robot can be analyzed based on the requirement of surgical workspace. Suppose that the surgical workspace is a cuboid, i.e., $x \in [-x_r, x_r]$, $y \in [-y_r, y_r]$ and $z \in [z_{rd}, z_{ru}]$, and $x_r \geq y_r$. The case of $x_r \leq y_r$ is similar, because the workspace is circular symmetry. The cross section of surgical workspace is a rectangle in the xz plane. Denote the vertexes as $\mathbf{V}_j$, $j = 1, \ldots, 4$, and the middle point of $\mathbf{V}_3\mathbf{V}_4$ as $\mathbf{V}_0$. Based on the geometric property of boundary, the cuboid is in the workspace of continuum robot by providing that $\mathbf{V}_0$, $\mathbf{V}_1$, ..., $\mathbf{V}_4$ are all in the workspace. If the robot length is very long, the robot will easily go out of the channel with a small rotation angle θ. Therefore, the length should be minimized to improve the dexterity of the continuum robot. The dimensional synthesis can be described as the following optimization problem:

$$\min\ l_1 + l_2 + l_3$$

$$\text{s.t.} \begin{cases} T\left(\phi_{1j},\theta_{1j},l_1,\ldots,\phi_{3j},\theta_{3j},l_3\right)\mathbf{d}^e = \mathbf{V}_j, & j = 0,\ldots,4 \\ -\frac{\pi}{2} \leq \theta_{ij} \leq \frac{\pi}{2}, & l_i > 0, \quad i = 1,2,3, \quad j = 0,\ldots,4 \end{cases} \tag{6}$$

where $\mathbf{T} = \mathbf{T}_1\mathbf{T}_2\mathbf{T}_3$ is the transformation from the base coordinate frame F_b to the frame F_e of the end effector, and $\mathbf{d}^e$ represents the end point of the end effector in the frame F_e. Therefore, the length of each segment can be determined through optimization.

Visual servo system design

After the mechanical structure of continuum robot is developed, the next step is to design an interactive system for assisting surgeon in controlling the robot in a user-friendly way.

It is assumed that camera can capture all the feature points on each segment of the continuum robot. Then, the configuration of segment can be determined by the feature point $(\mathbf{F}_i)$ on the top disk. Based on the projection principle, the image coordinate of feature point is

$$z_1^c[\mathbf{m}_1/f\,1]^{\mathrm{T}} = \mathbf{R}_{ce}\mathbf{R}_1\mathbf{F}_1 + \mathbf{R}_{ce}\mathbf{P}_1 + \mathbf{p}_{ce}. \tag{7}$$

where $\mathbf{m}_1$ is the image coordinate of feature point, f is the focal length, z^c is an arbitrary scale factor, and $\mathbf{R}_{ce}$ and $\mathbf{p}_{ce}$ are the extrinsic parameters of camera, and $\mathbf{R}_1$ and $\mathbf{P}_1$ are the rotation matrix and translation of the first segment, respectively. After eliminating z_1^c, a nonlinear equation is obtained:

$$\mathbf{m}_1 = \mathbf{g}(\phi_1,\theta_1), \tag{8}$$

The numerical solution of configuration (ϕ_1,θ_1) is calculated based on Eq. (8). The configuration of the end effector can be determined based on kinematics. On the other side, Eq. (8) can be applied to calculate the image Jacobian matrix

$$\begin{bmatrix} du_1 & dv_1 \end{bmatrix}^T = \mathbf{J}_1(\phi_1,\theta_1)\begin{bmatrix} d\phi_1 & d\theta_1 \end{bmatrix}^T. \tag{9}$$

where u_1 and v_1 are the coordinate in pixel. The image Jacobian matrix of the last two segments can also be derived by the above steps. The Jacobian matrix of the continuum robot can be calculated based on the kinematics. Then, one can design a controller based on Eq. (9).

Simulation experiment

To verify the design of the continuum robot, the continuum robot prototype was developed. The system consists of linear actuators, cables, pulleys, drivers, camera, digital pen, and digitizer tablet, as shown in Fig. 4. The superelastic nitinol rod was employed as the backbone of the continuum robot. Twelve cables passed through disks around the backbone. These cables were divided into three groups with respect to three segments. Each group has two pairs of cables, and each pair of cables is connected to a linear actuator. So each segment of continuum robot is just driven by two motors. Furthermore, each segment length can be varied by adjusting cable nodes in each group based on task requirement.

Fig. 4 The three-segment continuum robot prototype. It is driven by 12 cables and 6 linear actuators

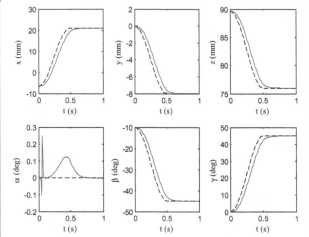

Fig. 5 Response of the control system. *Solid line* and *dotted line* represent the desire configuration and the response, respectively

To control the continuum robot, a PD controller was designed with visual feedback. The desired configuration was set as $y_d = \begin{bmatrix} 21 & -8 & 76 & 0° & -45° & 45° \end{bmatrix}$. The following function was employed for motion planning:

$$y = \begin{cases} y_s + \frac{(y_d - y_s)}{2} \left(\sin\left(\frac{\pi t}{t_d} - \frac{\pi}{2} \right) + 1 \right) & t \in [0, t_d] \\ y_d & t \in (t_d, \infty] \end{cases}.$$

(10)

where y_s is the initial pose. Here, proportional gain K_P was set as 1.5, and differential gain K_d was selected as 0.1. The response of the control system in simulation is shown in Fig. 5. One can find that the error converged to zero although the response had a little time delay.

Conclusion

This paper presents the design of a three-segment continuum robot. The approximate boundary of workspace

is formulated. The configuration of each segment is determined. In the future, the visual servo control system will be developed, and the controller will be improved.

Authors' contributions

A three-segment continuum robot is designed. The boundary of workspace is formulated. Moreover, a method for determining the configurations of each segment is proposed based on monocular vision. All authors read and approved the final manuscript.

Author details

[1] Department of Mechanical and Biomedical Engineering, City University of Hong Kong, Kowloon, Hong Kong, China. [2] Department of Mechanical and Automation Engineering, The Chinese University of Hong Kong, Shatin, China.

Acknowledgements

The work was also supported by a grant from Research Grants Council of the Hong Kong Special Administrative Region, China (Reference No. CUHK6/CRF/13G assigned to CityU).

Competing interests

The authors declare that they have no competing interests.

References

1. Jones BA, Walker ID. Kinematics for multisection continuum robots. IEEE Trans Rob. 2006;22(1):43–55.
2. Xu K, Simaan N. Actuation compensation for flexible surgical snake-like robots with redundant remote actuation. In: IEEE international conference on robotics and automation. 2006. p. 4148–53.
3. Simaan N, Xu K, Wei W, Kapoor A, Kazanzides P, Taylor R. Design and integration of a telerobotic system for minimally invasive surgery of the throat. Int J Robot Res. 2009;28:1134–53.
4. Dupont PE, Lock J, Itkowitz B, Butler E. Design and control of concentric-tube robots. IEEE Trans Robot. 2010;26(2):209–25.
5. Gravagne IA, Rahn CD, Walker ID. Large deflection dynamics and control for planar continuum robots. IEEE ASME Trans Mechatron. 2003;8(2):299–307.
6. Webster RJ III, Jones BA. Design and kinematic modeling of constant curvature continuum robots: a review. Int J Robot Res. 2010;29(13):1661–83.
7. Mahl T, Hildebrandt A, Sawodny O. A variable curvature continuum kinematics for kinematic control of the bionic handling assistant. IEEE Trans Robot. 2014;30(4):935–49.
8. Gravagne IA, Walker ID. Manipulability, force, and compliance analysis for planar continuum manipulators. IEEE Trans Robot Autom. 2002;18(3):263–73.
9. Rone WS, Ben-Tzvi P. Continuum robot dynamics utilizing the principle of virtual power. IEEE Trans Robot. 2014;30(1):275–87.
10. Rucker DC, Webster RJ III. Statics and dynamics of continuum robots with general tendon routing and external loading. IEEE Trans Robot. 2011;27(6):1033–44.
11. Braganza D, Dawson DM, Walker ID, Nath N. A neural network controller for continuum robots. IEEE Trans Robot. 2007;23(5):1270–7.
12. Yip MC, Camarillo DB. Model-less feedback control of continuum manipulators in constrained environments. IEEE Trans Robot. 2014;30(4):880–9.

17

Realization and swimming performance of the breaststroke by a swimming humanoid robot

Motomu Nakashima[1*] and Kosuke Kuwahara[2]

Abstract

In order to clarify the mechanics of human swimming, a full-body swimming humanoid robot called "SWUMANOID" was developed as an experimental platform for research about human swimming. SWUMANOID had a detailed human body shape, created using three-dimensional scanning and printing equipment, and was developed as an experimental model substituting for human subjects. Not only the appearance but also the methodology to realize various swimming strokes was considered. In order to reproduce complicated swimming motions with high fidelity, 20 waterproof actuators were installed. The free swimming of the crawl stroke at a velocity of 0.24 m/s was realized in the previous study. However, it could not perform the breaststroke due to mechanical limitations. The objectives of this study were to realize the breaststroke for SWUMANOID by improving its lower limbs, and to investigate the swimming performance of the breaststroke experimentally. The lower body of SWUMANOID was fully redesigned, built, and connected to the upper body. The swimming motion of the breaststroke was created based on that of an actual swimmer. A free swimming experiment was conducted in a 25 m outdoor swimming pool. In addition, in order to discuss the experimental results in detail, the experiment was reproduced by the simulation. From the experiment, it was found that SWUMANOID could perform the breaststroke successfully. The swimming speed for the stroke cycle of 2.3 s was found to be 0.12 m/s. Since this swimming speed was considered low compared to that of the actual swimmer, the reason for the discrepancy was examined by simulation. From the simulation, it was found that one main reason for the low swimming speed was insufficient output power of the motors, especially for the knee and shoulder joints.

Keywords: Swimming, Sports engineering, Biomimetics, Breaststroke, Fluid forces, Humanoid robot

Background

In spite of such a long history of human swimming, its mechanics still have not been fully clarified since it is an extremely complicated phenomenon, in which a complex human body moves unsteadily with many degrees-of-freedom (DOF) in the three-dimensional water flow. For example, the hand path in the crawl stroke depicts a distorted ellipse when viewed from the side in absolute space [1], showing that the hand does not push the water straight at a constant depth. Furthermore, the kinematics of the arm and hand during the underwater stroke is highly unsteady [2]. From this viewpoint, many attempts were made recently to quantify the unsteady fluid forces acting on a swimmer while swimming. The first approach was an experiment involving a human subject [3–5]. However, this method had problems with insufficient repeatability, physical fatigue of the subject, and difficulty in installing sensors on the subject. For this reason, some researchers have conducted experiments using physical models such as robots instead of human subjects. A lot of measuring experiments using physical models have been conducted to date [6–9], but there was no full-body experimental platform which could consider interactions between the many segments involved in normal swimming motions. Therefore, an analysis

*Correspondence: motomu@mei.titech.ac.jp
[1] Graduate School of Information Science and Engineering, Tokyo Institute of Technology, 2-12-1 Ookayama, Meguro, Tokyo 152-8552, Japan
Full list of author information is available at the end of the article

using physical models had been performed on an isolated segment and misleading conclusions could have been developed. To solve such problems, a full-body swimming humanoid robot was developed for research about human swimming by Chung and Nakashima [10]. The robot was named SWUMANOID. SWUMANOID had a detailed human-body shape, and was created using three-dimensional scanning and printing equipment since it was developed for the experimental model substituting for human subjects. The size of SWUMANOID was 1/2 scale of an actual swimmer. Not only the appearance but also the methodology to realize various swimming strokes was considered. In order to reproduce complicated swimming motions with high fidelity, 20 water-proofed actuators were installed. The free swimming of the crawl stroke at a velocity of 0.24 m/s was realized in the previous study [11].

Developing a swimming humanoid robot such as SWUMANOID is important for the following two reasons. First, it is expected that it will become an experimental platform for the research of human swimming. To date, many simulation studies about human swimming have been conducted [12–15]. Such simulation technique is very useful and powerful tool for analysis. However, simulation always needs validation and improvement by comparing with experimental results, since it is not an actual phenomenon after all. In order to conduct more accurate experiments for that purpose, more elaborate physical models, such as swimming humanoid robots, will be necessary. The second reason for developing a swimming humanoid robot is that it can be applied to robots for special tasks accompanying water environment, such as rescue robots in the sea and working robots around a pool in a nuclear plant. Developing a swimming humanoid robot and studying how it can swim will be useful for developing such robots in the future.

SWUMANOID had sufficient DOF to perform not only the crawl stroke, but also the back and butterfly strokes. However, it could not perform the swimming motion of the breaststroke due to the following three reasons: (1) the thigh joint only had one DOF, (2) the knee joint could not be as fully flexed as a human's, (3) and the ankle joint could not be as fully dorsi-flexed as a human's as well. These points for the lower limbs did not become problems when SWUMANOID performed the flutter kick for the crawl stroke, but did when it performed the breaststroke. Indeed, for the breaststroke, the thigh joints have to possess three DOF since the legs move in a complicated manner. The knee and ankle joints also have to be fully flexed and dorsi-flexed, respectively, for the recovery position of the legs.

The objectives of this study were to realize the breast-stroke for SWUMANOID by improving its lower limbs,

and to investigate the swimming performance of the breaststroke experimentally. In this paper, the improvement for the breaststroke, creation of swimming motion, and experimental and simulation methods are explained. Next, the experimental results are shown and discussed. Finally, the obtained findings are summarized.

Methods
Improvement for the breaststroke
As mentioned above, SWUMANOID, which was developed in the previous study [10, 11], did not have sufficient DOF in the lower body for the breaststroke. Therefore, the lower body of SWUMANOID was completely re-designed for the present study. The overview and DOF configuration of the designed lower body are shown in Fig. 1. Hip joint 2 (yaw) and hip joint 3 (roll) were newly introduced. Since hip joint 2 was located in the same position as hip joint 1, and there was no sufficient space for the motor, a belt pulley mechanism was introduced to this joint, as shown in Fig. 2a. The motor for this joint was installed in the thigh (the red part in Fig. 2a). With respect to the knee joint, the new joint had one degree-of-freedom, which was the same as the previously developed lower body. However, a much broader range of motion, especially in the flexing direction, was necessary for the breaststroke than for the crawl stroke. Therefore, the shank part was also re-designed to secure a sufficient flexion angle of 135°. For the ankle joint, a four-bar linkage mechanism was used, as shown in Fig. 2b. In particular, a large dorsi-flexion angle (90°) was secured to ensure the recovery position for the foot in the breaststroke. The outer cases of the robot were built by rapid prototyping

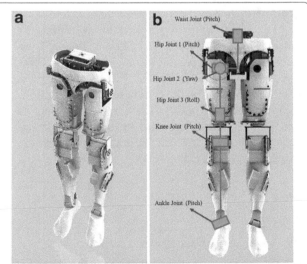

Fig. 1 Overview and DOF configuration of the designed lower body. **a** Overview, **b** DOF configuration

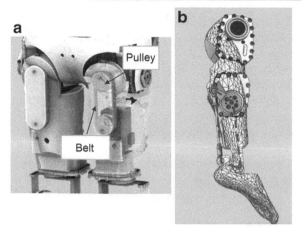

Fig. 2 Belt pulley mechanism for Hip joint 2 and four-bar linkage mechanism for the ankle joint. **a** Belt pulley mechanism, **b** four-bar linkage mechanism

Table 1 Specifications of improved SWUMANOID

Items	Specifications
Size	(H) 925 mm (W) 270 mm (D) 119 mm
Weight	7.4 kg
Actuators (dynamixel: Robotis Corp.)	RX28: 2.5 Nm (at 14.8 V) MX28: 3.1 Nm (at 14.8 V) MX28: 7.3 Nm (at 14.8 V)
DOF	Total: 24 DOFs Arm: 2 arms × 6 DOF Waist: 2 DOF Leg: 2 legs × 3 DOF
Controller	CM700 (Robotis Corp.)
Battery	Li-Po 14.8 V 1550 mhA × 2
Communication	ZigBee module ZIG-110A

Dynamixel MX64 (Robotis inc., Seoul) was used for Waist joint, hip joint 1, hip joint 2, and knee joint, while dynamixel MX28 was used for hip joint 3 and ankle joint. CM700 (Robotis inc.) and ZIG-110A (Robotis inc.) were used for the controller of the robot and wireless communication

with acrylic resin. The shapes of the cases were determined based on the scanned body data of an elite male swimmer.

Electrical parts and other specifications

The built lower body was connected to the previously developed upper body, as shown in Fig. 3. The specifications of the improved SWUMANOID are shown in Table 1. The stature became 925 mm. With respect to the actuators of the lower body, dynamixel MX64 (Robotis inc., Seoul) was used for waist joint, hip joint 1, hip joint

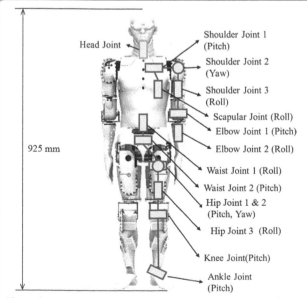

Fig. 3 Overview of the whole robot. The built *lower body* was connected to the previously developed *upper body*

2, and knee joint, while dynamixel MX28 was used for hip joint 3 and ankle joint. For the controller of the robot and wireless communication, CM700 (Robotis inc.) and ZIG-110A (Robotis inc.) were used, respectively.

The parameters of the body segments are shown in Table 2. The weight and volume of each body segment were measured. In order to adjust the total buoyancy, weights were installed in the head and abdomen parts. As a result, the total specific gravity became slightly smaller than 1, as shown in Table 2.

Creation of swimming motion

The overview of creating swimming motion is shown in Fig. 4. The swimming motion of the breaststroke in a previous simulation study by Nakashima [13] was used for the present study. This motion was created for the swimming human simulation model "SWUM" [12], based on the motion of an actual competitive swimmer. SWUM is a simulation model for analyses of human swimming, in which a swimmer's body is represented as a series of 21 truncated elliptic cones, and the rigid body dynamics and unsteady fluid forces are considered. By SWUM, it is possible to compute the fluid forces acting on a swimmer as well as joint torques, inputting the joint motion which is represented as multiple body positions in multiple time steps. This joint motion is defined using the original body reference coordinate system from SWUM, which is different from the relative coordinate system used for the robot. In addition, the arrangement of the joints and degrees-of-freedom for the robot and SWUM were different from each other. Therefore, the target position and orientation of body segments were first calculated in SWUM by forward kinematics. The joint angles of the

Table 2 Parameters of body segments

Segment	Weight [g]	Added weight [g]	Volume [cm³]	Specific gravity
Hip	426		410	1.04
Thigh × 2	1562		1460	1.07
Shank × 2	954		748	1.28
Foot × 2	106		200	0.53
Head and neck	368	335	679	1.02
Upper arm × 2	660		579	1.14
Elbow × 2	256		215	1.19
Forearm × 2	300		306	0.98
Breast × 2	1039		1514	0.7
Waist	838	583	1460	0.97
Whole upper body	3461	918	4753	0.92
Whole lower body	3048		2818	1.08
Total	7427		7571	0.98

The weight and volume of each body segment were measured. In order to adjust the total buoyancy, weights were installed in the head and abdomen parts. As a result, the total specific gravity became slightly smaller than 1

robot were then calculated by inverse kinematics, considering the motions of the scapular parts. The details of this procedure were described in the previous study [10]. Once the joint angles were obtained, they were put into CAD software, and the motions were checked visually. Finally, the motions were confirmed in a test on land. The

obtained swimming motion of breaststroke displayed in the CAD software is shown in Fig. 5. The swimming motion was represented as 18 step motions in this case.

Experimental method

A free swimming experiment was conducted in a 25 m outdoor swimming pool. Two measuring tapes were placed on the poolside and on the water surface in the pool along with the swimming direction of the robot, as shown in Fig. 6. The swimming movement was filmed by two cameras, one on land and one in the water. From the images filmed by the camera on land, the swimming speed of the robot was calculated.

Simulation method

In order to discuss the experimental results in detail, the experiment was reproduced by the simulation using SWUM. The simulation model of SWUMANOID was constructed, as shown in Fig. 7. Since the body segments in SWUM were represented as truncated elliptic cones, the lengths and radii of the segments were determined based on the geometry of SWUMANOID.

Results and discussion
Experimental results

The swimming motion of SWUMANOID in the experiment is shown in Fig. 8. The stroke cycle was 2.3 s in this trial. This value was determined based on the stroke cycle of the actual swimmer, whose swimming motion was utilized for that of SWUMANOID in the present study,

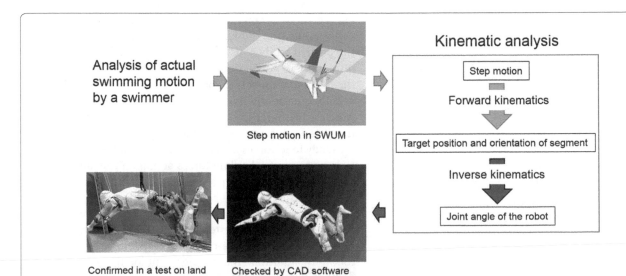

Fig. 4 Overview of creating the swimming motion. The target position and orientation of the body segments were first calculated in SWUM by forward kinematics. The joint angles of the robot were then calculated by inverse kinematics, considering the motions of the scapular parts. Once the joint angles were obtained, they were put into CAD software, and the motions were checked visually. Finally the motions were confirmed in a test on land

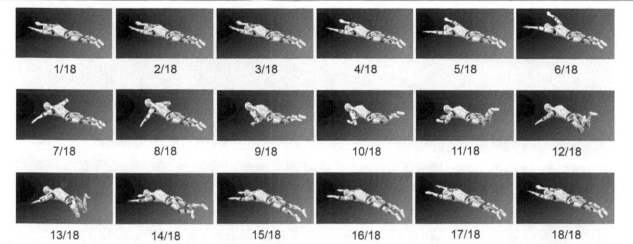

Fig. 5 Obtained swimming motion of breaststroke displayed in CAD software. The swimming motion was represented as an 18-step motion in this case

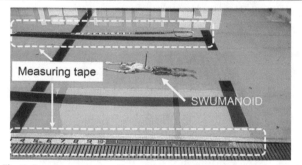

Fig. 6 Experimental setup. Two measuring tapes were placed on the poolside and on the water surface in the pool along with the swimming direction of the robot. The swimming movement was filmed by two cameras, one on land and one in the water. From the images filmed by the camera on land, the swimming speed of the robot was calculated

as well as the results of a preliminary experiment. In the preliminary experiment, the stroke cycle of 2.13 s, which was the one for the actual swimmer, was tested as the first trial. However, it was found that the output power of the actuators were apparently insufficient so that the robot could not perform the programmed motion at all. Therefore the slightly longer stroke cycle of 2.3 s was chosen as the main trial. It was qualitatively confirmed that SWUMANOID could perform the breaststroke in the experiment successfully. The experimental result of swimming speed is shown in Fig. 9. Note that this result was the change of average speed in one stroke cycle. It was not possible to measure the instantaneous speed by the present experimental system due to lack of the accuracy. Since the fluctuation of the speed in one stroke cycle

is seen in the actual breaststroke swimming and therefore is very important, it will be necessary to measure it by an improved experimental system in the future study. As shown in Fig. 9, the swimming speed almost converged to a constant value as time passed, and the average value was 0.12 m/s. The normalized stroke length, which is a non-dimensional propulsive distance in one stroke cycle and is defined by the product of the swimming speed and the stroke cycle divided by the stature, became 0.298 in this experiment. On the other hand, the normalized stroke length for the actual swimmer, whose swimming motion was utilized for that of SWUMANOID in the present study, was reported as 1.358 in the previous study [13] Note that the stroke cycle was 2.13 s in this case. It means that the swimming performance of SWUMANOID was almost one-fourth of that of an actual swimmer. One possible main reason of this discrepancy was the insufficient output power of the actuators. The target and measured values of the knee and shoulder joint angles for one stroke cycle are shown in Fig. 10. The shoulder joint angles are in the direction of flexion/extension. It was found that the measured angles (red) could not track the target angles (blue) both for the knee and shoulder when the angular velocities became large. The effect of this discrepancy in the joint angles is discussed by the simulation in the next section. Note that the measured angles for other joints were found to sufficiently track the target angles.

In addition, the trial of stroke cycle of 3.27 s was conducted. The average swimming speed was 0.11 m/s in this case. This value was only slightly lower than 0.12 m/s in the case of 2.3 s. Conversely, 0.12 m/s of 2.3 s was only slightly higher than 0.11 m/s of 3.27 s although the stroke

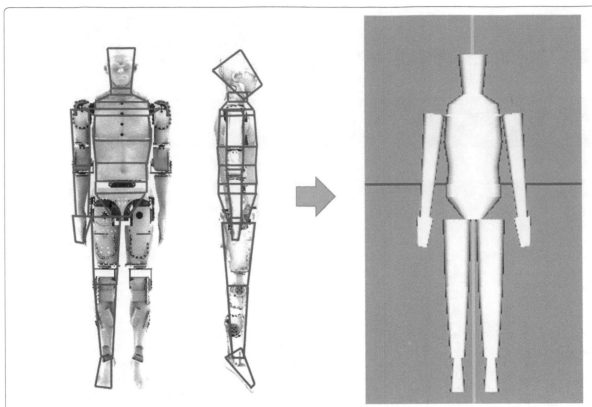

Fig. 7 Constructed simulation model of SWUMANOID. Since the body segments in SWUM were represented as *truncated elliptic cones*, the lengths and radii of the segments were determined based on the geometry of SWUMANOID

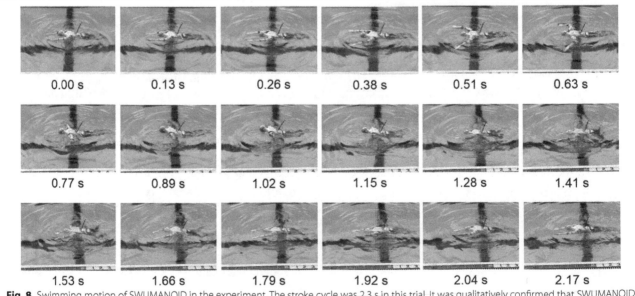

0.00 s	0.13 s	0.26 s	0.38 s	0.51 s	0.63 s
0.77 s	0.89 s	1.02 s	1.15 s	1.28 s	1.41 s
1.53 s	1.66 s	1.79 s	1.92 s	2.04 s	2.17 s

Fig. 8 Swimming motion of SWUMANOID in the experiment. The stroke cycle was 2.3 s in this trial. It was qualitatively confirmed that SWUMANOID could perform the breaststroke in the experiment as well

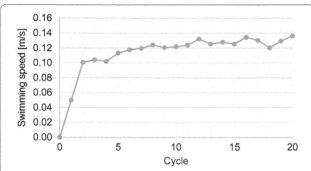

Fig. 9 Experimental result of swimming speed. The stroke cycle was 2.3 s in this trial. It almost converged to a constant value as time passed, and finally it reached 0.12 m/s

cycle became 70 %. Ideally, swimming speed is inversely proportional to stroke cycle. Therefore, from the result of 3.27 s, the average swimming speed of 2.3 s was expected to be 0.11/0.7 = 0.15 m/s. Since 0.11 m/s of 2.3 s in the experiment was certainly lower than 0.15 m/s, the insufficient output power of the actuators were suspected from this result as well.

Examination by simulation

In order to examine the effect of discrepancies in the joint angles, simulations in four conditions shown in Table 3 were carried out. In condition 1, the measured knee and shoulder joint angles were put into the simulation. That is, this simulation reproduced the experimental

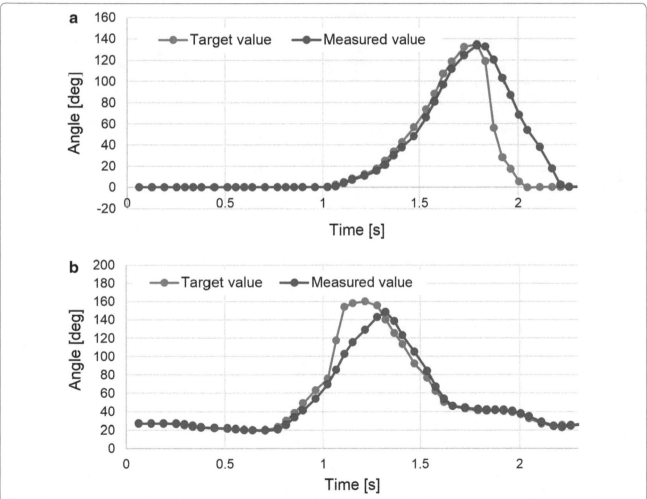

Fig. 10 Target and measured values of the knee and shoulder joint angles for one stroke cycle. **a** Knee joint angle, **b** shoulder joint angle. The shoulder joint angles are in the direction of flexion/extension. It was found that the measured angles could not track the target angles both for the knee and shoulder when the angular velocities became large

Table 3 Simulation results of swimming speeds for four conditions

Condition	Shoulder joints	Knee joints	Swimming speed [m/s]	Increase amount [%]
1	Measured	Measured	0.118	0
2	Target	Measured	0.162	37
3	Measured	Target	0.166	41
4	Target	Target	0.21	78

In condition 1, the measured knee and shoulder joint angles were put into the simulation. In condition 2, the target values of the joint angles for the shoulders were put into the simulation instead of the measured values. In condition 3, the target values for the knees were put into the simulation. In condition 4, the target values both for the shoulders and knees were put into the simulation

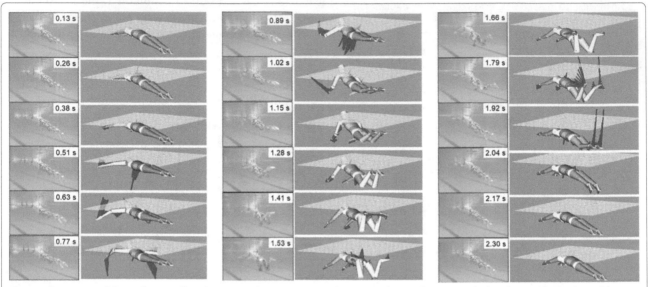

Fig. 11 Comparison of the underwater filmed images in the experiment and images in the simulation reproducing the experimental condition (condition 1). It was confirmed that the behavior of SWUMANOID in the simulation was consistent with that in the experiment

condition. Note that the target angles were put into the simulation for all the other joints since the measured angles for the other joints were found to sufficiently track the target angles. In condition 2, the target values of the joint angles for the shoulders were put into the simulation instead of the measured values. In condition 3, the target values for the knees were put into the simulation. In condition 4, the target values both for the shoulders and knees were put into the simulation. Comparison of the underwater filmed images in the experiment and images in the simulation reproducing the experimental condition (condition 1) is shown in Fig. 11. It was visually confirmed that the behavior of SWUMANOID in the simulation was consistent with that in the experiment. Note that one of the fluid force coefficients in the simulation model was adjusted so that the swimming speed in the simulation of condition 1 became equal to that in the experiment. The simulation results of swimming speeds for the four conditions are shown in Table 3. The increases in amounts of swimming speed against the

condition 1 are also shown in the table. It was found that the swimming speed increased by 37 and 41 % by changing the joint angles from the measured values to the target ones for the shoulders and knees, respectively. It was also found that the swimming speed increased for 78 % by changing the joint angles both for the shoulders and knees. The fluid forces acting on the hand, forearm and foot for one stroke cycle in conditions 1 and 4 are shown in Fig. 12. From Fig. 12a, it was found that the fluid forces acting on the hand and forearm in condition 1 had two negative peaks. From Fig. 12b, it was found that the fluid forces acting on the foot in condition 1 did not have a sharp positive peak at $t^* = 0.7$, unlike those in condition 4. These drawbacks in condition 1 resulted a swimming speed lower than that in condition 4.

The maximum joint torques of the shoulder and knee were calculated for condition 4 by the simulation. These were 0.84 and 2.37 Nm, respectively. Indeed, these were sufficiently smaller than 7.3 Nm in Table 1 for MX-64, which was used for the shoulder and knee joints.

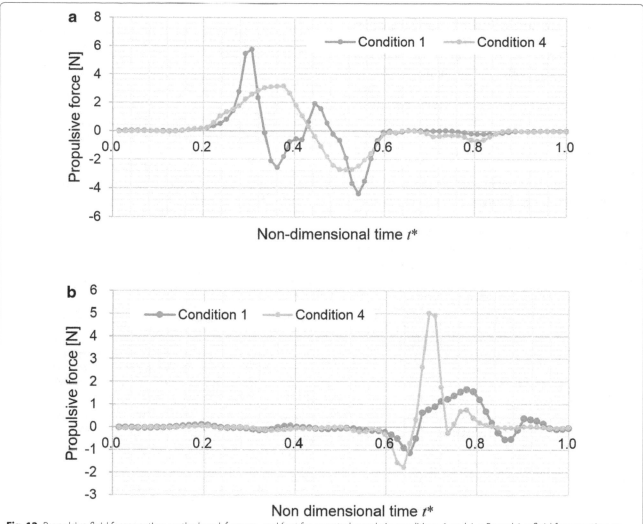

Fig. 12 Propulsive fluid forces acting on the hand, forearm, and foot for one stroke cycle in conditions 1 and 4. **a** Propulsive fluid forces acting on the right hand and forearm, **b** propulsive fluid forces acting on the right foot

However, 7.3 Nm was the stall torque, which was the maximum torque without rotation. It suggests that there is a possibility to improve the actuator performance by selecting more appropriate reduction ratio for the motor gears.

From above results, it was found that the insufficient output powers of the motors were one of the main reasons for the low swimming speed. Therefore, it is expected that the swimming speed will increase largely if the motors have sufficient output powers to realize the target joint motions. However, the swimming speed of 0.210 m/s (condition 4) in Table 3 means the normalized stroke length of 0.522. It is still much lower than 1.358 of an actual swimmer. One reason for this discrepancy may be that the difference in the actual joint angles and measured ones. The joint angles in the experiment were measured

by the internal function of the motors, and therefore mechanical errors such as backlash of the joints were not taken into account. Therefore, it was possible that the actual joint angles had some differences from the measured values. Another possible reason is that the target joint angles themselves were not sufficiently well-considered. Although they were determined based on the actual values of an actual swimmer, they had to be modified for SWUMANOID due to the limitation of the degrees-of-freedom as well as the range of motion of SWUMANOID. For example, in Fig. 12a, the fluid force acting on the hand and forearm still had a large negative peak at $t^* = 0.5$–0.6. This means the recovery motion, in which the hand moves forward, still was not sufficiently good. If such problems in the swimming motion are all solved by modification, the swimming speed may increase more.

Conclusions

In the present study, the swimming humanoid robot SWUMANOID was improved to perform the breaststroke. The lower body was fully redesigned, built and connected to the upper body. The swimming motion of the breaststroke was created based on that of an actual swimmer. From the experiment, it was found that SWUMANOID could perform the swimming motion of breaststroke successfully. The swimming speed for the stroke cycle of 2.3 s was found to be 0.12 m/s. From the examination by simulation, it was found that one of the main reasons for the low swimming speed was insufficient output power of the motors, especially for the knee and shoulder joints.

As the future tasks, the actuation system for the knee and shoulder has to be improved by some methods, such as providing instantaneous large current to the motors for the power peak timings, redesign of the motors or introducing subsidiary active/passive actuators.

Authors' contributions

MN provided the basic ideas of the overall system, and KK designed the robots and the overall system. All of the experiments were performed by MN and KK. All authors joined the discussions for this research. All authors read and approved the final manuscript.

Author details

[1] Graduate School of Information Science and Engineering, Tokyo Institute of Technology, 2-12-1 Ookayama, Meguro, Tokyo 152-8552, Japan. [2] Graduate School of Science and Engineering, Tokyo Institute of Technology, 2-12-1 Ookayama, Meguro, Tokyo 152-8552, Japan.

Acknowledgements

This work was supported by JSPS KAKENHI Grant Number 26282174.

Competing interests

The authors declare that they have no competing interests.

References

1. Maglischo EW (2003) Swimming fastest. Hum Kinet: 97
2. Toussaint HM, Van Den Berg C, Beek WJ (2002) Pumped-up propulsion during front crawl swimming. Med Sci Sports Exerc 34(2):314–319
3. Hollander AP, De Groot G, van Ingen Schenau GJ, Toussaint HM, De Best H, Peeters W, Meulemans A, Schreurs AW (1986) Measurement of active drag during crawl arm stroke swimming. J Sports Sci 4(1):21–30
4. Kolmogorov SV, Duplishcheva OA (1992) Active drag, useful mechanical power output and hydrodynamic force coefficient in different swimming strokes at maximal velocity. J Biomech 25(3):311–318
5. Takagi H, Sanders R (2002) Measurement of propulsion by the hand during competitive swimming. In: Ujihashi S, Haake SJ (eds) The engineering of sport 4. Blackwell Publishing, Oxford, pp 631–637
6. Lauder MA, Dabnichki P (2005) Estimating propulsive forces-sink or swim? J Biomech 38(10):1984–1990
7. Sidelnik NO, Young BW (2006) Optimising the freestyle swimming stroke: the effect of finger spread. Sports Eng 9(3):129–135
8. Nakashima M, Takahashi A (2012) Clarification of unsteady fluid forces acting on limbs in swimming using an underwater robot arm (development of an underwater robot arm and measurement of fluid forces). J Fluid Sci Tech 7(1):100–113
9. Nakashima M, Takahashi A (2012) Clarification of unsteady fluid forces acting on limbs in swimming using an underwater robot arm (2nd report, modeling of fluid force using experimental results). J Fluid Sci Tech 7(1):114–128
10. Chung C, Nakashima M (2013) Development of a swimming humanoid robot for research of human swimming. J Aero Aqua Bio-Mech 3(1):109–117
11. Chung C, Nakashima M (2013) Free swimming of the swimming humanoid robot for the crawl stroke. J Aero Aqua Bio-Mech 3(1):118–126
12. Nakashima M, Satou K, Miura Y (2007) Development of swimming human simulation model considering rigid body dynamics and unsteady fluid force for whole body. J Fluid Sci Tech 2(1):56–67
13. Nakashima M (2007) Analysis of breast, back and butterfly strokes by the swimming human simulation model SWUM. In: Kato N, Kamimura S (eds) Bio-mechanisms of swimming and flying -fluid dynamics, biomimetic robots, and sports science. Springer, Tokyo, pp 361–372
14. Nakashima M (2010) Modeling and simulation of human swimming. J Aero Aqua Bio-Mech 1(1):11–17
15. Takagi H, Nakashima M, Sato Y, Matsuuchi K, Sanders R (2015) Numerical and experimental investigations of human swimming motions. J Sports Sci. doi:10.1080/02640414.2015.1123284

18

Design of a passive, iso-elastic upper limb exoskeleton for gravity compensation

Ruprecht Altenburger[*], Daniel Scherly and Konrad S. Stadler

Abstract

An additional mechanical mechanism for a passive parallelogram-based exoskeleton arm-support is presented. It consists of several levers and joints and an attached extension coil spring. The additional mechanism has two favourable features. On the one hand it exhibits an almost iso-elastic behaviour whereby the lifting force of the mechanism is constant for a wide working range. Secondly, the value of the supporting force can be varied by a simple linear movement of a supporting joint. Furthermore a standard tension spring can be used to gain the desired behavior. The additional mechanism is a 4-link mechanism affixed to one end of the spring within the parallelogram arm-support. It has several geometrical parameters which influence the overall behaviour. A standard optimisation routine with constraints on the parameters is used to find an optimal set of geometrical parameters. Based on the optimized geometrical parameters a prototype was constructed and tested. It is a lightweight wearable system, with a weight of 1.9 kg. Detailed experiments reveal a difference between measured and calculated forces. These variations can be explained by a 60 % higher pre load force of the tension spring and a geometrical offset in the construction.

Keywords: Exoskeleton, Iso-elastic, Assistive device, Passive, Gravity compensation, Design optimisation

Background

Passive gravity compensation systems have existed for a long time [1] and many applications have evolved over the last century. One of the most well known applications is the parallel beam and spring systems used to balance lamp shades [2–4]. These systems were adapted to balance TV screens [5] and to reduce holding and actuation torques in robot arms [6].

In recent years, exoskeletons have gained significant attention in the research and development community [10] mainly for medical rehabilitation [11–13], medical assistance [14] and military applications [15]. Most exoskeleton developers focus on using electrically driven motors to support arm movement, ambulation or to carry objects on the back. Manipulation of heavy goods as seen in industrial applications are however rarely addressed. The main obstacles in these applications are the restricted power availability and the weight and volume of the exoskeleton.

Spring based systems have the advantage that no electrical power is needed, which reduces the weight and volume of the exoskeleton. Similar passive concepts do exist. However, due to the additional load, systems with counterweights are too heavy. In addition, springs with zero free length (i.e. ideal springs) compared to non-zero free length springs are costly or require an increased complexity [7]. For simplicity reasons, the solution focuses on using non-zero free length springs.

Requirementes

The design of this passive gravity compensation system targets the use as an upper limb (arm) exoskeleton. This arm exoskeleton should support the lifting of objects up to 7.5 kg. For this passive exoskeleton, no external power source is wanted, i.e. no sensors or powered actuators can be used. The gravity compensating force needs to be generated by spring forces only. Other main system characteristics are a minimal weight and a wearable, robust and compact design. The following requirements should be met:

1. minimum supporting force: 40 N

*Correspondence: ruprecht.altenburger@zhaw.ch
Institute of Mechatronic Systems, School of Engineering, Zurich University of Applied Science, Technikumstrasse 5, Winterthur, Switzerland

2. maximum supporting force: 120 N (these correspond to 4.5 kg of users arm weight plus 7.5 kg of external load)

3. adaptable to all supporting forces with a simple mechanism

4. close to constant gravity compensating force over a wide operating range (i.e. iso-elasticity) for all supporting forces

5. the vertical range of the compensation should cover at least ±45° from a horizontal position

6. feasible and manufacturable geometric dimensions

7. minimal friction and lightweight design (≤4.5 kg) to minimize the energy required to carry, move and accelerate the exoskeleton

8. based on commercially available springs.

Basic parallelogram—iso-elasticity

The early lamp poising systems consist of a parallelogram structure with a spring, Fig. 1. The parallelogram shown in Fig. 1—called the *standard parallelogram*—is actuated by an extension coil spring fixed diagonally within the parallelogram between points A and B. An external load acts on the right bar in the z-direction denoted by force F_z. Using vector addition, the resulting tension and compression forces in the bars a, b and c are related to their geometric lengths, i.e.

$$\frac{F_a}{a} = \frac{F_b}{b} = \frac{F_c}{c}. \tag{1}$$

In equilibrium, the force F_b corresponds to the external load force, hence

$$F_b = -F_z \tag{2}$$

and the force in F_c corresponds to the restoring force generated by the spring. Assuming that the spring is an ideal spring, the force is

$$F_c = F_{spring} = k\,c, \tag{3}$$

where k is the spring constant. Under this assumption, the lifting force F_b is

$$F_b = \frac{b}{c} F_{spring} = \frac{b}{c} k\,c = k\,b = \text{const.} \tag{4}$$

For an ideal spring, the lifting force F_b is therefore independent of the spring length c and hence the lifting force is constant for any value of φ. This means that no external force or torque is required to balance the weight in any orientation of the mechanism within its workspace for a predefined load force F_z. This characteristic is often referred to as "iso-elasticity" [1].

A second feature of this parallelogram spring system is that dynamic movements from one side cause no movements on the other of the parallelogram as Eq. (4) is independent of φ and if frictionless joints are assumed. This feature is utilised in stabilisation applications for motion picture cameras [8]. It is also found in support systems for workers for lifting tasks and as a balancer and stabiliser for heavy tools [9].

Effects of non-ideal springs and friction

In reality it is difficult to produce close-to-ideal springs. Figure 2 shows a selection of springs produced by *Durovis AG, Switzerland* (http://www.durovis.ch) compared to the ideal spring (zero free length) characteristics. Since a real spring always has a fixed free length l_0, a tension spring with a high pretension F_0 is needed to reproduce the ideal spring characteristic indicated by the solid line with endpoints in Fig. 2. Therefore, the restoring force needs to be described as

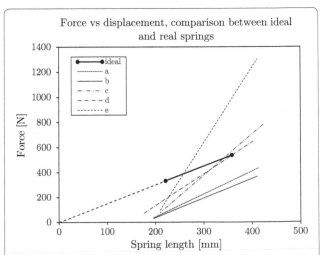

Fig. 2 Ideal spring characteristic for iso-elastic behaviour (ideal) compared to multiple real spring curves (**a** *Art Nr. 22/2/2*, **b** *Art Nr. 22/8/2*, **c** *Art Nr. 25/4/3*, **d** *Art Nr. 25/5/2*, **e** *Art Nr. 26/4/1*) Source: *Durovis AG, Switzerland*

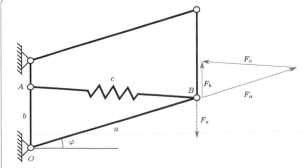

Fig. 1 Standard gravity compensation by parallelogram layout with extension coil spring

$$F'_{\text{spring}} = k\,(c - l_0) + F_0. \tag{5}$$

Preload values are typically only 3–8 % of the desired values as shown in Fig. 2. These realistic values for the spring parameters lead to a strong non-constant behaviour of the supporting force of the mechanism with forces that are too low at upper- and too high at lower positions. Again the lifting force F_b needs to be

$$F_b = \frac{b}{c}\,F'_{\text{spring}} = \frac{b}{c}\,[\,k\,(c - l_0) + F_0\,] \tag{6}$$

$$= \frac{b}{c}(F_0 - k\,l_0) + k\,b, \tag{7}$$

where the variable length c can be substituted with the geometrial condition

$$c = \sqrt{a^2 + b^2 - 2\,ab\,\sin(\varphi)}. \tag{8}$$

Hence, for a realistic spring, the system will always equilibrate at a specific angle φ for a given load F_z (see Fig. 3, solid line marked b). Obviously, there is no iso-elastic behaviour. The original lamp poising systems still worked nicely because the joints were not frictionless. For the motion picture camera stabilization however, the friction in the joints is unwanted because it weakens the effect of decoupling the motion between the person and the carried camera attached to different sides of the parallelogram structure.

Adaptability to load changes

For the ideal spring case, the parallelogram can be used to support different loads F_z and exhibit iso-elastic behaviour. This can be achieved by changing the length b in Eq. (4). The same approach can be used for a system with a real spring. In Fig. 3, the equilibrating position φ is shown versus the corresponding load F_z for bar length b, and for bar lengths $b_{80\%}$ and $b_{60\%}$, which are 80 and 60 % of length b, respectively.

For shorter bar lengths, the difference between maximal and minimal force reduces. Hence, increasing the

desired load compensation means that the system moves further away from an iso-elastic behaviour and therefore further away from the desired behaviour.

Two main options exist to manipulate Eqs. (7) and (8) to meet iso-elasticity and adaptability requirements. These are adjustable spring constants or a novel geometry, respectively. In [16], variable stiffness springs are used, which allows adjustment of the point of equilibration. The design is based on a nut with pitch equal to the spring. By changing the location of the nut along the length of the spring, sections of the spring are inactivated. This is a viable solution if the adjustment can be made while the spring is not elongated. For variable loaded springs (i.e. variable elongated springs), the pitch changes and therefore the adjustment of the stiffness is not easily possible.

In the next section, a geometrical change is proposed and optimised, which addresses the above-mentioned requirements.

Extension by a 4-link mechanism

If we consider the behaviour of the standard parallelogram with realistic values for the spring, the spring forces are typically too low for small extensions and too high for large extensions of the spring. The basic idea for the additional mechanism is to move point A in Fig. 1 (fixture of the spring) in a way that the spring is tensioned more in upper positions and less in lower positions relative to the standard parallelogram. This is achieved by the additional 4-link mechanism shown in Fig. 4. It depicts the mechanism in the two extreme positions, upper and lower. The additional bars are denoted y, z and r. The lengths x, y and z are constant whereas the length of bar r can be

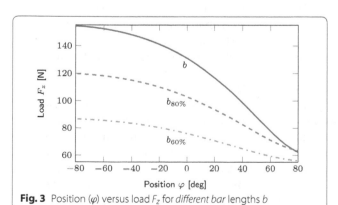

Fig. 3 Position (φ) versus load F_z for *different bar* lengths b

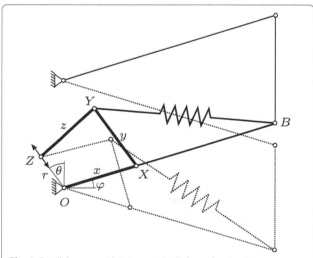

Fig. 4 Parallelogram with integrated 4-link mechanism in upper (*solid*) and lower (*dotted*) positions

adjusted along the axis $\overline{OZ}$. The spring is connected at point Y and it can be seen that the quadrilateral spanned by points $OXYZ$ "flattens" out when moving from the upper to the lower position, which reduces the effective distance between points Y and B compared to the case when point Y would stay at the same location. In addition, the reduced distance between O and Y has the same effect as reducing length b in the standard parallelogram from Fig. 1.

Angles and lengths are defined according to Fig. 5. The load force F_z acts on the end effector at point B. By taking the sum of moments around the origin O, as shown in Fig. 6, the resultant force F_z is as follows:

$$F_z\, a\, \cos(\varphi) + F_y\, x\, \sin(\alpha) - F_c\, a\, \sin(\beta) = 0 \qquad (9)$$

$$F_z = \frac{F_c\, a\, \sin(\beta) - F_y\, x\, \sin(\alpha)}{a\, \cos(\varphi)} \qquad (10)$$

where F_c and F_y are the spring force and the force in bar y, respectively. Using the geometrical arguments illustrated by Fig. 5, the force F_y is:

$$F_y = \frac{F_c\, \sin(\gamma + \alpha - \beta)}{\sin(\gamma)}. \qquad (11)$$

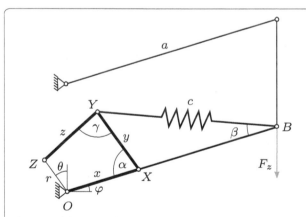

Fig. 5 Parallelogram with integrated 4-link mechanism showing the main variables and parameters

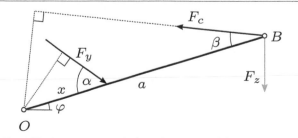

Fig. 6 The forces acting on the lower beam around the origin O

By substituting F_y by Eq. (11) in Eq. (10), the final resultant force as a function of spring force F_c and internal angles is:

$$F_z = F_c \left[\frac{\sin(\beta)}{\cos(\varphi)} - \frac{x\, \sin(\gamma + \alpha - \beta)\, \sin(\alpha)}{a\, \sin(\gamma)\, \cos(\varphi)} \right]. \qquad (12)$$

Hence, the force F_z can be described in terms of the spring force F_c—a real spring characteristic is assumed according to Eq. (7)—and geometrical relations which depend on the lengths of the four bars x, y, z and r. In Eq. (12) the angles α, β and γ can be algebraically replaced by the bar lengths. The expression is however omitted here for the sake of compactness.

Optimisation of the geometry

Qualitatively, the 4-link mechanism provides the desired behaviour. However, the design question remains as to how the dimensions of three bar lengths x, y and z (according to Fig. 5) need be chosen to provide a reasonable iso-elastic behaviour. For this purpose, a design optimisation problem is formulated. For practical reasons, it was decided to move point Z on a straight line starting at origin O to vary the supporting force F_z. The pitch angle θ is introduced as an additional parameter to be optimised. The length r should vary from 5 mm (40 N support) to approximately 45 mm (120 N support). The working range of the parallelogram was set to a range of $\varphi = -45° \ldots + 45°$.

To reach an iso-elastic behaviour over the working range, the curves in Fig. 3 need to collapse to horizontal lines. A simple optimisation criteria is therefore to use the difference between the maximally and minimally generated forces F_z within the operating range. Therefore, the optimisation reads as follows

$$\min_{x,y,z,\theta} R(x, y, z, \theta \,|\, r, \varphi) \qquad (13)$$

with

$$R(x, y, z, \theta \,|\, r, \varphi) = \max(F_z(\varphi \,|\, r)) - \min(F_z(\varphi \,|\, r)).$$

The notation $F_z(\varphi \,|\, r)$ means: the function $F_z(\varphi)$ at a given value r. In the present form an optimisation would favour geometrical parameters that lead to small absolute values for the lifting force. Thus an extra penalty term is added to the objective function R which measures the maximum distance of forces at $r = 45$ mm to the desired maximum force of 120 N.

In principle the three parameters of the spring (spring constant, length and pretension) could also be values to be optimised. But since the manufacturing process strongly restricts these values, it was decided to select a spring and calculate the values above for this particular spring. The values of the spring were chosen from energetic considerations like:

$$(l_{\max} - l_0) \cdot \frac{F_{\max} - F_0}{2} > m\,g\,h, \tag{14}$$

which means that the total energy in the maximum tensioned spring must be greater than the total lifting energy of the mass.

For manufacturing reasons the geometrical parameters x, y, z and θ are constrained to reasonable values. Their limits were given as

$$x, y, z \in [40\,\text{mm}, 100\,\text{mm}],$$
$$\theta \in [10°, 80°].$$

The resulting force $F(\varphi, r)$ can be calculated according to Eq. (12).

The optimisation is done using MATLAB *(The Math-Works, Inc.)* and the `fmincon` function of the optimisation toolbox. The used algorithm uses interior-point approach. The function $F(\varphi, r)$ was computed on a discrete grid of φ and r values.

The optimised geometrical parameters are $x = 58.2\,\text{mm}$, $y = 72.9\,\text{mm}$, $z = 94.3\,\text{mm}$ and $\theta = 51.4°$. Note that all values lie inside the given bounds. Figure 7 shows the resulting force values at the end effector for different angles φ and different settings of the point Z by varying distance r from 5 … 45 mm. For small lifting forces (small values of r) nearly perfect iso-elastic behaviour can be reached. The maximum force at 120 N shows a variation of ± 3.2 N for positions $\varphi = -45° \ldots 45°$.

If we consider the optimisation with respect to the four parameters x, y, z, θ one has to ask whether these four parameters give a proper description of the optimisation problem. Figure 8 shows a stacked plot of the optimisation function $R(x, y, z, \theta | r, \varphi)$ under variation of the parameters x, y, z. The angle θ is kept at the optimised value of 51.4°. For illustration purposes the z dimension is quantised. The figure shows only a very slight variation of R if we move diagonally through the parameter space shown by the dark blue areas on each surface plot. Note that the objective function has very large values at outer regions of the shown parameter space. For this reason the color bar has a nonlinear scale at the outer end. The optimisation routine has to find an optimum in this "flat valley", but there does seem to be a global optimum. However, there are many other possible combinations of the geometrical parameters with similar iso-elastic behaviour to that of the optimised parameters shown above.

First prototype

Figure 9 shows the prototype based on the 4-link mechanism. In the lower left, indicated by R, the adjustment for moving the supporting point Z can be seen. It is a nut running along a spindle when lever L is turned. By that, the desired lifting force can be adjusted.

The total weight of one parallelogram is 1.9 kg. As in other applications, two parallelogram segments are connected in series to provide a reasonable range of operation. The forearm of the user is attached to the end of the second segment using a cuff. This specific design supports

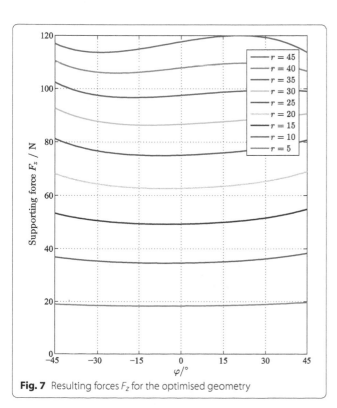

Fig. 7 Resulting forces F_z for the optimised geometry

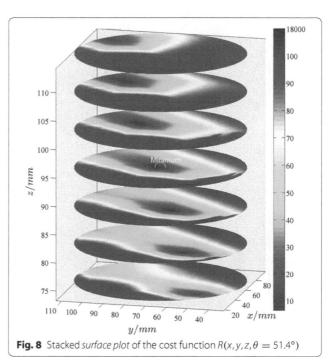

Fig. 8 Stacked *surface plot* of the cost function $R(x, y, z, \theta = 51.4°)$

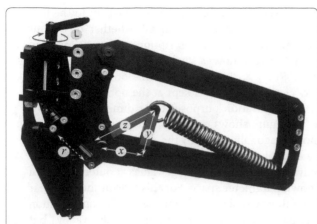

Fig. 9 Image of the prototype. The lengths x, y and z are highlighted. By rotating lever L the lower end of beam z can be positioned along the range r to adjust for different external loads F_z

up to 120 N at the location of the cuff. Considering that an arm consists of approximately 5 % of the human bodyweight, this system can therefore support an external load up to the required 7.5 kg. This is at the upper limit for most people when the load is held in the hand. A tool mounted directly to the system could weigh up to 12 kg. Ball bearings were used to reduce friction in the joints.

Experimental verification

Tests were carried out to verify the behavior of the developed system. Using different weights and a spring balance, the effective lifting force at 7 different angles φ were

measured. The weights correspond to 20, 50, 70, 90 N and since there is a static friction of approximately ± 5 N, the measurements were done in an upward and a downward direction. The force was applied with a hand held spring scale. A weight of 7 N was added which is the contribution to the weight of the arm itself. Figure 10 on the left shows results of the theoretical curve and the measurements. For each weight and angle the two squares show the upward and downward direction of the measurement. Especially for higher loads there is a significant discrepancy between calculated and measured forces. The supporting force tends to higher values at larger angles φ.

The reason for this discrepancy was found in a considerably difference of the spring data and also a geometrical effect which comes from the attachment of the spring to the construction (a small offset of a bolt). The spring characteristics was validated on a tensile testing machine. The values of the spring characteristics in the optimisation was taken from the datasheet an was: length $l_0 = 178.75$ mm, preload force: $F_0 = 115$ N, spring rate: $c = 11.15$ N/mm. The measured values were found as: length $l_0 = 178.75$ mm, preload force: $F_0 = 191.9$ N, spring rate: $c = 10.9$ N/mm.

Figure 10 on the right shows the measurements and calculated values with these corrected spring parameters. These values will be used in future for a new optimization for a second prototype. The first prototype was assembled to a supporting system (2×2 parallelograms for two arms) and show very positive results in tests (see Fig. 11).

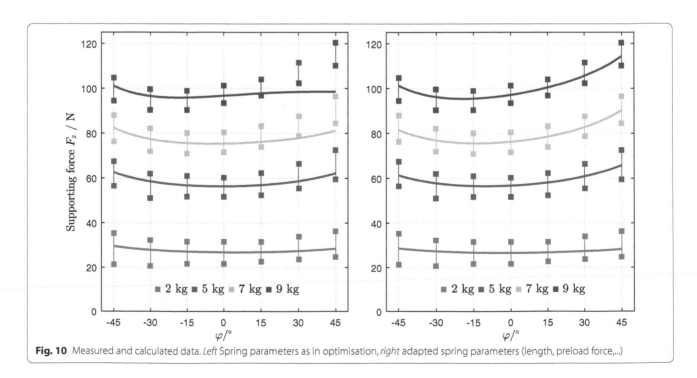

Fig. 10 Measured and calculated data. *Left* Spring parameters as in optimisation, *right* adapted spring parameters (length, preload force,...)

Fig. 11 Assembled parallelogram fixed at a carrying system

Conclusion and outlook

The resulting design exhibits close to iso-elastic behaviour over a wide operating range and is adjustable for loads in the range of 40–120 N. This makes the mechanism attractive for a lightweight exoskeleton arm, which is fully passive but still sufficiently powerful to support significant weight. The same mechanism can be used to balance an object such as a tool. Compared to electrically powered exoskeleton arms, the supporting force is continuously present and cannot be switched on or off depending on whether an object is being carried or not. The design is ideal for supporting the user's posture and for lower load weights.

The design optimisation approach provided an efficient framework for selecting the best parameters. Future development priority lies in further reducing the weight and designing the exoskeleton arm to be more suitable for specific tasks.

Authors' contributions
RA proposed the extended gravity compensation mechanism and supervised the prototype design and contributed to the parameter optimisation. DS developed the prototype and contributed to the parameter optimisation. KSS contributed to the background information. All authors contributed equally to paper preparation and revision. All authors read and approved the final manuscript.

Acknowledgements
This mechanism was developed within the project Robo-Mate. The Robo-Mate project has received funding from the European Union's Seventh Framework Programme for research, technological development and demonstration under Grant Agreement No. 608979.

Competing interests
The authors declare that they have no competing interests.

References
1. Lu Q, Ortega C, Ma O (2011) Passive gravity compensation mechanisms: technologies and applications. Recent Patent Eng 5:32–44
2. Carwardine G (1932) Improvements of elastic force mechanisms. UK Patent Specification 379,680
3. Jacobsen J (1962) Adjusting means for a lamp structure. US Patent 3,041,060
4. French MJ, Widden MB (2000) The spring-and-lever balancing mechanism, george carwardine and the anglepoise lamp. Proc Inst Mech Eng, Part C 214:501–508
5. Leporati RA (1968) Counterpoising or equipoising mechanism. US Patent 3,409,261
6. Chalfoun J, Bidard C, Keller D, Perrot Y, Piolain G (2007) Design and flexible modeling of a long reach articulated carrier for inspection. In: IEEE/RSJ international conference on intelligent robots and systems, pp 4013–4019
7. Agrawal A, Agrawal SK (2005) Design of gravity balancing leg orthosis using non-zero free length springs. Mech Mach Theory 40(6):693–709. doi:10.1016/j.mechmachtheory.2004.11.002 Accessed 26 Jan 2016
8. Brown GW (1979) Equipment support system. US Patent 4,156,512
9. Goldman D, Kelly H, Pagliery J (2014) 36 Coolest gadgets of 2014. CNN-Money. http://money.cnn.com/gallery/technology/innovationnation/2014/12/10/coolest-gadgets-2014/24.html
10. Pons JL (ed) (2008) Wearable robots: biomechatronic exoskeletons. Wiley, Hoboken
11. Rosen J, Brand M, Fuchs MB, Arcan M (2001) A myosignal-based powered exoskeleton system. IEEE Trans Syst Man Cybernet Part A Syst Humans 31(3):210–222
12. Nef T, Mihelj M, Kiefer G, Perndl C, Muller R, Riener R (2007) ARMin-exoskeleton for arm therapy in stroke patients. In: Proceedings of the 10th international conference on rehabilitation robotics (ICORR), pp 68–74
13. Banala SK, Kim SH, Agrawal SK, Scholz JP (2009) Robot assisted gait training with active leg exoskeleton (ALEX). IEEE Trans Neural Syst Rehabil Eng 17(1):2–8
14. Kong K, Jeon D (2006) Design and control of an exoskeleton for the elderly and patients. IEEE/ASME Trans Mech 11(4):428–432
15. Kazerooni H, Steger R (2006) The berkeley lower extremity exoskeleton. Trans ASME 128:14–25
16. Dorsser WDV, Barents R, Wisse BM, Schenk M, Herder JL (2008) Energy-free adjustment of gravity equilibrators by adjusting the spring stiffness. Proc Inst Mech Eng Part C 222(9):1839–1846
17. Byrd RH, Gilbert JC, Nocedal J (2000) A trust region method based on interior point techniques for nonlinear programming. Math Program 89(1):149–185

Permissions

All chapters in this book were first published by Springer; hereby published with permission under the Creative Commons Attribution License or equivalent. Every chapter published in this book has been scrutinized by our experts. Their significance has been extensively debated. The topics covered herein carry significant findings which will fuel the growth of the discipline. They may even be implemented as practical applications or may be referred to as a beginning point for another development.

The contributors of this book come from diverse backgrounds, making this book a truly international effort. This book will bring forth new frontiers with its revolutionizing research information and detailed analysis of the nascent developments around the world.

We would like to thank all the contributing authors for lending their expertise to make the book truly unique. They have played a crucial role in the development of this book. Without their invaluable contributions this book wouldn't have been possible. They have made vital efforts to compile up to date information on the varied aspects of this subject to make this book a valuable addition to the collection of many professionals and students.

This book was conceptualized with the vision of imparting up-to-date information and advanced data in this field. To ensure the same, a matchless editorial board was set up. Every individual on the board went through rigorous rounds of assessment to prove their worth. After which they invested a large part of their time researching and compiling the most relevant data for our readers.

The editorial board has been involved in producing this book since its inception. They have spent rigorous hours researching and exploring the diverse topics which have resulted in the successful publishing of this book. They have passed on their knowledge of decades through this book. To expedite this challenging task, the publisher supported the team at every step. A small team of assistant editors was also appointed to further simplify the editing procedure and attain best results for the readers.

Apart from the editorial board, the designing team has also invested a significant amount of their time in understanding the subject and creating the most relevant covers. They scrutinized every image to scout for the most suitable representation of the subject and create an appropriate cover for the book.

The publishing team has been an ardent support to the editorial, designing and production team. Their endless efforts to recruit the best for this project, has resulted in the accomplishment of this book. They are a veteran in the field of academics and their pool of knowledge is as vast as their experience in printing. Their expertise and guidance has proved useful at every step. Their uncompromising quality standards have made this book an exceptional effort. Their encouragement from time to time has been an inspiration for everyone.

The publisher and the editorial board hope that this book will prove to be a valuable piece of knowledge for researchers, students, practitioners and scholars across the globe.

List of Contributors

Haifei Zhu, Yisheng Guan, Shengjun Chen and Manjia Su
Biomimetic and Intelligent Robotics Lab (BIRL), School of Electro-mechanical Engineering, Guangdong University of Technology, Hi-education Mega Center, Guangzhou 510006, China

Hong Zhang
Biomimetic and Intelligent Robotics Lab (BIRL), School of Electro-mechanical Engineering, Guangdong University of Technology, Hi-education Mega Center, Guangzhou 510006, China
Department of Computing Science, University of Alberta, Edmonton, AB T6G 2E8, Canada

Bo Ding, Huaimin Wang, Zedong Fan, Pengfei Zhang and Hui Liu
College of Computer, National University of Defense Technology, Changsha, Hunan, China

Bruce E. Saunders
University of Bath, 54 Ballance Street, Bath, UK

Mannam Naga Praveen Babu and P. Krishnankutty
Department of Ocean Engineering, Indian Institute of Technology Madras, Chennai 600 036, India

J. M. Mallikarjuna
Department of Mechanical Engineering, Indian Institute of Technology Madras, Chennai 600 036, India

Hong Liu, Dapeng Yang, Shaowei Fan and Hegao Cai
State Key Laboratory of Robotics and System, Harbin Institute of Technology, HIT Science Park, No.2 Yikuang Street, Nangang District, P.O. Box 3039, 150080 Harbin, People's Republic of China

Yang Cao, Satoshi Miura, Yo Kobayashi, Shigeki Sugano and Masakatsu G. Fujie
1 Graduate School of Creative Science and Engineering, Waseda University, 3-4-1 Ohkubo, Shinjuku-ku, Tokyo 169-8555, Japan

Quanquan Liu
Graduate School of Advanced Science and Engineering, Waseda University, 2-2 Wakamatsu-cho, Shinjuku-ku, Tokyo 162-8480, Japan

Kazuya Kawamura
Graduate School and Faculty of Engineering, Chiba University, 1-33 Yayoi-cho, Inage-ku, Chiba 263-8522, Japan

Hiroki Matsuoka and Takefumi Kanda
Okayama University, 3-1-1, Tsushima-naka, Kita-ku, Okayama 700-8530, Japan

Koichi Suzumori
Tokyo Institute of Technology, 2-12-1-I1-60, Ookayama, Meguro-ku, Tokyo 152-8550, Japan

Tomohiro Suzuki and Satoshi Miura
The Graduate School of Science and Engineering, Waseda University, Tokyo, Japan

Yo Kobayashi and Masakatsu G. Fujie
The Faculty of Science and Engineering, Waseda University, Tokyo, Japan

Shin-yo Muto, Yukihiro Nakamura, Hideaki Iwamoto and Tatsuaki Ito
NTT Service Evolution Laboratories, 1-1 Hikarinooka, Yokosuka-Shi, Kanagawa 239-0847, Japan

Manabu Okamoto
NTT Media Intelligence Laboratories, 1-1 Hikarinooka, Yokosuka-Shi, Kanagawa 239-0847, Japan

Taku Nakamura and Akio Yamamoto
Department of Presicion Engineering, The University of Tokyo, Hongo 7-3-1, Tokyo 113-8656, Japan

Tran Vu Minh
School of Mechanical Engineering, Hanoi University of Science and Technology, Hanoi, Vietnam

Nguyen Manh Linh
Graduate School of Engineering and Science, Shibaura Institute of Technology, Saitama 337-8570, Japan

Xinkai Chen
Department of Electronic and Information Systems, Shibaura Institute of Technology, Saitama 337-8570, Japan

Makoto Sasaki and Kohei Onishi
Graduate School of Engineering, Iwate University, Morioka, Iwate, Japan

Dimitar Stefanov
School of Science and Technology, Middlesex University, London, UK

Katsuhiro Kamata
Pattern Art Laboratory Co., Ltd., Hanamaki, Iwate, Japan

Atsushi Nakayama
Department of Intelligent Systems Engineering, Ichinoseki National College of Technology, Ichinoseki, Iwate, Japan

Masahiro Yoshikawa
Graduate School of Information Science, Nara Institute of Science and Technology, Nara, Japan

Goro Obinata
Department of Robotics Science and Engineering, Chubu University, Kasugai, Japan

Yuichi Ambe and Fumitoshi Matsuno
Department of Mechanical Engineering and Science, Kyoto University, Katsura, Nishikyo-ku, Kyoto, Japan

Huei Ee Yap and Shuji Hashimoto
Department of Applied Physics, Graduate School of Advanced Science and Engineering, Waseda University, Tokyo, Japan

Satoshi Kitano , Atsushi Horigome and Gen Endo
Department of Mechanical and Aerospace Engineering, Tokyo Institute of Technology, Meguro-ku, Ookayama 2-12-1, 152-8550 Tokyo, Japan

Shigeo Hirose
Hibot Corp., 5-9-15 Kitashinagawa, Shinagawa-ku, 141-0001 Tokyo, Japan

Bo Ouyang and Dong Sun
Department of Mechanical and Biomedical Engineering, City University of Hong Kong, Kowloon, Hong Kong, China

Yunhui Liu
Department of Mechanical and Automation Engineering, The Chinese University of Hong Kong, Shatin, China

Motomu Nakashima
Graduate School of Information Science and Engineering, Tokyo Institute of Technology, 2-12-1 Ookayama, Meguro, Tokyo 152-8552, Japan

Kosuke Kuwahara
Graduate School of Science and Engineering, Tokyo Institute of Technology, 2-12-1 Ookayama, Meguro, Tokyo 152-8552, Japan

Ruprecht Altenburger, Daniel Scherly and Konrad S. Stadler
Institute of Mechatronic Systems, School of Engineering, Zurich University of Applied Science, Technikumstrasse 5, Winterthur, Switzerland

Index